REDEFINING EFFICIENCY

TECHNOLOGY
AND THE
ENVIRONMENT

JEFFREY STINE AND
JOEL TARR

SERIES EDITORS

HUGH S. GORMAN

REDEFINING EFFICIENCY

POLLUTION CONCERNS, REGULATORY MECHANISMS, AND TECHNOLOGICAL CHANGE IN THE U.S. PETROLEUM INDUSTRY

THE UNIVERSITY OF AKRON PRESS

AKRON, OHIO

LIBRARY OF CONGRESS CATALOGING-IN-PUBLICATION DATA

Gorman, Hugh S. (Hugh Scott), 1957–
    Redefining efficiency : pollution concerns, regulatory mechanisms, and
technological change in the U.S. petroleum industry / Hugh S. Gorman.—1st ed.
      p. cm. — (Technology and the environment)
    Includes bibliographical references and index.
    ISBN 1-884836-74-7 (hardcover) — ISBN 1-884836-75-5 (pbk.)
    1. Petrolueum industry and trade—Environmental aspects—United States.
2. Industrial efficiency—United States. 3. Environmental protection—United
States—History. 4. Pollution—Law and legislation—United States. I. Title.
II. Technology and the environment (Akron, Ohio)
TD195.P4 G67 2001
363.73'5765—dc21

                                 2001002140

To my father

Contents

List of Figures

List of Tables

Preface

How has industry's response to pollution concerns changed over the course of the twentieth century? It is an important question to ask if we are to get beyond the stereotype of industrial firms being negligent in the years before systematic regulation and resistant in the years after. Although there is some truth to this characterization, it is just the skeleton of a much richer story. Putting a body of evidence on this skeleton not only brings that story to life but also provides new insights into the interaction of pollution concerns, regulatory mechanisms, and technological changes.

Here, I trace the response of a single industry, the U.S. petroleum industry, to pollution-related concerns over the course of the twentieth century. Few other industries can serve as a better window through which to view the changes that have occurred. Indeed, as Daniel Yergin has made clear in *The Prize*, petroleum flows through the veins of industrial society and the history of the twentieth century is intimately intertwined with efforts to control that flow. Yergin, of course, was interested in global politics, not pollution. But one can also use the petroleum industry to examine the history of pollution control.

In tracing pollution-related concerns associated with the extraction, transportation, and refining of petroleum, I asked the following questions:

- What petroleum-related pollution concerns did people articulate at the beginning of the twentieth century and how did those concerns change over time?
- How did decision makers in the petroleum industry respond to early pollution concerns in terms of legal arguments, changes in industrial practice, and technological innovation? How did general patterns associated with their responses change over time?
- Over the course of the twentieth century, what technological and regulatory changes have influenced industry's response to pollution concerns?

The petroleum industry, of course, is not a monolithic actor. It consists of many different firms and organizations performing a variety of activities, each involving different technological, economic, legal, political, and social constraints. Still, in answering such questions, a coherent story emerges. What takes shape is not a simple story of enforcement and compliance but a process of debate, challenge, and discovery, resulting in the development of an industrial ecology that would have been unimaginable at the beginning of the twentieth century.

The historical record available to address these questions is scattered but voluminous. Numerous secondary sources exist. Many insightful and exhaustive histories of the petroleum industry, individual oil companies, and industry leaders have been written, as have histories of the various industry sectors, oil-producing and oil-refining regions, and petroleum-related technologies. The industry's regulatory history—especially in regard to production rules, tariffs and quotas, and pricing regulations—is also well documented. Although these books do not focus on pollution-related issues, many make some reference to such issues. Government documents and government-sponsored reports that address petroleum-related pollution concerns also abound, usually related to one of the many controversies in which the industry has found itself embroiled over the course of a century. Petroleum-related treatises, textbooks, and trade magazines also mention pollution concerns and provide additional insight into changing public concerns and industry's response to those concerns.

The publicly accessible archival record is also interesting and informative. The corporate records of the Sun Oil Company, housed at the Hagley Museum and Library in Delaware, provide as complete a picture of an oil company as one could hope for. Furthermore, the private papers of industry executives can be found in archives scattered around the United States. In many cases, their papers contain extensive correspondence, minutes, and reports associated with their businesses, their positions in industry-related organizations, and their roles on various industry committees. Presidential libraries and the archival records of government agencies such as the U.S. Bureau of Mines and the U.S. Environmental Protection Agency also provide good access to relevant primary sources. Archived transcriptions of oral histories and various clipping files on petroleum-related environmental issues also proved useful. Finally, for insight into the details of the existing regulatory system, I examined the state-level wastewater and air emissions records of three refineries.

Funds for travel to the various archives have come from a variety of sources, including the National Science Foundation (NSF Grant 9422671), the Hagley Museum and Library, the Hoover Presidential Library Association, and a faculty development grant from Michigan Technological University. A second NSF grant (Grant SES-9817913) to research post-1970 changes associated with monitoring refinery effluents and emissions also contributed to the later sections of this book.

I specifically want to thank David Hounshell, Joel Tarr, and Jeffrey Stine, each of whom read the original manuscript closely and made numerous suggestions. I thank Edward Constant for providing me with insights into the subtleties of oil production and the culture of the industry. I also thank Christine Rosen and Christopher Sellers, who helped me to refine my thoughts by critiquing an article based on this research. Finally, I thank my wife, Bonnie Gorman, who with patience and good humor has continually provided the encouragement and support needed to stay focused on a project such as this.

Introduction

In the United States, public concerns with industrial pollution and efforts to address those concerns have a long history, dating back well into the nineteenth century.[1] Initially, those concerns hovered in the shadows of larger and more pressing issues associated with the industrialization and urbanization of an agricultural society. As railroads linked local economies together, waves of immigrants, entrepreneurs, and capital radiated outward from Boston, New York, Philadelphia, and Baltimore, establishing manufacturing clusters in cities throughout the Northeast and Midwest. In most urban centers, industrial pollution was just one of many issues to be addressed.

Concerns associated with domestic sanitation, the need for a disciplined work force, and crime prevention often spoke to civic leaders with more urgency.[2] By the late nineteenth century, though, civic leaders were pointing to smoke-choked industrial centers such as Pittsburgh and expressing hope that their cities could avoid that same fate. Only hope, and factors such as favorable weather patterns and access to an inexpensive source of clean coal, prevented air quality in many industrial cities from deteriorating as rapidly.[3] The quality of urban rivers, streams, and bays also deteriorated, with many becoming little more than sewage and navigation channels.

Only as the twentieth century dawned and as Progressive Era reformers sought to organize this new urban-industrial society did the problem of industrial pollution receive more attention. Still, no elegant solution presented itself. In the first two decades of the twentieth century, resource conservationists, industrial hygienists, sanitation engineers, biologists, and medical doctors all confronted the issue of industrial pollution at one time or another, each struggling for ways to reconcile the public's desire for clean air and clean water with industry's need to discharge pollution-causing wastes.[4]

Gradually, a rough guiding ethic for controlling industrial pollution emerged. This ethic—rooted in the same technocratic ideals of efficiency, utilitarian conservation, and the elimination of waste that guided the nation's natural resource policies—remained in place for another thirty years, rarely questioned by nonexperts and taken for granted by engineers and technical decision makers responsible for industrial practices. Unless a pollution control effort resulted in the recovery of valuable material or decreased the amount of money spent on damage and nuisance lawsuits, firms generally did not take direct action to reduce their emissions, effluents, leaks, spills, and other discharges. However, engineers and technical managers did not see their lack of aggressive action, beyond that dictated by pure economics, as a conspiracy of silence. On the contrary, they saw economic incentives to improve efficiency and reduce waste as tightly coupled with long-term efforts to fight pollution. From this perspective, engineers saw pollution-causing wastes as analogous to waste heat, a symptom of inefficiency. More than anything, pollution indicated that an industrial process had yet to be optimized. Pollution was, in their minds, a short-term problem that good engineering would eventually eliminate.[5]

This book is about the creation, temporary success, and ultimate failure of this first guiding ethic for controlling industrial pollution. It is also about the replacement of this guiding ethic with a second ethic that crystallized in the debates of the 1960s, took on a definite shape in the 1970s, and matured in the 1980s and 1990s.[6] Although much has been written about pollution control under this newer environmental ethic, which depends on explicitly defining environmental objectives and constructing policies to achieve those objectives, relatively little has been written about the "gospel of efficiency" as a pollution control ethic.[7] However, both ethics—and the transition from one to the other—must be considered if one is to capture the full response of industrial firms to pollution concerns over the course of the twentieth century.

First, both ethics are important to consider in making sense of the debates over industrial pollution that occurred in the 1960s. Those debates were triggered not only by the rising expectations of a growing middle class but also by the failure of one pollution control ethic and the need to construct another.[8] By the late 1950s, engineers and technical managers could no longer argue that steady increases in the efficiency of industrial operations would eventually eliminate pollution concerns. Despite tremendous increases in the efficiency of industrial operations, the quality of the nation's air and water continued to deteriorate. Furthermore, the scale of production continued to increase while incentives to secure additional reductions in pollution-causing wastes dwindled. A new approach was clearly needed.

Second, both ethics are important in any discussion of technological change related to pollution control. Under an efficiency-minded ethic of self-regulation, few people placed much emphasis on measuring and monitoring the flow of pollution-causing wastes into the environment. Neither did many people examine the movement of contaminants through the environment or study the effect of those contaminants on ecological processes and human health. After all, the assumption was that concerns with pollution-causing wastes would gradually be reduced as engineers increased the efficiency of industrial operations. The focus, therefore, remained on improving production processes.

Under an ethic that requires a society first to define and then to achieve environmental objectives, the focus shifts. The flow of contaminants into the shared environment becomes a transaction to be measured and monitored. Indeed, pollution control regulations have restructured the economics of pollution control. From the perspective of those who make business decisions, the notion of what it means to operate an "efficient" industrial operation has gradually been redefined to include environmental measures of performance.

Finally, both ethics are necessary to understand the development of the current regulatory regime. In the transition from one guiding ethic to another, older assumptions do not simply disappear. For example, efforts by some to use cost-benefit analyses to "compute" the worthiness of a specific environmental objective reflects the continued power of an efficiency ethic. While cost-benefit analyses may be useful in determining how best to achieve objectives once set, such analyses cannot determine which objectives are worthy and which are not. Reaching consensus on environmental objectives is inherently a political process. However, to people who continue to view efforts to control industrial pollution as a technical exercise in economic optimization, the use of cost-benefit analysis to judge environmental objectives will continue to carry significant weight. Examining the construction of the existing regulatory system with this transition from one guiding ethic to another in mind places many of the conflicts that occurred in a new light.

The Petroleum Industry As a Case Example

Any serious historical study of efforts to control industrial pollution requires one to trace how specific pollution concerns change over time. In this book, I trace pollution-related concerns associated with the extraction, transportation, and refining of petroleum over the course of the twentieth century. By focusing on a single industry—the petroleum industry—one can examine the history of pollution control in the United States in as comprehensive manner as possible.

Almost everybody is familiar with some aspect of the petroleum industry. Indeed, in the United States, the industry's main product is often taken for granted. As motorists pull into a gasoline station, many do not even look at the posted price or notice the name of the company selling the product. They simply pull up to a pump and set about the routine task of filling their tank. In five minutes or so, they have taken on approximately seventy pounds of fuel and are ready to resume their journey. As they accelerate into traffic and drops of fuel are ignited in their car's engine, energy absorbed from the sun millions of years ago propels them forward.

The technological system that keeps gasoline flowing is, of course, extremely sophisticated. Deepwater drilling platforms, oceangoing tankers, long-distance pipelines, and modern refineries are awesome in scope and complexity. Over the course of a century and a half, this system for extracting, transporting, and processing petroleum has evolved from a piecemeal collection of loosely interconnected components into an highly integrated global network. Whether one embraces or questions the energy-intensive and automobile-dependent society that this technology makes possible, one cannot fail to recognize the engineering skill that stands behind it.

The petroleum industry is also an inherently messy business. Petroleum is often found in environmentally sensitive, not to mention culturally and politically sensitive, areas.[9] The movement of oil, gas, and briny water through underground formations cannot be precisely controlled, and these fluids sometimes find unexpected paths through which to flow. Oceangoing tankers move incredible quantities of petroleum over long distances, and one spill is enough to threaten a local ecosystem. Those same tankers, when empty, must routinely be deballasted and cleaned, providing another potential path for pollution-causing discharges. Underground pipelines carry even larger quantities of petroleum and petroleum products, and leaks from those pipelines are difficult to prevent and detect. All petroleum must also pass through a refinery capable of separating, manipulating, and cleaning the various components of crude oil, and managing the wastes from such a facility is a major task.

Given the enormous potential for pollution-causing effluents, emissions, leaks, and spills to escape, this system of extraction, transportation, and processing is surprisingly clean. As one might expect, this was not always the case. In the early twentieth century, a significant percentage of the material that flowed into this system never made it to market. If one includes the petroleum that oil producers left abandoned underground, stripped of viscosity-reducing natural gas and too thick to flow through underground formations, only about 15 percent of the petroleum discovered reached the market as fuels and other products. Of the petroleum that producers actually extracted, the percentage of material lost to the environment was also quite high, approaching 20 percent. Today, the loss from field to filling station is far less, well under 1 percent.

What changes in technology and industrial practices made this reduction in pollution-causing discharges possible? What determined the pace of change? What changes were made under an efficiency-minded ethic of self-regulation? What changes were made after systematic regulations were in place? What conflicts, concerns, and debates remain? By raising and answering such questions in the context of shifting pollution control ethics, this book captures the interaction of publicly articulated pollution-related concerns, actions taken to address those concerns, and technological changes affecting the release of contaminants in a more comprehensive manner than would otherwise be possible.

Using the petroleum industry as the focus of study also ensures broad coverage because a wide range of pollution-related issues emerge in the course of extracting, transporting, and processing petroleum. In addition, the petroleum industry has been around since the mid–nineteenth century, allowing one to see the response of mature firms to pollution concerns over a long period of time. There is also an extensive public record to examine. Over the course of the century, one finds the petroleum industry at the center of many pollution-related and environment-centered debates, conflicts, trends, and events. From the Oil Pollution Act of 1924 and the early conflict over leaded gasoline to the blowout at Santa Barbara and the Trans-Alaska pipeline, high-profile debates have forced decision makers in the oil industry to publicly articulate their positions on various issues. In addition, as producers of energy, oil companies also have had to weigh in on issues related to resource conservation and on issues in which environmental quality is pitted against industrial development.

Finally, the industry's main trade organization, the American Petroleum Institute, played a significant role in the debates that occurred as the pollution control legislation passed by Congress was transformed into enforceable regulations. All industries had to respond to the new legislation passed by Congress, but the wide range of pollution control issues associated with the production, transportation, and refining of petroleum forced the oil industry to respond on numerous fronts. For example, all six criteria pollutants for which the U.S. Environmental Protection Agencies set ambient air quality standards—lead, sulfur dioxide, carbon monoxide, nitrogen oxides, ozone, and particulates—were associated with burning of petroleum products in one way or another. Oil spill legislation and legislation protecting groundwater also involved the petroleum industry more than most.

Today, as scientists examine the seriousness of global warming, oil companies, along with coal producers, find themselves at the center of still another debate. Most petroleum ends up being burned as fuel, with water vapor and carbon dioxide given off as the products of combustion. For most of the twentieth century, people saw these products of combustion as perfectly natural. Today, however, with the accumulation of carbon dioxide in the atmosphere and the associated potential for global climate change, the burning of fossil fuels has increasingly come under attack. The world's response to the threat of global warming from burning fossil fuels will be one more step in the long process of integrating environmental objectives into economic decisions affecting industrial practices.

Pollution Concerns, Regulatory Mechanisms, and Technological Development

To what extent are innovations that make a technological system more compatible with natural ecosystems and socially defined environmental objectives similar to other technological innovations? Are environmental innovations fundamentally the same as innovations that improve the performance of a product or allow a firm to produce a product less expensively? What are the implications of any differences that exist?

Clearly, from a historical perspective, differences do exist between pollution-related innovations and innovations that improve the performance of a product or a manufacturer's ability to produce products. Technological development typically depends on someone being able to gain an advantage from the dissemination of an innovation. Whether one is talking about weapons, textiles, automobiles, chemicals, electric power generation, aircraft, electronics, satellites, or software, innovations propagate when they are embraced by producers and consumers in a position to benefit from those innovations. In the absence of government-enforced regulations, though, producers only benefit from pollution control efforts when they can get a return on investment—mainly through savings in materials or fewer damage suits—greater than that available elsewhere. Hence, few technological system builders paid much attention to pollution control.[10]

Government-enforced regulations, of course, have restructured the economics of pollution control. Theoretically, firms manufacturing similar products now compete in their ability to satisfy these regulations. To what extent have firms started developing the technology that allows them to comply with pollution control regulations less expensively than competitors? If incentives to advance technology in this way have not emerged, why haven't they? What are the barriers preventing this change? In examining the industry's response to the emerging regulatory system, this book addresses such questions.

Advances in technology associated with monitoring flows of contaminants and characterizing their effect on environmental quality are also relevant. Regulatory agencies can only do their job if they have adequate information about these contaminants and their effects on ecological processes and human health. In the first half of the century, firms operating industrial facilities had little incentive to monitor discharges or seriously study the effects of those discharges. When engineers at a facility did undertake such research, firms had little incentive to make the results public unless that information proved favorable. What important changes relative to measuring, monitoring, and characterizing effects have since taken place? What role has the industry played in encouraging or resisting the production of environmental knowledge? A shift in the control over the production of such knowledge is an important part of the transition that took place as industrial self-regulation gave way to a new regulatory regime.

What follows, then, is a story that captures the initial validation of an efficiency-minded pollution control ethic, the temporary success and eventual collapse of that ethic, and the transition to a new ethic rooted in achieving socially determined environmental objectives. To highlight these changes, this book is divided into three parts, each focusing on a different period and highlighting the petroleum-related pollution concerns that emerge in that period. Each part also examines the petroleum industry's response, technological or otherwise, to those concerns.

Specifically, the first five chapters focus on what engineers in the early twentieth century considered to be state-of-the-art technology in each sector of the petroleum industry and indicates how the technological changes they championed either added to or addressed pollution concerns. In general, during this period, incentives to use resources more efficiently overlapped with efforts to reduce pollution-causing discharges. Hence, industry made a case for self-regulation based on efforts to become economically and technologically more efficient. In the 1920s, federal legislators tacitly accepted as sufficient this strategy for reducing industrial pollution. This acceptance of industrial self-regulation in the area of pollution control established a pattern that lasted for several decades.

Next, the focus shifts to the period between 1925 and 1955. In this period, engineers and technical managers in the petroleum industry validated the notion of self-regulation by reducing the quantity of pollution-causing discharges released for every barrel of oil produced. Quite simply, in the early part of this period, their efforts to increase the efficiency with which firms extracted, transported, and refined petroleum also reduced the quantity of leaks, spills, emissions, and effluents being discharged. However, by the 1950s, efforts to improve the efficiency with which firms transformed crude oil into inexpensive gasoline no longer overlapped with efforts to reduce pollution-causing discharges. Arguments rooted in an efficiency ethic had lost their potency.

In the final set of chapters, I examine the transition to government-enforced regulations designed to accomplish specific environmental objectives. As in the first two sections, many of the petroleum-related incidents, debates, and conflicts examined—such as those triggered by the smog problem in Los Angeles, the wreck of the *Torrey Canyon*, the blowout at Santa Barbara, the Trans-Alaska pipeline, debates over the sulfur content of fuel oil, and problems with injection wells—have significance that go beyond the petroleum industry.[11] In the end, a regulatory system emerges that encourages firms to incorporate environmental performance into what it means to operate an efficient industrial operation.

Early Pollution Concerns and the Direction of Technological Change

ONE   # Pollution Concerns Articulated

AS THE NINETEENTH CENTURY turned toward the twentieth, the long-term future of the oil industry looked bleak. In the United States, electric lights had become a fixture in most urban areas, leaving older methods of illumination in the shadows. Hence, the oil industry's main product—kerosene for use in lamps—no longer shined as brightly as it once did. Even the future of transportation appeared to depend on electricity. What else could replace the network of trolleys that animated urban areas? True, electric generating plants required fuel, but oil burned as fuel could never be priced much higher than coal, and this fact dampened its potential as a valuable product.

The company that had long dominated the kerosene-based petroleum industry, Standard Oil, was also coming under attack. For almost a quarter century, John D. Rockefeller had used the power of Standard Oil to crush competitors. In 1906, journalist Ida Tarbell focused the nation's attention on these practices, and the resulting glare of publicity triggered a federal investigation. Five years later, after a long and exhausting trial, federal courts gave Standard Oil six months to divide its assets in such a way that no company or committee of stockholders could coordinate the actions of all the various businesses.[1] As a result, Standard Oil divided into companies such as Standard Oil of New Jersey (later Exxon), Standard Oil of California (later Chevron), Standard Oil of New York (later Mobil), and Standard Oil of Indiana (later Amoco). The industry was by no means dead. Oil consumption was still rising. However, the industry's glory days appeared to lie in the past, not the future.

By the end of World War I, the picture had changed dramatically. The sounds and smells of automobiles—pinging engines, poorly muffled exhausts, and gasoline vapors—emanated from towns across the United States. Entrepreneurs in search of new oil fields fanned out across the nation, and companies constructed refineries larger and more sophisticated than ever before. Companies also expanded the system of pipelines, tankers, and storage facilities necessary to transport petroleum from one place to another. In the end, the gasoline-based petroleum industry that emerged was bigger and more powerful than the kerosene-based industry had ever been.

By any measure, this system for extracting, transporting, and refining petroleum was a technological marvel. Wells scattered throughout the nation continuously extracted flammable fluid from formations deep underground. Pipelines and tankers then transported this fluid great distances to refineries, where heat and pressure transformed it into an array of salable fuels and chemical products. However, the network of wells, pipes, tanks, tankers, process vessels, and valves through which petroleum moved allowed significant quantities of oil and other pollution-causing discharges to escape.

In the years after World War I, these discharges of oil and oily wastes became a national concern. By 1920, slicks of oily scum covered the surface of most crowded harbors, soaking into wooden docks and mixing with debris trapped by pilings and other structures (fig. 1.1). Residents of coastal towns, especially those near major ports, also saw oily wastes and tarry residue foul their beaches, routinely preventing people from enjoying the ocean. Furthermore, people realized that the problem was getting worse, and numerous voices articulated concern.

Strictly speaking, oil pollution was nothing new. The problem had been around for a half century, with oily wastes often turning pockets of

Figure 1.1. In waterways such as the Schuylkill River, shown in this 1917 view of docks operated by the Atlantic Refining Company's refinery, pollution due to oily discharges grew worse in the period after World War I. (Library of Congress LC-US-Z62-226A)

land and water into ecological wastelands.[2] Residents in early refining centers certainly were no strangers to oily scum and oil slicks. Even G. S. Davison, president of Gulf Refining in the 1920s, recalled the days when "as a boy swimming in the Allegheny River" his skin quickly got "coated with the floating tar seeping or discharged from a neighborhood petroleum refinery."[3] Although most refineries had reduced their waste discharges in the intervening years, the amount of petroleum being produced and consumed in the United States had grown by an order of magnitude. Dramatic increases in the manufacturing of gasoline, fuel oil, lubricating oils, and asphalt resulted in domestic oil production rising from about 60 million barrels per year in 1900 to approximately 600 million barrels per year in the early 1920s. More and more of this petroleum, in one form or another, eventually ended up in rivers and harbors.

People concerned about fisheries certainly expressed alarm. In a letter to the U.S. House Committee on Rivers and Harbors, an official with the

Texas Game, Fish, and Oyster Commission, J. G. Burr, noted that he had seen "veritable streams of waste oil going into the bays from loading stations." He attributed a 90 percent decline in oyster production around Galveston to this oil and urged strong action against oil polluters.[4] Researchers with the U.S. Bureau of Fisheries expressed similar concerns. In a report titled "Danger To Fisheries From Oil and Tar Pollution," bureau researcher J. S. Gutsell succinctly summarized how waste oil affected fisheries. According to Gutsell, some of the oil floated on the surface, which not only killed eggs and delicate larvae but also interfered with the process by which water absorbed oxygen. Some oil also emulsified in the water column or sank to the bottom. This oil, he explained, killed the organisms on which fish and clams depended for food.[5]

Sportsmen belonging to organizations such as the American Game Association also expressed concern. For example, in 1921, after one oil slick left thousands of ducks dead and dying in Rhode Island's Narragansett Bay, the Game Association issued a bulletin describing how oily wastes affected waterfowl. According to the author of that bulletin, any bird that landed on oil-polluted water soon died due to its feathers becoming matted and ineffective. He also reported that oil leaking into "ducking" waters from a pipeline in Montana had killed numerous birds. Finally, the author described ducks he had personally seen in California and Kansas that were unable to fly due to contact with small amounts of oily wastes.[6]

Members of the Waterway League of America also registered complaints, with their main concern being the way in which oil slicks in urban harbors fouled their boats.[7] People who lived in resort areas began complaining about oil washing onto their beaches, leaving a tarry film on everything it touched. Some of this material had the consistency of sticky chewing gum, which bathers could remove only by scrubbing with gasoline. Members of the Coney Island Board of Trade complained that this residue hurt their business by keeping people away from the beaches.[8] Resort towns all along the Jersey coast faced similar problems, as did those in the vicinity of Boston, Baltimore, Norfolk, Miami, and Galveston.[9]

Insurance companies also expressed concern. In 1920, a fire in lumber piles along a Baltimore pier suddenly intensified when the oil-soaked pier ignited. Oil floating in the harbor caught fire, and the flames spread rapidly to an adjacent pier, destroying twelve ships. Following that incident, the National Board of Fire Underwriters began investigating the condition of harbors throughout the United States.[10] Reports from most cities indicated that oily scum and slicks were commonplace. Another fire, this one in New Orleans, reinforced the assessment. There, a conflagration that destroyed two wharves and several large ships began after a workman dropped a red-hot rivet into the water. The rivet ignited a pool of fuel oil trapped by debris, and the resulting inferno quickly consumed the wharves and destroyed the ships.

Where was the oil coming from? The New York Department of Health attempted to answer this question for the area around New York City. Inspectors familiar with disposal practices knew that many industrial plants discharged oily effluents and tarry residues into public bodies of water, with oil refineries and plants manufacturing gas from coal being the most obvious sources. They also knew that employees at service garages, as well as many individuals, poured used engine oil into sewers and drains. Storm water could also wash oil from roads into streams or storm sewers that eventually emptied into harbors.

In order to pin down the source of oily wastes, investigators from the Health Department visited sixty-seven industrial plants. After summarizing the effluents being discharged by each plant, they concluded that those plants could not have been responsible for the most serious oil slicks.[11] Investigators then concentrated on sources in New Jersey. Again, after a visit to each plant and an inspection of the harbor by motorboat, they concluded that municipal sewers and industrial outlets did not discharge enough oil to account for the slicks that were plainly evident. They also concluded that accidents—such as storage tanks overflowing, underwater pipelines being ripped out by anchors, or ships losing fuel oil in collisions—routinely occurred but did not represent the main source of oil.

Investigators finally traced the largest quantity of oily waste to a fleet of tankers delivering heavy fuel oil to distribution facilities. After unloading their cargo, these ships pumped water into their tanks for ballast. Then, after returning to a refinery for more cargo, crews pumped the ballast water overboard, along with any oily residue mixed in with the water.[12] Given that the heavy fuel oil being delivered was little more than crude oil with the gasoline and kerosene skimmed off, a significant amount of oily residue stuck to the sides of tanks, and this residue eventually ended up in the harbor.

Later conversations with shipping companies also indicated that all oil-burning steamships contributed to the problem. Typically, steamships crossing the Atlantic purchased their fuel in U.S. ports, where oil was plentiful and inexpensive. On their return trip from Europe, these ships filled their empty fuel tanks with water for ballast. After safely reaching the Atlantic seaboard, they discharged their oily ballast water directly into the harbor. Tankers carrying crude oil from the Gulf of Mexico to the Atlantic seaboard also contributed to the problem, except that those ships loaded ballast water on the Atlantic coast and discharged the resulting oily mixture on the Gulf coast.

What was the response to this problem? In New York City, the supervisor of the port, Captain Roy Smith, began fining ships and facilities for discharging oil into the harbor.[13] Smith's authority to do so came from an 1888 federal law that applied only to New York Harbor. Unfortunately, by placing pressure on ships, Smith simply shifted the problem from one location to another. Steamship captains concerned about being fined generally responded by flushing their tanks before entering the harbor. Hence, more residue ended up washing onto bathing beaches. Furthermore, given that a single motorboat with a top speed of six miles per hour served as Smith's main enforcement tool, violators could easily avoid detection.[14]

Authorities in most other port cities had even less authority and resources than Captain Smith. A federal law passed in 1899, commonly known as the Rivers and Harbors Act, did prohibit the disposal of refuse in navigable waters, but the Army Corps of Engineers, the enforcing body, applied that law only to refuse that impeded navigation. Furthermore, the Rivers and Harbors Act did not explicitly prohibit oily discharges or other effluents.[15] Hence, when port supervisors in other cities attempted to take action, the operators of industrial facilities often challenged their authority.

For example, in 1920, when Colonel Benjamin Allin, director of the Port of Houston, took action against the Galena Signal Oil Company for discharging oily effluent into the Houston ship channel, W. W. Moore, Galena's attorney, immediately called for Allin's dismissal. Allin exhibited, in Moore's words, a "persistent disposition to unnecessarily harass the management of various industries along the ship channel and to interfere with and hamper their legitimate activities."[16] In this case, the mayor of Houston and Houston's city council both supported Allin's efforts. The mayor, in his response to Moore, noted that "I have personally inspected the outlets from the refineries into the Channel and have seen with my own eyes large quantities of oil flowing into the Channel. By this, I do not mean the milky discharge, but black oil in quantities sufficient to cover the water to the side of the outlet for a considerable distance."[17] The mayor's support may have helped Allin keep his job as port director, but Allin backed off in his attempts to prevent such discharges.[18]

Although the severity and source of the pollution problem varied from port to port, authorities everywhere took interest. However, the oily discharges continued. For example, on May 23, 1921, crews repairing a pipeline that carried crude oil under the Hudson River accidentally released a significant slug of the volatile fluid into the water. Given the familiar sight of an oil sheen on the river, the crew apparently felt no need to clean up their mess. Burning cinders happened to ignite the oil, and the resulting fire quickly spread. Among other things, the fire consumed an historic, wooden New York Naval ship, the *Granite State,* and destroyed the dock where a yacht carrying President Warren Harding had landed earlier in the day. The yacht itself, anchored several hundred feet away from the dock at the time of the fire, escaped damage.[19]

Had the oil not caught on fire, nobody would have given the pipeline leak much attention. Had the fire not destroyed the *Granite State* and threatened the president's yacht, the fire probably would not have made front-page news. As it happened, the fire did make the front page and became one more piece of evidence supporting those calling for some action. Indeed, one month after that incident, New Jersey congressman William Appleby introduced a bill to prohibit the discharge of any kind of oily waste into the nation's streams and harbors. Given the failure of a similar bill introduced the previous year, Appleby took pains to publicize and line up considerable support for his proposal.[20] The resulting attention triggered a flurry of related bills and a series of congressional hearings on water pollution.

Congressional Hearings on Oil Pollution

In the United States, no serious debates over industrial pollution at the national level had occurred before, and the bills introduced to address the oil problem reflected this inexperience.[21] Participants, all of whom recognized oil pollution as a concern, could not imagine the tangle of issues that would soon emerge. Any hope that one of the proposed bills would prove acceptable soon evaporated.

The first wave of bills differed mainly in how they specified the types of discharges to be prohibited and in how they justified the prohibition. For example, Representative Clay Briggs of Texas proposed simply extending the phrase "any refuse matter" in the 1899 Rivers and Harbors Act to include "any oil, fuel oil, oil sludge, or any other oil refuse."[22] The existing law already empowered the secretary of war to take action against any establishment that discharged navigation-impeding refuse into public waters. Briggs saw prohibiting oily discharges on the grounds that it impeded navigation as sufficient. Appleby's bill, in contrast, did not tie the prohibition on oily discharges to navigation. Another bill, introduced by Representative Benjamin Rosenbloom of West Virginia, specified discharges of "free acid, acid waste or acid-forming material" as also being unlawful.[23] In the way of penalties, all bills kept, with some variations, the same general language as the 1899 law. That law allowed the secretary of war to impose a fine between $500 and $2,500 and to imprison responsible individuals for up to a year.

When the House Committee on Rivers and Harbors held its first series of hearings on the proposed bills, participants raised a wide range of issues, many of which had not been anticipated by Appleby and the authors of the other bills. Legislators soon discovered that addressing concerns associated with industrial pollution required more than simply passing a law that prohibited discharges into public waters. Some discussion of technology, goals, and philosophical approach was also needed if they were to create a practical piece of legislation.

Participants at the first hearing agreed that a problem existed and had to be addressed. Representative Appleby opened by introducing David Neuberger, a prominent citizen of New York, who read a statement signed by members of the Interstate Committee on the Prevention of Pollution of

Coast Waters and Beaches. This committee, specifically created to lobby for pollution control legislation, consisted of approximately two hundred civic leaders from the New York and New Jersey area. Their main interest lay in preserving the attractiveness of beaches in shore towns.

Neuberger's statement of support emphasized respecting certain standards of environmental quality. Although nobody at the time would have used the phrase "environmental quality," his intention was clear. He asserted that industry should simply respect the rights and interests of others and advocated tracing each line of pollution to its source and eliminating it by "legal remedies rigidly applied."[24] Similarly, Dr. James Marshall, representing the Ocean Grove Camp Association, noted that people from all over the nation visited the shore for recreation and invigoration, and that the oily scum threatened to cut them off from the recuperative powers of the sea. Marshall pleaded for Congress to keep the Atlantic coast pure and safe for the people who visited there.[25]

The discussion moved to more practical issues of implementation when two representatives of the American Petroleum Institute (API), a lobbying organization recently established to represent the interests of the petroleum industry, presented statements. Van H. Manning, the API's director of research and a former director of the U.S. Bureau of Mines, argued that oil should not be singled out as a special type of effluent. After all, other forms of industrial waste caused more harm to fisheries than the oil discharged by refineries, an assertion he backed up with data from a Bureau of Fisheries report. The other API spokesperson—J. H. Hayes, an attorney for Standard Oil of New Jersey—raised the issue of intention. What if a plant discharged oil unintentionally? What if a pump leaked at a loading station and a few barrels of oil ended up in the water? Should someone go to prison for that?[26]

Other speakers also raised similar concerns about the details of the various bills. Samuel Taylor of Pittsburgh, noting that large quantities of acidic water drained out of coal mines in western Pennsylvania, made a case for exempting those discharges from any legislation.[27] Similarly, a representative for the Manufacturers Association of Connecticut observed that one of the bills, as written, would allow an industrial plant connected to a municipal sewer to discharge pollution-causing wastes while a plant not connected would be prohibited from doing so. Was that, he asked, the intent?[28] Another speaker pointed out that Appleby's bill, as written, treated one drop of oil spilled from a motorboat in the same way as thousands of gallons of oily ballast discharged from a tanker. Should not the bill define pollution more precisely?[29] A spokesperson for the Manufacturing Chemists Association wondered why the various bills specified the War Department as the enforcing agency. Would it not make more sense to specify the Commerce Department?[30]

Others argued about the accuracy of various statements. For example, early in the hearing, a spokesperson for the Waterway League of America testified that oyster beds around New York City, once extremely productive, had been destroyed by oil pollution.[31] Another speaker, William Gibbs, chair of the Technical Committee of the American Steamship Association, challenged that assertion. Gibbs agreed that oil-burning steamships did cause problems but noted that most of the oyster beds in and around the city were destroyed by metropolitan sewage, not oily wastes.[32] Representative Appleby also submitted a report by a biologist, Thurlow Nelson of Rutgers, that identified dyestuffs and chemical wastes as being responsible for major damage to oyster beds.[33]

Despite concerns associated with the specifics of a bill or the accuracy of various assertions, most speakers still expressed support for taking action to reduce the pollution problem. By the end of the first hearing, dozens of statements, letters, and reports had been entered into the record, all explicitly supporting the goal of keeping rivers and harbors free of contaminants. Indeed, at this stage in the debate, no one challenged the general notion that oily discharges should be prohibited. That level of agreement would soon change.

As one might expect, many supporters of strong action justified their support not only in terms of environmental quality but also in economic terms. For example, speakers representing fire insurance companies framed their concerns in terms of economics. So did speakers representing various chambers of commerce and speakers representing the fish and oyster industry. Even Neuberger, the spokesperson for the Interstate Committee on the Prevention of Pollution, justified his support in economic terms. He noted that the assessed value of real estate in coastal towns had increased dramatically over the last generation, with the assessments in Deal, New Jersey, jumping from $60 thousand to $8 million. The oil nuisance, he suggested, threatened the value of that property.[34]

Justifying pollution control in terms of economics opened a rhetorical door that supporters of strong legislation would find difficult to close. Perhaps it is more accurate to say that this door was already open, and that people in favor of strong action initially saw no reason to close it. After all, justifying one's position in economic terms made perfect sense. For Neuberger and his colleagues, the strategy certainly proved successful in attracting the support of civic leaders from all over the country. This support eventually allowed the group to expand their geographic base, giving rise to the League of Atlantic Coast Seaboard Municipalities Against Oil Pollution and, later, the National Coast Anti Pollution League.[35] Although these groups remained elite organizations without broad participation, they had influence under the association-based model of policy making championed by the influential secretary of commerce, Herbert Hoover.[36] And Hoover, in his role as commerce secretary, found himself at the center of the oil pollution debate.

As head of the Department of Commerce, Hoover, who had been a successful mining engineer before becoming a public figure, had responsibility over several agencies and bureaus related to commercial activity, including the U.S. Steamship Inspection Service, the Bureau of Navigation, the Bureau of Standards, and the U.S. Bureau of Fisheries. The antipollution organizations rightfully looked upon Hoover as an ally. He previously had spoken out forcefully against the damage being done to fisheries by pollution-causing discharges. However, when the Rivers and Harbors Committee requested Hoover to make a statement, which he did in early 1922, he rooted his comments in the principle of utilitarian conservation—the notion that society should optimize the value extracted from a set of resources so as to do the most good for the most people. In pursuing this line of thought, he demonstrated that arguments based on the economics of pollution control were a double-edged sword.

In his statement to the committee, Hoover noted that by passing too strong a pollution bill "we might be doing in some localities infinitely more damage to industry and to commerce, imposing infinitely more cost upon municipalities than the value of any fishery that could be established would warrant." No one, Hoover asserted, could justify the costs associated with reestablishing productive fisheries in the New York Harbor. In other areas of the country, he acknowledged, the cost of pollution control would clearly be worth the effort. When asked if he meant that the benefits of pollution control should be weighed against its costs, Hoover answered in the affirmative.[37]

Nobody challenged Hoover's way of thinking or asked from whose perspective these costs and benefits were being computed. Undoubtedly, everybody assumed that one could determine the pattern of use that resulted in the greatest benefit to society. An assumption also seemed to be that this optimal pattern of use would result in an acceptable level of environmental quality. This line of thought shifted the direction of debate. Speakers at the earlier hearings seemed to assume that everybody intuitively understood that certain measures of environmental quality were worth sustaining and that Congress should pass laws to accomplish those objectives. The economic argument in favor of strong pollution control legislation was simply icing on the cake. Hoover, though, explicitly moved the debate in the direction of costs and benefits, away from an ethic rooted in the notion of environmental quality and toward an ethic rooted in the efficient use of resources.

Speakers concerned about the future of industry moved readily in the direction of costs and benefits. A representative from Massachusetts pointed out that the Rivers and Harbors Act of 1899 applied to all navigable waters and that, according to past interpretations by the court, any stream on which one could float a log for the purpose of moving timber qualified as a navigable body of water. Hence, technically, the proposed amendment would prohibit mills in New England from discharging waste liquids into streams and canals that ran by their facilities. This prohibition worried him because, if enforced, it meant the certain closure of these plants.[38] A representative from Connecticut expressed the same concern, noting that although he cared about pure water and the fate of the oyster industry, he feared that Connecticut, which was "a hive of industry," would be put out of business.[39]

Advocates of strong pollution control legislation also proved willing to justify their position using economics-based arguments. Representative Rosenbloom of West Virginia, who introduced the bill prohibiting the discharge of acids, certainly appeared confident in his ability to argue in those terms. He began by enumerating the great benefits of acid-free water. Although he could not quantify all benefits, such as the value of people being able to catch fish from a river, he translated others—such as having to spend less money maintaining locks and dams damaged by corrosion—into dollars and cents.[40] Then, after making the observation that everybody in the room would support his bill "if the cost of it would not be greater than the benefit derived," he introduced a witness to detail the costs.[41]

Unfortunately for Rosenbloom, his opponents were also prepared to argue in terms of costs and benefits. A spokesperson from the Pennsylvania Department of Labor and Industry asserted that companies would end up spending $5 million on lime just to save the $57 thousand that the government spent maintaining locks and dams.[42] The vice president of a coal company even pointed to the benefits of not taking action. Citing an article published by the U.S. Geological Survey, he claimed that acid mine drainage actually improved water quality by killing bacteria associated with raw sewage.[43] Another member of the House argued that lawmakers in West Virginia had explicitly eliminated laws protecting fish because they already decided that "coal properties and sawmills and pulp mills and other industries of like kind and character were of more value than a few fish."[44] Finally, witnesses representing mining and manufacturing companies successfully challenged the expertise of Rosenbloom's witness. They did so by demonstrating that this witness, an inspector who investigated stream pollution in Ohio, had little knowledge of basic chemistry.[45] Much of the rest of the two-day hearing continued in this vein.

Three months later, at another hearing on oil pollution, this one held by the House Committee on Foreign Affairs, the logic of utilitarian conservation again rose to the surface. At that hearing, the deputy commissioner of the Bureau of Fisheries, Dr. H. Moore, reinforced Hoover's general focus on costs and benefits. After explaining the effect of oil on fisheries in some detail, Moore was asked whether he thought all pollution-causing discharges should simply be prohibited. Moore responded by saying that prohibitive action only made sense "if it could done without working an injury greater than that which is currently wrought by the pollution."[46]

Applying the logic of utilitarian conservation to pollution concerns also shifted attention away from enforcement mechanisms. If the debate had proceeded in its original direction, participants would have had to spend more time reaching consensus on specific environmental objectives and on enforcement-related issues. After all, it was naive to think that a complete prohibition on discharges could be effectively enforced without the law being more specific. Arriving at consensus on procedures, thresholds, and enforcement tools, however, promised to be a long and torturous ordeal.

Those who framed the problem in terms of optimizing the use of resources offered an alternative approach. Rather than passing what they referred to as "prohibitive" legislation, they encouraged Congress to take "constructive" steps.[47] By constructive steps, they meant giving firms the time to make industrial operations more efficient and the time to develop practical disposal methods. If Congress moved in this direction, they suggested, efforts to fight pollution would be far more cost effective and might even pay for themselves in the long term.

The best example of this hope that pollution control efforts could pay for themselves emerged in discussions of how to prevent oily discharges from ships. When speaking to the House Committee on Foreign Affairs, Commerce Secretary Hoover expressed confidence that some method of collecting the oily waste could be put in place, making the entire effort pay for itself through the recovery of the waste oil.[48] Hence, enforcement would not be a significant issue as shipowners would take action on their own. Later in the day, Eugene Moran, the head of a pollution prevention committee established by the supervisor of the New York Harbor, noted that New Jersey Standard (Standard Oil of New Jersey, predecessor to Exxon) had been experimenting with a shipboard device capable of separating oil from water. According to Moran, this device would pay for itself by allowing ships to reclaim their waste oil.[49] On the following day, a spokesperson for the Sharpless Manufacturing Company appeared before the committee and indicated that his company already manufactured oil-water separators. He could see no reason why ships could not install similar devices.[50]

The members of Congress embraced the hope that this technology would solve the problem of ships discharging oily ballast. Indeed, they passed a resolution calling for the president to convene an international maritime conference to determine an effective way to use such technology in preventing the pollution of navigable waters.[51] The presidential committee charged with organizing the conference wisely decided to prepare by assembling as much data as they could about oily wastes and how to prevent those wastes from causing pollution. The U.S. Bureau of Mines performed the actual investigation with the help of the American Petroleum Institute and the American Steamship Association.[52] While these organizations collected data, Congress placed its consideration of pollution control legislation on hold.

Approximately a year later in September 1923, the Bureau of Mines released a preliminary report in which it documented numerous sources of oil pollution and methods of prevention. The authors of the report concluded that discharges from ships represented the most serious problem, but that no shipboard oil-water separator capable of reclaiming all oily residue existed. Eliminating oily discharges from ships would also require a system of barges to collect some wastes and facilities on shore to reclaim or dispose of those wastes. Other sources of oily discharges were also detailed and discussed.[53]

Legislators quickly introduced more pollution control bills.[54] These bills ranged from limited proposals to prohibit oily discharges only from oceangoing ships to strong legislation that would prohibit all types of pollution-causing discharges from industrial plants throughout the United States. In the hearings that followed, advocates of strong legislation found themselves on the defensive. Numerous points of contention emerged, with representatives of various trades groups steadfastly defending their interests.

At these hearings, David Neuberger, then the president of the National Coast Anti Pollution League, finally challenged the logic of costs and benefits. He noted that when health officials in New York put pressure on the Vacuum Oil Company to reduce its oily discharges into the Genesee River, the company responded and the river returned to its "pristine" condition. "What they spent," Neuberger said, "I do not know. That is no consideration when it is a matter of public health and safety." Even here, though, he felt compelled to justify pollution control in terms of health and safety. In any event, his effort to reframe the problem was too little too late.[55]

In the end, legislators put great weight on Secretary Hoover's recommendation that Congress first address the specific problem of ships discharging oily effluent. Hoover reasoned that many chemicals threatened the health of fisheries but that oil pollution represented the most serious problem. In addition, based on the Bureau of Mines investigation, Hoover estimated that between 75 and 90 percent of all oily effluents came from ships. All other sources of oily discharges, including those from refineries, appeared to be less of a concern. Hence, according to Hoover, Congress could take a significant step forward by addressing the specific issue of oily discharges from ships.[56]

The strategy Hoover recommended served as the basis of the Oil Pollution Act of 1924, which simply prohibited ships from discharging any oily water within three miles from shore. Two years later, the United States did convene a preliminary international conference, but that conference did not lead to any treaty. It did, however, result in an informal agreement that all oceangoing tankers would flush their tanks far—fifty miles or more—from shore. As a result, the frequency with which oily residue washed onto beaches declined, and conditions in harbors improved enough to alleviate the most serious concerns, especially fire-related concerns. The push for strong federal pollution-control legislation, which turned contentious in the weeks before passage of the final legislation, quickly lost momentum.

Organizational Actors in the Pollution Control Debate

Two organizations that played an important role in the hearings leading up to the Oil Pollution Act of 1924, the U.S. Bureau of Mines and the American Petroleum Institute, are key to understanding the development of pollution control practices in the petroleum industry. More than any others, these organizations shaped the petroleum industry's response to pollution concerns. In these efforts, the Bureau of Mines served as a guardian of the same ethic in which Herbert Hoover rooted his fight against waste and inefficiency. The American Petroleum Institute, on the other hand, stood as the guardian of an industry.

In 1922, when Congress called for a systematic investigation into oil pollution, the U.S. Bureau of Mines emerged as the governmental agency responsible for performing the study. This choice was not simply a case of using available personnel. In terms of its mission, the Bureau of Mines was the ideal candidate. This agency, more than any other, had translated the Progressive Era ideal of resource conservation into an ethic that could be applied to industrial practices.

Congress originally established the Bureau of Mines in 1910 to conduct inquiries and investigations that would lead to safer and more efficient methods of mining, with unsafe conditions in coal mines being the prime target.[57] But accidents, and the waste of resources associated with accidents, represented only one kind of waste that could be studied and eliminated. One could—and the first director of the bureau, Joseph A. Holmes, did—frame the bureau's mission even broader. He pushed for and succeeded in getting Congress to extend the bureau's mission to conducting "inquiries and scientific and technological investigations concerning mining, and the preparation, treatment, and utilization of mineral substances, with a view to improving health conditions, and increasing the safety, efficiency, economic development, and conserving resources through the prevention of waste in the mining, quarrying, metallurgical, and other mineral industries."[58] The oil and gas industry was one of those "other mineral industries."

How did pollution control fit into this mission? To the extent that pollution affected the health and safety of workers, the link was clear. However, an indirect link also existed. Pollution-causing discharges represented a form of economic waste that conservation-minded leaders hoped to eliminate. Indeed, when Herbert Hoover became the nation's secretary of commerce in 1921 and focused attention on his campaign to eliminate wasteful practices, the editor of *Chemical and Metallurgical Engineering* devoted an entire issue to the subject of "waste in industry." In that issue, the editor credited Hoover with clearly seeing "the loss to the human race in goods and services due to waste and inefficiency" and included articles on the waste of human resources due to accidents, waste associated with lawsuits, waste due to the lack of standardization, and wastes associated with poor marketing.[59] Pollution was just one more symptom of waste and inefficiency.

Pollution, from this point of view, was something akin to waste heat in a poorly designed industrial process. This way of thinking suggested that improving the economic efficiency of an industrial operation would also decrease pollution-causing discharges associated with that operation. In the long term, then, at least according to this logic, engineers could best serve the public good by continually striving to increase the economic efficiency of an industrial operation. The Bureau of Mines translated this philosophy into practice.

In the petroleum industry, the Bureau of Mines executed its mission mainly through the work of several field stations, with the Petroleum Experiment Station in Bartlesville, Oklahoma, being one of the more important. This station, established in 1919, served as the home base for engineers and scientists hired to identify wasteful practices that prevented oil companies from extracting, transporting, and using the nation's supply of petroleum in the most efficient manner possible. In general, the bureau encouraged its engineers to first identify wasteful practices and then research ways in which those practices could be changed. Finally, through publications, presentations, and demonstrations, they were to disseminate their findings.[60]

Although investigating sources of pollution-causing oily discharges and assessing the technology available to prevent those discharges was not the bureau's main goal, it certainly fit the agency's mission. Hence, when Congress expressed interest in a study of oil pollution, the Bureau of Mines stepped forward. The agency also had an indirect link with the newly formed American Petroleum Institute: Van H. Manning, a former director of the bureau, had become that organization's director of research.

Formally chartered as a private organization in 1919, the American Petroleum Institute grew out of a committee that the federal government established in the years leading up to World War I. After the United States entered the war, this organization, then called the National Petroleum War Services Committee, worked closely, though not always without conflict, with the U.S. Bureau of Mines and the Federal Trade Commission to coordinate oil production. Rather than disband after the war, the members—all presidents and vice presidents of major oil companies—agreed to reconstitute as the American Petroleum Institute.

When government leaders such as Bernard Baruch first mobilized the nation's industries for war, the petroleum industry had no trade organization capable of representing the general interests of the industry.[61] The lack of any such organization, though, was easily explained. Before 1911, Standard Oil, which dominated the industry, coordinated its own affairs. At best, independent trade organizations filled smaller niches within the industry. Then, in the years immediately following Standard Oil's dissolution, executives of the various oil companies spawned from that dissolution, under court order not to coordinate their activities, hesitated to form trade organizations. Furthermore, the larger independent companies such as Sun Oil, Gulf Oil, and the Texas Company (later Texaco) still looked upon any Standard-related company with suspicion. Before the war, many independents would not think of sitting at the same table as somebody within the Standard fold.

The experience of meeting regularly during the war allayed many of the concerns and suspicions that various individuals held. By the time the war ended, most members on the war-inspired committee recognized the usefulness of an organization that could speak for the entire industry. Hence, they agreed to keep meeting as a trade organization, with their stated mission being to represent the petroleum industry in government-related matters, to promote trade in petroleum products, and to promote scientific and technological advances of mutual interest to its members.[62]

From the start, the directors of the American Petroleum Institute agreed that the new organization should not undertake anything that individual companies could do for themselves. Only projects that demanded an unusual degree of cooperation—such as lobbying, public relations, collecting statistics, and standardization—would be pursued. Furthermore, the directors of the API knew they had to demonstrate that the organization was not merely a reincarnation of Standard Oil. Given that Jersey Standard's chairman of the board, A. C. Bedford, had led the wartime committee and strongly supported the creation of a trade organization, such suspicions came easily. For this reason, Bedford purposely encouraged independents to take a larger role in the organization than they might have otherwise.[63]

To establish a research agenda, Bedford consulted with Van H. Manning, a former director of the Bureau of Mines. Manning suggested that researchers be funded to "point out the manufacturing losses, and indicate the necessary investigative work whereby those losses will possibly be reduced or eliminated." According to Manning, "those results should not be stored in the records of one company, but should be available to other manufacturers" so that "the greatest good can be accomplished and duplication minimized."[64]

But where did one drew the line between research of mutual interest and research that could be done by individual companies seeking a competitive advantage? Tensions emerged as the small staff hired to run the new institute struggled to determine where their efforts could be best spent. In the end, the API's main effort went into collecting statistics of common interest. Manning, hired as research director, received some funds to organize a program of fundamental research on the origin of oil, the composition of reservoir rocks, and the characteristics of anticlinal structures, but research projects related to the extraction, production, and refining of oil were scrapped.[65]

When the situation demanded, the directors of the API proved willing to call upon the vast pool of technical expertise available to them through their individual companies. For example, when oil companies operating railroad tank cars sought changes to government-enforced safety rules that they considered too strict, the API coordinated an effort to challenge those rules. Not only did industry engineers succeed in convincing the American Railway Association, the organization that maintained the rules, to make the desired changes, but the institute also expedited the implementation of those changes.[66]

The API proved to be somewhat slower in collecting data and presenting recommendations for changes its members did not support. In 1921, in the wake of a serious explosion involving a tank car carrying gasoline with a high vapor pressure, the Interstate Commerce Commission and the Bureau of Explosives proposed a significant change. They desired to reduce the maximum allowable vapor pressure of gasoline shipped in uninsulated tank cars from ten pounds per square inch to eight pounds per square inch. Such a change would have required the industry either to upgrade a number of tank cars or to ship only the less volatile gasoline.[67]

In this case, the API opposed the proposed change on the grounds that the device used by the Interstate Commerce Committee to measure vapor pressure was imprecise. Their concerns triggered an investigation by the Bureau of Mines, which eventually supported the API's position. Both organizations agreed that R. P. Anderson, an industry chemist, had developed a better measuring device than the one being used by government inspectors. The chief inspector for the Bureau of Explosives then proposed adopting the Anderson apparatus, but the API requested that it first be allowed to evaluate thoroughly the new device.

Two years later, the API's evaluation committee reported back. The committee recommended that a commission be formed to compare the new and old instruments. They desired to measure the vapor pressure of various gasolines throughout the country to determine both the differences between the two instruments and whether changes to existing regulations were necessary. In total, three years passed with no action taken on the proposed tank car modification. Although the inspectors with the Bureau of Explosives recognized the need for precise and accurate instruments, they also expressed concern that the API was merely delaying unwanted changes.

Hence, the API's access to expertise gave it important leverage in committees formed to perform technical studies. As long as the API could frame an issue as a technical problem—that is, capable of being solved by optimizing the use of material, money, and labor—one could be sure that the interests of petroleum companies would be well represented. Not surprisingly then, in 1922, when the Bureau of Mines received its charge to study the sources of oil pollution and to assess the technology available to eliminate those sources, the API quickly offered its assistance. And given the nature of the project, the API's access to technical expertise proved attractive to the Bureau of Mines.

T W O   # Concerns in the Oil Fields

WHILE CONGRESS DEBATED over what to do about oil-related pollution, other government leaders expressed concern about another oil-related issue that emerged in the early 1920s. Too much oil was being wasted. A significant proportion of the petroleum in newly discovered oil fields never made it to market as refined products. Given that economic growth and national security seemed to depend on securing a steady supply of liquid fuel, the waste of this valuable resource proved disconcerting, especially since many national leaders feared that domestic oil fields would run dry within ten or fifteen years.[1]

Both concerns—oil pollution and oil conservation—were indirectly linked, especially in the minds of efficiency-minded conservationists. Both issues had to do with preventing the waste of a valuable resource. Indeed, immediately after the Oil Pollution Act of 1924 placed closure on the debates over pollution control, Commerce Secretary Herbert Hoover and several other cabinet members established a Federal Oil Conservation Board to investigate the waste of oil. By conservation, they did not mean using less oil. Rather, they emphasized using that resource more efficiently, with as little waste as possible.

To appreciate their concerns, one must examine conditions in a typical field of the early 1920s. The search for new oil deposits intensified in the years after World War I, and wildcatters—entrepreneurs seeking oil in new locations—scoured the land for promising places to drill.[2] And for every exploratory well drilled, a half dozen more were drilled into known oil-producing formations. In 1920 alone, over 30,000 new wells were drilled, almost double the average of 16,800 holes drilled per year over the previous ten years.[3] The resulting glut of oil and the wasteful practices associated with getting that oil to market only heightened fears that the nation would soon exhaust its supply of this valuable resource.

The Discovery and Development of an Oil Field

The speed with which oil producers responded to the postwar demand for oil surprised nobody. After all, a substantial infrastructure for extracting, transporting, and refining petroleum had been in place for over a half century. In 1859, Edwin Drake, financed by investors seeking a steady source of "rock oil" to process and sell as an illuminating oil, successfully drilled a commercial oil well in western Pennsylvania, placing that region at the center of oil production for several decades.[4] As the demand for kerosene increased, the search for oil and, eventually, the production of oil spread to the surrounding states of Ohio, West Virginia, and New York. Entrepreneurs in search of more oil also moved further west and, in the mid-1880s, discovered important new fields along the border of Indiana and Ohio.

By the 1890s, a single network of companies, Standard Oil and its affiliates, dominated the kerosene-based industry in the United States.[5] In addition to controlling a significant amount of production, Standard Oil also wielded great power in the transportation and refining of petroleum. Indeed, some remote fields, such as those discovered in Kansas, remained undeveloped due to Standard Oil's lack of interest. In the absence of a local market, producers in those new fields could do little with the oil they lifted.

Around the turn of the century, the industry began to change. Gas and electric lamps grew more popular, and the demand for kerosene started leveling off. Therefore, in 1901, when a flood of kerosene-poor oil poured out of a field called Spindletop near Beaumont, Texas, Standard Oil showed little interest. In this case, independents had easy access to the Gulf of Mexico, and much of the free-flowing oil ended up as fuel for steamships and industrial boilers, strengthening the market for petroleum as an energy source. The demand for gasoline generated by the internal combustion engine also represented a new market. As the number of automobiles on the road increased, discoveries of major new fields, such as Oklahoma's Glenn Pool (1905) and Cushing (1912) fields, kept a supply of inexpensive gasoline flowing.[6]

With the dissolution of Standard Oil in 1911, the companies created from that monopoly found themselves competing with independents. This competition, along with significant technological advances, kept the price of gasoline low. By 1920, the petroleum industry had been transformed into a gasoline-based industry, with five states—Kansas, Oklahoma, California, Texas, and Louisiana—accounting for 86 percent of the nation's oil production.[7] Wildcatters in search of new fields focused their attention on these and neighboring states.

On April 22, 1920, one wildcatter drilling in El Dorado, Arkansas, a small cotton town located just north of the Louisiana border, received an encouraging sign while searching for oil. In drilling an exploratory well, his crew tapped a gas-bearing formation 2,200 feet below the surface. A tremendous flow of natural gas—about 30 million cubic feet per day, enough to satisfy the daily heating needs of a small northern town— roared uncontrollably from the hole.[8] Two days later, the company that drilled the well, the Constantin Oil and Refining Company, attempted to close off the hole, but the gas continued to escape through underground channels, surfacing in places up to one-half mile away and creating huge craters in the process.[9] At one location, a young man lighting a cigarette accidentally ignited the gas and caused an explosion that killed him and several other people.[10]

Although the fiery, gas-spewing holes attracted a lot of attention, few long-time residents got too excited about their economic prospects. Outsiders searching for oil had drilled gassers before, but nothing had ever come of their discoveries. Besides, a pipeline from the Caddo field in Louisiana had already saturated the local market for natural gas, and no economical method for storing the excess gas existed.[11] Therefore, the residents of El Dorado saw the wild well as a hazardous diversion but little more.

Scouts for oil companies, though, did take notice.[12] They knew that natural gas was simply a component of petroleum that vaporized easily. Scientists had long known that the dark, viscous mixture extracted from oil fields contained molecules of different sizes, each having a backbone of carbon to which hydrogen atoms and a smaller number of impurities, such as sulfur and nitrogen atoms, were attached. They also knew that molecules containing one carbon (methane, CH_4), two carbons (ethane, C_2H_6), three carbons (propane, C_3H_8), and four carbons (butane, C_4H_{10}) all vaporized into gases at common pressures and temperatures and formed the bulk of the gas that blew out of wells. Indeed, those who financed the Constantin well had secured the advice of a geologist, and further analysis of the formation suggested that conditions for finding oil were favorable.[13] Although stories of drillers who could locate oil on intuition alone would continue for years to come, the real story lay in the increasingly important role that geologists, scientific theories, and technological tools played in the search for new oil fields.[14]

By the 1920s, everybody working in oil fields knew that petroleum occupied tiny holes in porous formations, usually sandstone, far below the surface. Although they referred to these deposits of petroleum as oil reservoirs or oil pools, they knew that these reservoirs were not large open cavities. Rather, the oil-saturated rock contained tiny crevices through which the oil could flow and drain into wells. The size and interconnectivity of these microscopic cavities and small crevices determined much about the field's characteristics. Geologists knew that if the pores and crevices holding the oil were small or, even worse, poorly interconnected, not much oil would flow into a well. At the other extreme, smaller reservoirs in formations with large, highly interconnected fissures could prove to be far more prolific, at least in the short term. Such free-flowing fields sometimes attracted far more investment than they deserved, deceiving many into thinking that they had tapped into the edge of a much larger pool.[15] The gas-spewing well in El Dorado, Arkansas, while promising, did not necessarily mean that a huge oil field lay underground.

Geologists also had some idea as to why the gas and oil ended up where it did. [16] They speculated that the oil and gas, being less dense than water, rose through water-saturated porous formations until reaching a surface that could not be penetrated (fig. 2.1). Over eons, large pockets of oil and gas collected at high spots over salt domes and other such structures, with any undissolved gas rising to the highest level possible. The surrounding water—ancient, stagnant, and high in dissolved salts—trapped the oil in place.

Because geological forces could fold various layers of rock back upon themselves and create irregularly shaped reservoirs difficult to visualize, oil wells did not always pass through pockets of gas, oil, and salt water in the expected order. One might hit only gas or only briny water without any show of oil. Multiple layers of brine, gas, and oil could also confuse matters, making the task of figuring out what was where all the more difficult.

As to the specific origins of petroleum, most chemists and geologists

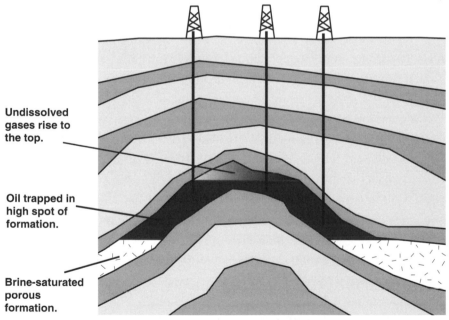

Figure 2.1. Petroleum deposits exist where oil and gas have risen to the top of a high spot in a brine-saturated porous formation. (Illustration by author)

believed the substance had organic origins, but precisely how the remains of tiny plants or animals were transformed into this soup of hydrocarbons remained a subject of debate.[17] Several leading geologists believed that the original organic material had settled, layer upon layer, in ancient bodies of shallow water and that, after a long period of decay and pressure, hydrocarbons had flowed, or were somehow washed, out of their original location. However, in the years after World War I, scientists were still presenting and disproving new hypotheses.[18]

In describing the petroleum industry for underwriters of fire insurance, one author dismissed such theories as unimportant and commented that "you may believe petroleum to have originated from organic or inorganic sources, as your fancy prefers, or believe both, if you are inclined against discrimination."[19] However, geologists interested in finding new petroleum reserves saw refinement of these theories as far more than an academic exercise. An accurate theory concerning the origins of petroleum could help determine where to drill.

An earlier theory, never well accepted, had been that petroleum came from the drippings of coal, which implied that one should look for oil near coal mines. Initially, this theory made sense. When entrepreneurs first became interested in the natural seeps of petroleum found west of the Alleghenies, they saw this dark mixture as a substitute for a popular illuminant known as coal oil, itself a recent replacement for whale oil and turpentine-based illuminants. After a prominent chemist confirmed that the "rock oil" of western Pennsylvania contained hydrocarbons that could be burned as an illuminant, entrepreneurs understood that by extracting this fluid directly, they could avoid the expense of mining, transporting, and processing coal to extract illuminating oil. Edwin Drake and his financial backers made no claims about petroleum being drippings from coal seams, but others did.[20] However, those theories soon ran counter to observations and never proved useful in finding new reserves.[21]

Another point of debate centered around the question of whether salt had been a necessary ingredient in the creation of oil.[22] Not only were the deep waters surrounding petroleum reservoirs high in various salts, but a number of oil fields had been discovered where the thrust of salt domes had created the necessary oil traps. But what was the connection between the salt domes, the saline waters surrounding oil-bearing formations, and the oil itself? Did that deep water represent the remains of ancient seas, or was it due to water leeching minerals out of various formations over geological time? At a 1920 meeting of the American Institute of Mining and Metallurgical Engineers, David White, chief geologist for the U.S. Geological Survey, reminded his audience that "the geological principles controlling the distribution of oil and gas have yet to be discovered" and argued against the theory that "salts in amounts premising marine or brackish water [were] essential as a catalyzer."[23]

Regardless of how oil was created, most people involved in the search for oil in the early 1920s understood that salt thrusts created high spots in which oil could collect, and they sought out geologists to help locate potential thrusts. As a guide to where one should drill, though, salt thrusts came in a distant second to a show of oil in an exploratory well. After someone discovered oil in an area, everybody realized that the best place to drill was as close as possible to the discovery well. The show of natural gas at El Dorado, while not as definite as an oil strike, was a close second.

Development under the Rule of Capture

Before drilling, entrepreneurs interested in El Dorado first had to secure drilling rights to whatever oil they hoped to extract. Knowing that oil lay below the ground meant nothing if one did not have the right to drill on the land above the oil. As soon as people identified the location of a potentially productive oil deposit, the race to secure drilling rights began. Hence, in the weeks following the discovery of gas at El Dorado, geologists and real estate people from established oil companies visited the town, first to survey the area and then to secure whatever leases looked promising.

The pattern of leasing in a field influenced the way the field developed. If a few companies secured most of the drilling rights in the area, the field could be developed in a fairly orderly fashion. Typically, this occurred when major companies proved willing to secure large leases before anybody actually discovered oil. However, if anybody discovered oil before such leases had been secured, a mad rush to obtain drilling rights followed, resulting in large tracts of land being subdivided into a number of

Figure 2.2. In oil fields divided into small parcels, such as in this California field developed around the turn of the century, producers rushed to extract oil before their neighbors. (Library of Congress LC-US-C62-16663)

small, but relatively expensive leases. In such cases, producers rushed to extract as much oil as possible as quickly as possible. Otherwise, neighbors might extract most of the oil before one's own wells even started producing (fig. 2.2).

In El Dorado, with independents securing two-thirds of the leases, the latter scenario of uncontrolled development unfolded and resulted in a pattern that the industry had seen time and time again.[24] In January 1921, after a well in a cotton field started spewing 3,000 to 10,000 barrels of oil per day, entrepreneurs hoping to get in on the action descended on the town. A parade of trucks carrying pipe, boilers, sheet metal, and other oil field goods streamed into the area, along with the derrick builders, tank assemblers, drillers, and roughnecks needed to transform a cotton field into an active oil field.[25]

As in most oil fields, the business relationship between those who owned the surface land, those who secured the rights to drill, those who drilled, and those who operated the completed well took on a number of permutations.[26] In all cases, the price to secure the drilling rights to a piece of land rose as the probability of finding oil under that land increased. For example, after the original gas well blew in but before the discovery of oil, a representative from the Houston Oil Company paid $4,000 and one-eighth in royalties for the right to drill on eighty acres. After the discovery of oil, he had to pay $30,000 in cash, $20,000 in oil, and one-eighth in royalties for only twenty acres located three-quarters of a mile from the producing well. Even so, the representative reported that "prices went sky high but we got in early in the game." [27] To pay for those leases, one can be sure that the owners of the Houston Oil Company planned to lift as much oil as they could in as short a time as possible.

To a large extent, the grab for land and the uncontrolled development that followed reflected both the fluid nature of petroleum and the "rule of capture," the legal system governing the extraction of oil. According to the rule of capture, a legal philosophy that originally had been applied to the rights of hunters, whoever extracted the oil owned it. Because oil flowed through the porous rock that contained it, the oil did not necessarily stay under one piece of land. Therefore, everybody with some financial interest in a well wanted to extract and sell as much oil as possible before all the oil was gone. In essence, the legal system treated oil as a wild animal.

As a result, most leaseholders drilled as many wells as they could to prevent others from draining oil away from their land. For example, in El Dorado, the field superintendent for Houston Oil grew concerned when the Gladys Bell Oil Company started drilling near the boundary of a lease owned by Houston Oil. The field superintendent soon contacted the home office to ask if he should drill any protective wells in return.[28]

In some fields, leaseholders constructed drilling rigs on every available square foot, a strategy that usually resulted in a chaotic oil glut and a drop in prices that made recouping one's investment difficult. And drilling was not inexpensive. At El Dorado, if everything went right, a drilling crew could complete a 2,200-foot hole for about $12,000, or just under $5 per foot. But in the early months of developing this field, most things did not go right. The inexperienced crews did not have the necessary skills, and few possessed much knowledge about the formation into which they were tapping. As a result, the initial wave of drilling proceeded slowly, raising the cost of the first wells to about $11 per foot.[29] Furthermore, crews occasionally lost control of a well, resulting in large quantities of gas and oil shooting freely out of the ground.

By the middle of March 1921, just two months after the initial oil strike, conditions had deteriorated to the point where the understaffed Arkansas Conservation Commission requested help from the U.S. Bureau of Mines.[30] At this time, petroleum engineers from the Bureau of Mines represented the vanguard of a knowledge-based, scientific approach to oil production. In this role, they served as gatherers of data, as missionaries to the unconverted, and as advisers to those advancing the state of the art. In El Dorado, they arrived to find a field in relative chaos.

Initially, the engineers from the Bureau of Mines could not do much to change conditions in the field. Even when a crew did bring a well into production without losing control, they often did not have enough storage to hold the oil (fig. 2.3). In many cases, operators simply ran their crude into large, open pools behind earthen dams, and as the more volatile fractions of this oil evaporated, an untimely spark could ignite an entire pond. And fire was not unusual.[31] Operators routinely flared whatever gas they could not use as fuel and, on occasion, the gas and oil spewing from a gusher caught fire, making El Dorado a thoroughly messy and sometimes dangerous place. By September, after over seven million barrels of crude oil had been shipped from the field, almost half as much had been lost to fire, spills, seepage, runoff, and evaporation.[32]

Initially, new arrivals to El Dorado were surprised to see so many companies drilling wells with rotary rigs. Although rotary technology had been used in oil fields for over twenty years—the famous Spindletop field in Texas had been drilled with rotary equipment in 1901—most contrac-

Figure 2.3. In developing fields lacking sufficient storage, operators ran significant amounts of crude oil into pits with earthen walls. (Library of Congress LC-US-Z62-11528)

tors in Texas and Oklahoma still preferred the less expensive method of impact drilling, especially when working a shallow field.[33] And El Dorado, with oil-saturated sandstone at 2,200 feet, almost qualified as a shallow field.[34]

In the older system of impact drilling, also known as cable tool drilling, an oscillating beam continually lifted and dropped a heavy chisel-like bit, which slowly chipped away at the bottom of the hole. Every few feet, the driller reeled up the impact bit and lowered down a tool for bailing out the crushed rock. However, at El Dorado, the hole had to pass through several loosely consolidated formations, which sometimes caved in after the bit moved on to deeper layers. If that happened, the loose mud and debris buried the bit, and a careless driller could cause a derrick to collapse under the strain of pulling out the buried tool. In some cases, abandoning such a well proved easier and less costly than recovering the bit and cleaning out the hole.[35] Pockets of high-pressure gas also posed problems for cable tool drillers, as no easy way existed to control the gas.

The more sophisticated and capital-intensive rotary method of drilling solved the problem of cave-ins and reduced the chances of a well blowing out of control. In the rotary method, drillers attached a rock-grinding bit to a string of pipe. A mechanical device on the floor of the rig rotated the entire pipestring, and this rotary motion caused the bit to grind up whatever rock lay at the bottom of the hole. Each time the bit chewed twenty to thirty feet deeper into the earth, the crew stopped drilling and screwed another length of pipe onto the top of the pipestring. The drilling then resumed for another twenty to thirty feet, which could take several hours or several days depending on the type of rock encountered at the bottom of the hole.[36]

An important feature of rotary drilling had to do with how the crew removed ground-up rock, or cuttings, from the bottom of the hole. To accomplish this task, they continuously pumped a mixture of water and clay, known as drilling fluid or "mud," inside the rotating pipestring. The mud flowed down the pipe, passed through openings in the rock bit, and traveled back up to the surface via the space between the wall of the hole and the outside of the drilling pipe. The cuttings rose with the drilling fluid. Then, at the surface, screens filtered the cuttings from the mud as it flowed into a settling pool. From there, pumps reinjected the fluid back into the hole (fig. 2.4).

The drilling fluid also served several other functions. For one thing, it kept the bit from overheating. The fluid also plastered the walls of the hole with a cake of mud, preventing loose formations from collapsing into the hole. Finally, the weight of the drilling fluid helped keep pockets of high-pressure oil or gas from blowing out of the well.

Although drilling holes with rotary equipment took less time than those drilled with cable tools, operating rotary equipment required a larger crew. In addition to other tasks, these crews had to pull the entire pipestring out of the ground whenever the rock bit had to be changed (fig. 2.5). In that process, the operator of the crownblock, the device from which drilling pipe hung, raised the pipestring two pipe lengths at a time or, if the rig's derrick were tall enough, three pipe lengths. Laborers on the

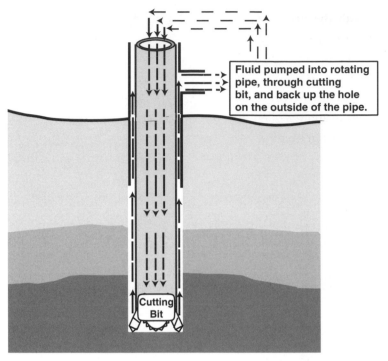

Fluid pumped into rotating pipe, through cutting bit, and back up the hole on the outside of the pipe.

Cutting Bit

Figure 2.4. In rotary drilling, a mudlike fluid is circulated down through the middle of a rotating pipe and back up the outside of the pipe. (Illustration by author)

rig floor then unscrewed the portion of pipe above ground and stored it upright against the derrick. After removing the entire pipestring, the crew attached a new cutting tool and reinserted the pipestring back into the hole, reassembling it as they did. Crews at El Dorado performed such "round trips" numerous times in the thirty days it took to drill an average hole in that field.[37]

The use of rotary drilling technology indirectly increased the prominence of trained geologists and engineers in an oil field. With cable tools, the driller in charge of a rig could easily spot oil in the bail of cuttings brought to the surface. In rotary drilling, the heavy column of drilling mud often prevented any oil from flowing into the hole, making it difficult to recognize when the rock-cutting bit reached an oil-bearing formation. Even when some oil did flow, a driller might not notice it in the drilling fluid. Furthermore, cuttings took a while to reach the surface. As a result, a crew could easily pass through an oil-producing formation unless someone with the necessary training analyzed the cuttings and made sense of the geological structures below.[38]

In 1921, the person with the necessary training was still likely to be called a geologist. However, a new profession was emerging. By the time El Dorado was discovered, several university programs had started training engineers to apply geological knowledge to oil production, and graduates of such programs typically referred to themselves as "petroleum engineers."[39] Regardless of what they were called, these experts realized that

Figure 2.5. In rotary drilling, crews have to remove the entire pipestring to change the bit. (Library of Congress LC-US-F33-12173-M3)

constructing a three-dimensional image of the formations below would allow them to better understand the behavior of the field.

However, in the early 1920s, nobody had an easy way to record and store information about the different layers through which a drill passed. Sophisticated methods of collecting and logging data, which would allow later engineers to plot information about each stratum, had not yet been introduced. Seismic techniques for mapping underground structures had not come into practical use either. At El Dorado, crews could take core samples, a technology that had been reduced to practice in Oklahoma two years before, but doing so consumed time and money. First, they had to replace the regular cutting bit with a special tool—a diamond-studded tool worked best—that could core out a formation without pulverizing it. Not only was the actual drilling slower, but the crew also had to retrieve the core every time they drilled the length of the tool.[40]

In practice, they collected most of their information by noting the layers of rock that drillers encountered. A good driller could tell something about a formation just by the rate of progress and, if nothing else, could record the depth at which the bit moved from one formation to another. Engineers with the Bureau of Mines encouraged operators to use this information to assemble a rough three-dimensional model of the field. They could do so by using different pegs to represent the various formations and by running string from peg to peg at the level of significant formations.[41]

The accuracy of the information supplied by drillers limited the usefulness of peg models. Although the Arkansas Conservation Commission required drillers to keep manual logs, the engineers from the Bureau of Mines noted that drilling contractors at El Dorado, in their rush to reach paying formations, kept poor notes, making the task of mapping out the field difficult. Many contractors agreed to drill to a certain depth for a fixed fee, and keeping accurate logs only slowed them down. In practice, unless someone systematically collected and analyzed cuttings from various holes and matched the depths at which layers of rock occurred, there was not much chance of forming an accurate, three-dimensional image of the oil field.[42]

Compounding the difficulty of underground mapping was another problem that engineers and geologists failed to appreciate at the time. Holes drilled with rotary equipment were not straight. Too much weight on the rock bit could cause the pipestring to deflect by a few degrees, especially if it encountered sloping hard rock. Such deflections resulted in large horizontal offsets. Many holes drifted so far off the vertical that the length of pipe required to reach the same producing sand varied widely, distorting any three-dimensional model that engineers and geologists attempted to create.

In the end, the most useful information came from people who drilled on the edges of a known field. Lacking any knowledge of what lay below, they risked coming up with a dry hole, that is, a well that failed to produce oil. Still, dry holes provided useful information about the boundaries of a field, encouraging some drillers to refer to them, euphemistically, as "geological successes," albeit successes nobody wanted. Much more desirable were exploratory wells that extended the known boundary of a reservoir, causing those working the field to describe the pool as "dipping west" or "headed level south."[43]

Casing, Cement, and Water Shutoff

Soon after their arrival in El Dorado, the engineers from the Bureau of Mines focused on how to prevent water from flooding the oil-bearing sand. Water—which could flow in from the edges of an oil pool, flow up from below, or drain from aquifers above the oil-bearing sands—presented a problem in all fields. If water entered the producing region, it could affect the flow of oil by turning the microscopic surfaces of oil-bearing rock water-wet rather than oil-wet.[44] Once that happened, oil tended not to flow as easily in that portion of the formation, creating a ready channel for water but not for oil. Then, not only did wells produce more water, but operators of those wells also had to separate and dispose of more water.

At El Dorado, engineers identified several water-related concerns. First, all holes passed through a freshwater aquifer that lay about two hundred feet below the surface. In the absence of the necessary precautions, this water could flow down the hole and mix in with the oil. Their concern was for the oil-producing formation, not the aquifer. Second, the main producing sand was relatively thin, less than ten feet in places. Hence, many crews drilled too far and into a layer of water below the oil. The engineers also expressed concern with edge water that inevitably flowed into the oil-bearing region of a formation as operators removed the oil.

To seal off the freshwater aquifer, the Arkansas Conservation Commission required crews to set a string of twelve-and-one-half-inch casing, basically steel pipe that served as a wall for the hole, to a depth just below the aquifer. Among other things, the casing prevented sections of the hole from caving in after crews completed the hole and pumped out all drilling mud. However, water from the aquifer could still flow down the space behind the casing. To prevent this from happening, the Conservation Commission required drilling contractors to fill the gap behind the casing with at least twenty-five sacks of cement.[45]

To install and cement this first string of casing, crews at El Dorado stopped drilling after the hole passed through the aquifer. Then, they pulled out the drilling pipe and inserted the casing, threading each segment together until the entire string reached the bottom. Next, to seal the gap between the outside of the casing and the exposed surface of the hole, they injected a cement slurry down the inside of the casing and forced the wet cement at the bottom back up the outside of the casing. In theory, after the mixture set, which took several days, the cement kept the fresh water from draining into the oil sands.

Although most operators recognized the importance of using cement to isolate water-bearing from oil-bearing formations and generally cooperated with the Arkansas Conservation Commission, not all crews in El Dorado, Arkansas, had the necessary knowledge and skill to cement a well properly. Neither casing nor cement had been used in the earliest oil fields. In western Pennsylvania, early oil producers had solved problems with water flowing down the hole in other ways. The first wave of drillers simply used doughnut-shaped bags of seed that expanded after coming in contact with water. If placed around the pipe, the expanded bags sealed off the top part of the hole from the bottom part and prevented water from flowing down the hole.[46] Later, when casing came into wider use, drillers set the bottom of the casing firmly on the caprock above the oil bearing formation and hoped that debris falling behind the casing would improve the seal.

The use of cement to seal the gap behind the casing first emerged in California. In 1902, operators in the newly discovered Lompoc oil field found that they could not prevent water from flowing into the producing formation. After some experimentation, an operator with the Union Oil Company successfully stopped the flow of water by setting casing directly in a slurry of cement.[47] As they gained experience, companies such as Union Oil found that cementing, while generally successful, sometimes failed to produce the expected results. To learn more about what was happening underground, Union Oil turned to mining engineers from Stanford University. These consultants, the first trickle of a stream of university-trained engineers that oil companies would eventually hire, confirmed suspicions that water leaking into an oil formation from one hole could affect a large area. To protect a lease fully, all wells in an area had to be cemented.[48]

Therefore, in California, Union Oil and other large companies pushed to make cementing a standard practice, but some independent operators who did not see cementing as worth the effort frustrated the campaign.[49] The established companies then lobbied for legislation to make cementing mandatory. By 1915, the State Mining Bureau of California required drillers of each new oil well to demonstrate that they had successfully isolated the oil-bearing sand from other formations. Given that poorly plugged abandoned holes also could serve as conduits between formations, the bureau also defined procedures for plugging those holes. Established oil producers, and the emerging cadre of university-trained geologists and petroleum engineers who were replacing drillers as the chief technical personnel in oil fields, heartily backed enforcement of these regulations.[50]

The use of cementing as a way to seal casing did not spread to oil fields outside California until after World War I. One individual who helped to disseminate this practice, Erle P. Halliburton, started out as an employee of Almond A. Perkins, the founder of an innovative well-cementing service in California. Though innovative, the Perkins Company had not been particularly successful because most companies in California preferred to cement their own wells. After Perkins refused Halliburton's request to start a branch division, Halliburton simply quit and started his own operation in Texas. There, Halliburton faced a different situation than Perkins. In Texas, few drilling contractors had any expertise at cementing. But just as few even bothered with the task.[51]

Halliburton's break came in 1920 with the development of the Hewitt field in Carter County, Oklahoma. There, the oil sands dipped steeply, and no easily recognizable formation warned drillers when they were approaching the oil-bearing sand. Hence, many crews drilled too far, requiring them to plug the bottom of the hole with cement if they were to prevent water intrusion from below. When Halliburton moved his equipment to this developing field, he found continuous demand for his services, especially after engineers with the U.S. Bureau of Mines recommended the use of cement. By 1924, his company operated a fleet of twenty trucks. However, when oil was discovered in El Dorado, Arkansas, Halliburton, whose name would eventually become synonymous with cementing, had expanded his operation to a newly discovered field near Duncan, Oklahoma, but no further. Contractors in fields such as El Dorado simply did not have the same level of expertise available to them, and their lack of skill often resulted in poor cement jobs.[52]

Although crews at El Dorado had little experience with cementing, their lack of skill did not necessarily slow them down. Inspectors from the Arkansas Conservation Commission rarely verified whether cement jobs actually held. Just making sure that crews used cement to seal their casings proved difficult enough. Hence, as soon as the cement set, crews at El Dorado resumed drilling without knowing whether the freshwater aquifer had been successfully isolated. Using a slightly narrower bit than before, one that fit through the first string of casing, they simply drilled though whatever cement remained at the bottom and continued their downward course.

Drilling the rest of the hole involved additional complications. Before reaching the El Dorado oil sands, drillers often passed through a layer of natural gas under higher than normal pressures. In anticipation of such pockets, crews typically increased the density of the circulating mud, so that the added weight would counteract the upward pressure of any gas encountered. However, they did not want to adjust the mud too far in advance as the heavier fluid made drilling more difficult. Crews also had to watch out for situations in which the drilling fluid suddenly flowed into fissures and porous formations, so-called thief sands. If too much escaped, the remaining drilling fluid might not be able to contain gas or oil under high pressure. Such runaway conditions, if noticed early enough, perhaps through a decrease of mud in the settling pool, could be contained. If the crew failed to notice such a situation, high-pressure gas and oil could break through the mud, causing a gusher or, as the condition would be called in later years, a blowout.[53]

When a driller safely reached the caprock above the oil sands, the Arkansas Conservation Commission required that a second string of casing be installed. Once installed, this string of casing ran from the top of the hole, through the inside of the first string of casing, and down to the bottom of the well. Then, the crew sealed the gap behind this inner string of casing with sixty sacks of cement and waited ten days.[54] Ideally, after the cement set, the contractor had a well that ended just above the oil-producing formation and isolated all fluids in their original formations. The only thing left to do, either by the drilling contractor or a service company, was to tap into the oil-bearing sands and prepare the well for production, complete with a "christmas tree" of valves at the wellhead.

The Problem of Flush Production

Initially, operators in El Dorado had no need to install any pumping mechanisms in their oil wells. Most operators simply lowered a string of production pipe inside the casing to the producing area, along with a packing device to prevent any oil from flowing outside the pipe. As the oil drained from the sandstone and flowed into the bottom of the well, natural pressure forced the gaseous oil up through the production pipe, past any open valves at the wellhead, and into storage. In some fields, operators attempted to increase the flow of oil by exploding nitroglycerin in the hole.[55] Other methods of loosening a formation would be developed in later years, but crews at El Dorado had little need for such techniques. Keeping the fluid in the ground proved to be the more difficult problem.

Geologists knew that in some fields the natural lifting force came from hydrostatic pressure on the oil.[56] Edge water pressurized by channels reaching the surface could lift the oil all the way to the top of the hole.

However, in a formation isolated from the surface, such as the oil-bearing sands at El Dorado, geologists suspected that the primary lifting force came from the pressure of natural gas, which could be quite high in the early stages of production. They reasoned that large amounts of natural gas remained dissolved in the oil and caused the reservoirs, when first tapped, to act like a carbonated drink that had been opened too rapidly. Liquid in the container simply gushed out and bubbled over.[57] In the early days of El Dorado, this gushing force propelled the oil out of wells with little difficulty.

Operators referred to this initial period in which a field produced oil with ease as a period of flush production, and the engineers from the Bureau of Mines saw efforts by operators to capture this production as contributing to the field's rapid development. Nobody knew how long flush production would last, and so all attempted to take advantage of this free-flowing oil for as long as they could, even if much of the oil ended up being lost to seepage, evaporation, or fire. The bureau's engineers discouraged this practice and strongly encouraged operators to choke back on their production. Doing so, they told operators, not only saved oil and gas that would otherwise be wasted but also prevented high-speed sand particles from wearing away valves and other equipment. But many operators, believing that a strong initial flow permanently etched favorable underground channels in the oil-producing formation, often kept production flowing at full bore.

The escaping natural gas also concerned the engineers from the Bureau of Mines. Although the Arkansas Conservation Commission passed a regulation requiring operators to shut down wells that produced only gas, many operators attempted to "blow a well in to production" by letting gas flow until some oil showed.[58] Then, because no method of storing this natural gas existed, most of these producers simply flared whatever gas they could not use as fuel. The bureau's engineers argued that this rapid removal of natural gas would deplete the field's lifting energy, just as if the reservoir were a carbonated drink going flat. Eventually, the internal pressure of the field would not even be enough to lift the oil to the surface. When that time came, operators would have to let the oil rise as far as it could and then pump it the rest of the way.

Although most engineers did not yet know it, dissolved gases also made the oil less viscous, helping it to flow through underground formation more easily.[59] As the amount of dissolved gases decreased, the oil became thicker and flowed more slowly through the tiny crevices in the sandstone. At some point, even pumping would not help because, despite the sandstone still being saturated, little oil would flow from the sandstone into the well.

In addition to wreaking havoc with the flow of oil, the practice of flaring natural gas also represented the direct loss of two valuable commodities—the natural gas itself and any gasoline vapors, mostly hydrocarbons with five and six carbons, dissolved in the gas. Most operators knew that they could recover this casinghead gasoline by running the natural gas through a special compression or absorption plant, but building such a plant required time, specialized knowledge, and capital—scarce resources in a field such as El Dorado.[60] Some plants would eventually be built to service the field but, in the meantime, operators continued to vent huge amounts of gas to the atmosphere. A few creative operators piped the natural gas to the bottom of their storage tanks and allowed it to bubble up through their stored oil. The oil, they explained, absorbed the gasoline vapors out of the natural gas. The bureau's engineers warned these operators to be cautious and pointed out that if the gas were drier than expected, it might actually absorb lighter fractions from the tank and reduce the value of the oil.[61]

Even if an oil field had adequate storage, operators still faced challenges in keeping a lease clean. Representatives from pipeline companies only accepted oil from which most of the salt water and sediment had been removed. Specifically, most pipeline companies would not accept any crude that contained over 3 percent water or too much sediment. As a result, oil field operators had to dispose of any salt water, oily emulsions, and oil-soaked dirt and debris they separated from their petroleum. In many fields, oil field operators often paid little attention to this disposal task (fig. 2.6).

Because the oil produced at El Dorado contained significant amounts of salt water, from 25 to 50 percent by volume, operators attempted to remove that water as inexpensively and quickly as possible. To remove some

Figure 2.6. In the early twentieth century, oil producers such as those operating these wells in the Salt Creek field near Casper, Wyoming, often discharged gas, salt water, and other wastes indiscriminately. (Library of Congress LC-US-Z62-124120)

of this water, along with any gas being produced, most operators in the field first passed their crude through a system of risers and bleeder pipes. Water sprayed out of these roughly constructed devices and ran along the ground until seeping into the soil or reaching a stream. A significant amount of oil also escaped with the salt water.[62]

Not all of the water could be removed in this manner. The churning of the oil and water as the two fluids flowed to the surface produced emulsions, intimate mixtures of water and oil in which a thin layer of one material encased fine droplets of the other. Such emulsions resisted easy separation, forcing operators to experiment with one of several techniques, including the use of heat, chemicals, electrical currents, and centrifuges. All methods raised the cost of production by three to ten cents per barrel, making the task of breaking emulsions a significant operating expense.[63]

Salt water, along with some dirt and emulsions, also settled to the bottom of storage tanks. To dispose of the water, operators simply opened a valve near the bottom of the tank and allowed the water to run free, closing the valve only when oil began to flow instead of water. Every once in a while, laborers with shovels also had to remove the emulsions and bottom sediment, known as "b. s.," and dump it wherever convenient. If the oil content was high enough, operators often set the sludge on fire. Otherwise they simply let it accumulate.[64] In such cases, heavy rains also washed much of that waste, along with any drilling mud left in slush pits, into nearby streams. Keeping an oil field lease clean was, for many operators, a luxury they could not afford.

In Pennsylvania fields, disposal of salt water had never been a major problem because it was lifted in relatively small amounts, did not have a high concentration of dissolved salts, and could be quickly diluted by heavy rains that occurred in all seasons.[65] However, as oil fields matured in dry areas—including portions of Kansas, Oklahoma, California, and Texas—operators lifting large quantities of very salty water learned quickly that they could not run this briny waste directly into slow-moving streams without some risk of being sued for damages by farmers and ranchers.[66] Some of the water being lifted had a salt concentration several times higher than seawater, and one barrel of this brine could make about 15,000 gallons of fresh water too salty for human consumption.[67]

Furthermore, where problems existed, those problems grew worse over time. As operators extracted more oil, the deep salt water surrounding that oil encroached on producing wells and forced operators to lift an ever-increasing percentage of water, eventually more water than oil. The streams and soil that received this wastewater gradually grew saltier, making complaints even more likely. If the concentration of dissolved salts in a stream rose too high, the water could not be used to irrigate crops or to supply livestock with drinking water. In sufficient amounts, salt water running across farmland and seeping into the soil also damaged crops.

Damage suits filed by farmers and ranchers encouraged many of the larger oil producers to pay more attention to releases of oil and salt water than they would have otherwise. According to one industry observer, the damage suits associated with the pollution of streams "were expensive to fight and very annoying. Practically, every time that a cow or horse was found dead near these streams, the oil companies were blamed. It was alleged that that death occurred through the drinking of waters polluted by waste oil."[68] In many areas, salt water proved to be more troublesome than oily wastes.

The lawsuit that Bessie Edwards of Oklahoma brought against the Pulaski Oil Company was typical. Edwards owned a farm adjacent to one of Pulaski Oil's wells, and salt water released from Pulaski's separation tanks flowed diagonally across her property, saturating a strip of land. When crops failed to grow on this strip of land, Edwards sued for damages and the court awarded her $800. Pulaski Oil accepted its liability but argued over the amount. Given that land in the area sold for about $50 per acre, the attorneys representing Pulaski Oil could see no reason why damages should be more than several hundred dollars. After all, they argued, even though the farm was large, the salt-saturated strip did not amount to more than two or three acres. In this case, the court ruled that the unproductive strip had affected the value of the entire farm and upheld the $800 award.[69]

The $800 penalty undoubtedly encouraged the Pulaski Oil Company to be more careful in the future. Perhaps the company began paying Edwards a fee to run a drainage ditch or pipe along the edge of her property. Perhaps the well had been a poor producer to begin with, encouraging the Pulaski Oil Company to plug its well sooner than otherwise. In any event, in 1923, when the Oklahoma Supreme Court upheld the damage award, few people saw any need to treat problems associated with the disposal of oil field brines in a more systematic fashion.

Waste, Pollution, and Oil Production

The engineers from the Bureau of Mines often talked about reducing waste but did so in terms of the conservation of resources, not pollution. To a large extent, though, efforts to increase the efficiency of operations overlapped with efforts to reduce pollution-causing discharges. Indeed, until oil producers extracted oil more efficiently, little hope for reducing pollution existed.

In general, engineers with the Bureau of Mines were more concerned with the waste occurring underground than with the oil spilled, consumed by fire, or lost through evaporation on the surface. In most fields, producers recovered less than 15 percent of all oil underground.[70] The remaining oil stayed underground, stripped of its viscosity-lowering natural gas and trapped behind pockets of water-wet sand. The chief petroleum technologist with the Bureau of Mines, A. W. Ambrose, in an address to members of the American Institute of Metallurgical and Mining Engineers, could say without fear of being disputed that "the underground losses of oil exceed by hundreds of thousands of barrels all the oil that has been lost in storage, transportation, or refining. The quantity lost is, of course, indeterminate; but when it is considered that the contents of an entire field have been excluded from recovery by invading waters, some idea of the amount wasted may be gained. . . . Conservation of the oil, therefore, should start before it is brought to the surface rather than after it is placed in storage tanks."[71]

In addition, most conservation-minded petroleum engineers and geologists agreed that operators, if they cooperated, could drain one square mile of an oil field with about sixteen wells rather than the hundreds they often used.[72] Drilling more wells than necessary added to the cost of lifting oil without increasing the amount being extracted.

To efficiency-minded conservationists, the rule of capture represented the major barrier to cooperation. As long as producers felt compelled to extract oil as fast as possible, little could be done to manage a reservoir. Hence, the well-managed field that Ambrose and other conservation-minded engineers envisioned—one in which nobody drilled more wells than necessary, wasted natural gas, or allowed water to encroach on producing wells—remained a distant hope. In most fields, leaseholders continued to drill more wells than necessary so as to extract oil as fast as possible.

Unregulated production also complicated efforts to control the amount of salt water being lifted, separated, and disposed of. Although engineers from the Bureau of Mines were more concerned with the effect of this salt water on the dynamics of a field, their efforts to control the flow of this water underground overlapped with efforts to address the disposal problem aboveground. Until operators cooperated in managing the extraction of oil from the reservoir, little could be done to address the pollution problem.

Not everybody saw uncontrolled production as a problem. If nothing else, uncontrolled production resulted in cheap oil, and oil companies with a heavy investment in refining liked cheap oil. In addition, many oil producers objected to any legislation that might prevent them from extracting as much oil as possible from their leases. Some operators had been fortunate enough to secure small leases in the middle of a productive field and did not want to see any limits placed on their ability to extract oil. Others had been fortunate enough to maintain high-production wells without paying attention to the dynamics of the larger field, and they found arguments based on the beneficial role of natural gas and the hazards of encroaching water unconvincing.

Even the heads of integrated oil companies feared potential government intrusion more than wasted oil. They were also concerned with antitrust legislation and charges of collusion. After all, by controlling the manner and pace of production, one could also control the supply. Given that Senator Robert LaFollette of Wisconsin had just embarked on a major investigation into the price of gasoline—an investigation that would eventually examine the control of refining patents—those people who called for controls on production seemed at odds with the current antimonopoly sentiment.[73]

Hence, by the early 1920s, the waste of oil and gas due to poor production practices had emerged as an important subject of discussion. Increasingly, as engineers and geologists learned more about the movement of oil and water and more about the role of natural gas in maintaining reservoir pressure, voices calling for controls on the rule of capture grew louder. Most oil-producing states already had some experience with conservation-inspired laws and regulations, such as statutes associated with the cementing and plugging of wells. In 1915, to prevent dramatic price drops associated with gluts, Oklahoma even passed a statute that limited production to market demand. However, this statute quickly proved unenforceable.[74]

Though relevant, concerns associated with the release of pollution-causing discharges in oil fields, such as oily wastes and salt water, re-

Table 2.1. Pollution- and Waste-Related Concerns in the Oil Production Sector, Ca. 1920

Concerns	Incentives to Change	Barriers to Change
Venting of natural gas to the atmosphere	Operators in a field could significantly increase the amount of oil extracted if they reduced the amount of gas escaping.	The unregulated rule of capture discouraged operators from cooperating in efforts to prevent natural gas from escaping.
Fluids migrating from one strata to another	Keeping oil from escaping and keeping water out of the producing formation allowed more oil to be extracted.	Using cement to isolate formations was costly and time consuming. The rush to extract oil under the unregulated rule of capture often discouraged its use.
Vapor losses from storage tanks	Preventing vapors from escaping would reduce losses of gasoline due to evaporation.	Many producers were not aware of how much gasoline was being lost to evaporation, even in fields with adequate storage.
Seepage losses from earthen pits	Pipeline companies paid lower prices for oil stored in earthen pits.	In flush fields, the unregulated rule of capture encouraged operators to lift as much as oil as possible, even without adequate storage.
Fires	Preventing fires would save oil, save resources devoted to fighting fires, and make oil fields a safer place.	Conditions in fields developed under the unregulated rule of capture made fires difficult to prevent.
Storage tank failures	The loss of oil, the cost of tanks, and the cost of potential damage suits encouraged companies to prevent such failures.	Few professional design standards or governmental regulations existed.
Blowouts	No operator wanted to lose control of a well as it could lead to a cave-in and force abandonment of the well.	The technology to prevent blowouts in high-pressure fields was poor and the rush to extract oil discouraged careful drilling.
Disposal of bottom sediment, drilling muds, and other oily wastes	The cost of damage suits brought by farmers and ranchers when rain washed oily wastes into streams encouraged some producers to be careful.	No clear disposal practices were established and conditions in fields developed under the unregulated rule of capture discouraged good housekeeping practices.
Disposal of oil field brines	Runoff that damaged crops or contaminated water supplies could result in damage suits, providing some incentive to change.	No easy way to dispose of the salt water existed, and the rush to extract oil often resulted in operators lifting a greater percentage of water.

mained secondary. In the short term, damage suits appeared to be a sufficient mechanism for addressing specific pollution-related concerns. Furthermore, as table 2.1 suggests, little could be done to systematically address pollution concerns in the nation's oil fields as long as the production of oil remained unregulated. In practice, the development of most new oil fields proceeded in the same chaotic fashion as it had in the field in El Dorado, Arkansas.

THREE   Keeping Oil in the Pipelines

MOST PETROLEUM IN THE UNITED STATES being transported
from oil fields to refineries moved through pipelines. The extent of the
U.S. oil pipeline system by 1920—over 25,000 miles of interstate trunk line
and 11,000 miles of gathering lines—testified to its significance.[1] However,
from the perspective of those who wished to conserve the nation's oil supply, this system had a major problem. It leaked.

Below the surface, corrosion attacked all pipelines. Short of encasing
every inch of pipe in asphalt or cement, oil companies found they could
do little to prevent corrosion and the leaks that resulted from corrosion.
Losses by evaporation were also significant. At each pumping station
along a pipeline, all oil flowed into storage tanks before being pumped
into the next section of the line. The pumping equipment simply was not
coordinated enough to move oil continuously through the entire line. As a
result, the crude oil frequently came into contact with air, and significant
quantities of the most volatile fraction, gasoline, evaporated. Tank farms
owned by pipeline companies also suffered from significant vapor losses.

Numerous incentives to improve the performance of the nation's
pipeline system existed. First, to pipeline companies, the escape of petro-

leum through leaks and evaporation represented the loss of a valuable product. Second, the cost of repairing and replacing pipelines encouraged companies to investigate ways to protect their lines from corrosion in a more systematic fashion. Damage and nuisance suits associated with pipeline leaks provided additional incentives to make engineering improvements. Finally, fire insurance organizations successfully lobbied many cities to pass statutes requiring oil companies to meet higher standards when designing tank farms and pipeline stations. With these incentives already in place, few people saw any need for pollution-control regulations aimed specifically at preventing the loss of petroleum from pipelines.

In the short term, though, efforts to prevent leaks and vapor losses took a back seat to moving oil. Hence, when the oil field in El Dorado, Arkansas, came on line in 1921, the main focus was on moving oil from the field to refineries, not preventing losses. And doing so meant extending the existing pipeline network to reach that field as quickly as possible.

Connecting to the Pipeline System

In the early months of production at El Dorado, before long-distance pipelines reached that field, all oil had to be shipped to refineries by tank car. Fortunately for producers, two railroads already served the town, placing all leases within easy reach of a hastily constructed loading station. Indeed, the field's proximity to the railroad made El Dorado all the more attractive to early investors, allowing them to install small-diameter gathering lines to carry oil from individual leases to loading racks capable of filling about four hundred tank cars each day. By rail, the oil then moved to refineries in Louisiana, Missouri, and Kentucky.[2]

As it happened, use of the railroad gave way to the less expensive method of transporting oil by pipeline sooner than expected. A productive oil field with pipeline service lay approximately fifty miles south of El Dorado, just across the Louisiana border. Two companies operating pipelines there, Standard Oil of Louisiana and the Louisiana Oil Refining Company, quickly made plans to extend their pipeline to El Dorado. A third company, the Shreveport–El Dorado Pipeline Company, started work on an entirely new line that would connect El Dorado to refineries in Shreveport, Louisiana.[3]

By July 1921, barely five months after oil started flowing in El Dorado, all three pipelines to the field were in place and pumping. Combined, the three lines could transport about 60,000 barrels each day, roughly as much as could be handled by the rail-based loading stations. In its first two days of operation, the Shreveport–El Dorado Pipeline Company purchased 38,000 barrels of crude, of which 26,000 barrels went to filling the eighty miles of empty pipe that stretched to Shreveport.[4]

In addition to transporting oil from one point to another, pipeline companies also served as the industry's bookkeeper. Each transfer of oil from a producing lease to a pipeline company was a controlled event. When that transaction occurred, a representative from the pipeline company called a gauger first sampled crude from the middle of the operator's transfer tank. Using portable equipment, the gauger then measured the oil's specific gravity and water content. Then, because no reliable instruments for measuring the flow of oil existed, the gauger computed the amount transferred by noting the depth of the oil in the tank before and after opening the valve to the pipeline. Finally, the representative gave the operator a "run ticket" that recorded the amount and quality of crude transferred. Because most royalties were based on how much oil had been produced, all transfers of oil had to be recorded, even if the same company owned the lease and the pipeline. Indeed, these tickets served as the official record of all oil produced on that lease, transforming the fluid mineral into a documented asset.[5] In the rapidly developing El Dorado field, these run tickets brought much-needed order to a field otherwise enmeshed in chaos.[6]

Pipeline systems also served as a fairly sophisticated scheduling and switching network, with operators tracking and routing the oil in batches as it moved through the system and directing it to storage tanks, trunk lines, and branch lines with relative precision and little mixing between batches. To accomplish this task, pipeline companies had to maintain a sufficient amount of storage. Otherwise, companies would have had to match the amount of oil entering the pipeline system with the demand by refineries more closely than they did.

Therefore, when production at El Dorado proved reliable, each pipeline company also made plans to construct enough new storage to hold about five days of throughput. Large, vertically integrated oil companies with leases in the field also maintained significant amounts of storage. Hence, by the time the three pipelines began pumping crude from El Dorado, over forty tanks, each with a capacity of 55,000 barrels, had been constructed.[7] Standard of Louisiana also constructed another tank farm consisting of fifty-two 55,000-barrel tanks south of El Dorado in Haynesville, Louisiana.[8]

In theory, oil producers could sell their oil to anybody who wished to buy it. If the producer had a buyer, the pipeline company simply performed a gathering and transportation service for a fee. Independent operators, of course, desired the best price possible for their crude, which often meant holding the oil until they found a suitable buyer. Few operators, however, could afford the necessary storage. Hence, most oil producing states required pipeline companies to accept crude oil even if the producer had not yet found a buyer. Then, while the oil sat in the pipeline company's tank farm, the producer had thirty days to find a buyer. After that grace period, operators had to sell their oil directly to the pipeline company at the company's posted price—typically the lowest price around—or be billed for storage space.

Historically, those who controlled the system for transporting crude oil from production fields to refineries exerted considerable leverage over producers. Indeed, the success of Standard Oil before its breakup in 1911 sprang from its control over transportation. Only a few independent companies, such as Gulf Oil, Sun Oil, and the Texas Company, succeeded in bypassing Standard Oil's transportation network, which played an important role in these companies developing as vertically integrated firms.

The importance of controlling transportation became obvious soon after entrepreneurs drilled the first commercial oil wells in western Pennsylvania. Initially, producers shipped most of their crude petroleum in wooden barrels, with horse teams hauling those barrels to a creek that flowed into the Allegheny River.[9] When the creek did not have enough water to float barges, oil shippers arranged for upstream dam owners to flood the creek by releasing water in an orderly sequence.[10] The barges eventually reached the Allegheny, allowing them to transport their cargo to small refining facilities in Pittsburgh.

This system of transporting crude, which remained in place for the duration of the Civil War, quickly became the source of many headaches. First, barrels were expensive and in short supply. Second, they leaked. Even when a barrel made it to a refinery intact, which many did not, much of the crude leaked out during the journey. Third, teamsters in the region charged what producers considered outrageous prices to haul the barrels a short distance. When one considers that some of the early wells along Oil Creek produced several thousand barrels of oil each day, and that no more than five or six barrels, each weighing about three hundred pounds, could be transported by the average wagon, one begins to appreciate the chokehold that teamsters exerted on the rest of the industry.[11]

Furthermore, the entire system was awkward. Sealing, lifting, and stacking the barrels consumed much time and labor. And once loaded, wagons did not always make it where they were going. Heavily rutted roads frequently broke the axles of the overloaded wagons. When wet, those same roads turned into wheel-swallowing mud, preventing wagons from moving until they could be unloaded and pulled free. Continually dripping oil turned sections of road into a paste that never dried, and dead horses and abandoned barrels lined the sides of such roads. The artificial flooding of creeks took a further toll. Some barges smashed into each another. Some overturned. Others were left grounded before reaching deeper water. In 1864, one flood resulted in a loss of 20,000 to 30,000 barrels of oil.[12] According to one chronicler, such accidents resulted in large amounts of oil floating out to the Allegheny and down to Pittsburgh, "to the intense disgust of the worthy burghers of all river cities that obtained their water supply from said streams."[13]

Soon after the Civil War, oil buyers laid the foundation for a transportation system that involved piping crude from producing leases to railroad loading stations. One of the first oil buyers to do so profitably, Samuel Van Syckel of Pithole City, Pennsylvania, constructed a five-mile-long pipeline out of two-inch-diameter pipe.[14] The pipeline emptied its oil into storage tanks mounted on railroad flatcars. From there, the oil moved to refineries by rail. Other oil buyers soon followed suit and built similar pipelines, putting thousands of teamsters out of work. Van Syckel's line alone, capable of delivering about 2,000 barrels per day, replaced about 300 full-time teamsters. As a result, tension between teamsters and pipeline owners ran high, forcing owners to hire armed guards to protect their pipelines from sabotage.[15]

This combination of small-diameter gathering lines and tank railcars revolutionized the growing industry by allowing potential refiners to locate their facilities in places other than Pittsburgh. Indeed, entrepreneurs in Cleveland, Ohio, and Erie, Pennsylvania, both close to the oil fields of western Pennsylvania by rail, quickly constructed a number of small refineries. By 1866, these cities boasted a total of sixty refineries, surpassing Pittsburgh's total of fifty-eight. Oil companies on the East Coast operated another twenty refineries. Several years later, the number of refineries in Erie and Cleveland had climbed to eighty while the number in Pittsburgh dropped to about forty.[16]

In the late 1860s, John D. Rockefeller, a commission merchant by training and the co-owner of a refinery, set about consolidating the Cleveland refining industry. Rockefeller recognized that the owners of many early refineries ran their plants casually as side businesses and had no special expertise or ties to the industry. Most could refine no more than one or two hundred barrels in a day.[17] He reasoned that, after struggling to make a profit, they would jump at the chance to recover some of their investment. In most cases, Rockefeller was right. After gathering over 10 percent of the nation's refining capacity, by far the largest block of control at that time, Rockefeller and his associates negotiated with several railroads to haul oil at a discount, even to the point of receiving a percentage of the oil-shipping fees that competitors paid. Intelligence received about the shipments of competitors also allowed Rockefeller to determine the strategic value of other refineries and proved almost as valuable as the rebate.[18]

Increasingly intricate rebate schemes followed, in part driven by the monopolistic hopes of railroad owners to move into the refining business.[19] In the end, Rockefeller benefited more than the railroads, and by the late 1870s, Standard Oil and its affiliates controlled over 90 percent of the nation's refining capacity. Now dominating both the transportation and refining sectors, Rockefeller put producers on the defensive. He effectively eliminated any outlet for oil other than refineries in the Standard fold.[20]

People talked about constructing a long-distance pipeline to bypass the railroads and to loosen Standard's growing hold on the industry, but nobody financed such an effort until 1874. Then, Dr. David Hostetter, an entrepreneur and the manufacturer of the popular Hostetter's Stomach Bitters, set about constructing a three-inch-diameter pipeline from producing fields in Butler County, Pennsylvania, directly to refineries in Pittsburgh.[21] In addition to potentially undermining Standard Oil's control, this pipeline, designed to transport 3,500 barrels per day, also threatened the oil-hauling business of the Pennsylvania Railroad. Hence, executives of that railroad refused to let the line cross their right-of-way.[22] When Hostetter's pipeline crew crossed the railroad's right-of-way by running the pipe in a gully under a small railroad bridge, employees of the railroad immediately ripped out that segment of pipe.

Hostetter then leased the incomplete line to another group of financiers, who put the pipeline in operation as the Columbia Conduit Company. These entrepreneurs crossed the railroad's right-of-way using tank wagons. On one side of the railroad, crews emptied the pipeline into the tank wagons. They then crossed the railroad's right-of-way by public highway and pumped the oil back into the pipeline.[23] Although they faced some additional problems—in the first winter, pipe laid the previous summer contracted and a number of joints pulled loose—the company managed to deliver oil to Pittsburgh less expensively than those who shipped by rail.[24]

Standard Oil countered by purchasing all the main refineries in Pittsburgh, effectively eliminating any outlet for the crude being transported by Columbia Conduit. Columbia Conduit then began loading its oil onto barges at Pittsburgh and floating that oil to Huntington, West Virginia. From there, the company transported most of its oil to the East Coast over the Chesapeake and Ohio Railroad. Standard Oil then offered to buy the pipeline, which it did in 1877.

Ironically, the owners of Columbia Conduit used their profits to finance the Tide Water Pipe Line Company, which ambitiously set about constructing a crude-oil pipeline over the Allegheny Mountains to carry oil directly to refineries on the East Coast.[25] In planning the route for this line, the new company worked swiftly and secretly, attempting to stay one step ahead of both the Pennsylvania Railroad and Standard Oil. In the end, Tide Water Pipe Line Company managed to secure the necessary right-of-way. Once again, though, the pipeline company had to guard against acts of sabotage. As before, changes in the weather also damaged the line. This time pipe laid in cold weather expanded in the summer, pushing some joints out of the ground and causing a twisting, writhing mess.[26] And again, when the line went into operation, which it did in 1879, Standard Oil put pressure on the company by purchasing most of the refineries that purchased crude carried by the pipeline.[27] Standard also constructed a competing line, effectively neutralizing Tide Water as a serious threat.[28]

Railroad executives continued to challenge the growing use of long-distance pipelines, but their efforts proved futile. In one case, the Pennsylvania Railroad unsuccessfully attempted to prevent its former ally Standard Oil from laying a pipeline in Newark Bay. They argued that leakage from the pipe would adversely affect oyster beds in the area.[29] And early pipelines did leak. For example, after the Standard-controlled National Oil Pipe Line Company built its first line across Pennsylvania, farmers there routinely complained about breaks and leaks, often in the hope of receiving compensation. "In every instance of such breakage," read one such complaint, "the land was rendered completely useless for farming or agricultural purposes, as it was impossible to raise anything on."[30]

All the while, companies affiliated with Standard Oil continued to expand their pipeline systems. By the mid-1880s, a gang of twenty-eight men could lay about 150 lengths of 18-foot pipe, approximately a half mile, each day.[31] With multiple crews, companies could complete a line one hundred miles long in about a month. The pipeline network controlled by subsidiaries of Standard Oil grew so rapidly that by the time Congress passed the Sherman Antitrust Act in 1890—an act that had been inspired in part by Standard's monopolistic deals with the railroads—railroads no longer carried much of the crude produced in the United States.[32]

As the pipeline network grew, producers lobbied hard for legislation that would force pipeline companies to operate as common carriers. Although the federal government and state legislators passed laws to this effect, most notably the Hepburn Act of 1906, all efforts fell short of their intent. For one thing, pipelines constructed by Standard-controlled affiliates mainly went to tank farms and refineries controlled by other companies in the Standard Oil fold. Because no sympathetic buyers could be found at the other end of most lines, few independents could expect much advantage in using a Standard Oil line as a common carrier.

In some cases, Standard Oil prevented a pipeline being designated as a common carrier by taking advantage of loopholes in the law. For example, in 1909, when constructing a long-distance pipeline running from Oklahoma to the Gulf Coast, Standard Oil divided the line into three segments and assigned a different company to operate each segment: in Oklahoma, the Oklahoma Pipeline Company; in Arkansas, the Prairie Oil and Gas Company; and in Louisiana, the Standard Oil Company of Louisiana.[33] The short Arkansas segment of this line, which cut across the corner of the state and had no internal connections, served as a convenient foil to common-carrier legislation. By crossing state borders, the Arkansas line allowed the segments in Oklahoma and Louisiana to remain entirely in their own states and prevented them from being subject to common-carrier regulations.

Other precautions were also taken. When securing land for the line's right-of-way, the company's real estate department made no use of the power of eminent domain. Doing so also would have automatically made the line a common carrier in Louisiana. Instead, the company secured permission directly from all the necessary landowners. And where the line crossed under public highways, local lawyers were hired to secure the necessary ordinances.[34] Other loopholes also prevented the common carrier legislation from being effective. For example, even if declared a common carrier, a pipeline company could make its minimum shipment too large for most small companies. Hence, even after the federal government dissolved Standard Oil in 1911, the question of how to effectively regulate pipeline companies remained unanswered.[35]

Louisiana Standard did run into a small legal and political problem after building its line to El Dorado. Actually building the pipeline proved relatively easy. A year earlier, the company had completed an eight-inch branch line to an oil field that lay just below the Arkansas border near Homer. Moreover, the company already had plans to add another eight-inch line to Homer. Extending the pipeline another fifty miles to El Dorado posed few difficulties. However, crossing into Arkansas transformed Louisiana Standard's pipeline system into a common carrier by virtue of being engaged in interstate commerce.

As it turned out, this change in status mattered. In the same year that oil was discovered in El Dorado, a young politician from Louisiana, Huey Long, began lobbying for legislation that would allow the state to regulate Standard of Louisiana more closely.[36] Long, who once lost money when the company abruptly canceled an offer to purchase certain oil leases, failed in his initial attempts to access Standard's records.[37] The extension into El Dorado, and the pipeline's subsequent designation as a common carrier, gave Long more ammunition. However, rather than release information about the entire company to Long, who had since become chairman of the state's Public Service Commission, Standard of Louisiana reorganized its pipeline department into a separate company and provided only the records associated with that portion of the company, effectively thwarting Long.[38]

Despite unresolved regulatory issues, the logistics of building a pipeline —securing a right-of-way, moving pipe and equipment to remote areas, and laying the pipeline—had become more or less routine by the time crude started flowing from the El Dorado field. In general, when planning a line, engineers attempted to secure the most level and accessible route possible, with the fewest and least difficult stream crossings. They then determined the best location of pumping stations, selected equipment for those stations, and specified the type of pipe to use. Construction crews took over from there. Although each new line required a major effort in planning, companies could draw on the skills of engineering teams, real estate departments, and constructions crews experienced in this line of work.

At the time, all oil was pumped through pipelines in segments. The first pumping station, located in the oil field, forced the crude oil through twenty to forty miles of pipeline and into a holding tank at the next station. The second station then pumped the oil from its holding tank and through another twenty to forty miles of pipeline, and so on until the oil reached its destination. Hence, for all practical purposes, each segment of the pipeline operated independently of the others. Individual segments could continue to operate even if other stations stopped pumping, at least for a while, until the station's storage tanks were empty or until the tanks at the next station were full.[39]

How much oil could be transported through the pipeline depended, of course, on the diameter of the line. An eight-inch-diameter pipe could move about 20,000 barrels per day, with the exact amount depending on pumping pressures and the type of crude being transported. Theoretically, a twelve-inch line could deliver more than twice that of an eight-inch line because capacity increased as the square of any change in diameter. However, practical considerations reduced the amount of oil that a twelve-inch pipe could carry. Steel pipes over six inches in diameter had to be lap-welded, a process in which plate steel was rolled into the shape of pipe and then fused at the edges. At pressures necessary to pump oil, lap-welded pipes larger than eight inches in diameter tended to split at the seams.[40] Seamless pipe, manufactured by piercing a hot billet and using centrifugal force to form the pipe, was stronger and more reliable but available only in small diameters and relatively short lengths.[41]

Hence, many pipeline designers favored eight-inch pipe, which represented the best mix of capacity, strength, availability, and cost. When necessary, they ran two lines in parallel rather than use a single large pipe.[42] For example, Standard's trunk line from Tulsa to Baton Rouge, laid in 1909, had been an eight-inch line. In 1913, engineers increased its capacity by adding another eight-inch line in parallel. The company added a twelve-inch line in 1915, but to avoid splitting the twelve-inch pipe with high pumping pressures, the company ran smaller pipes in parallel for the first fifteen miles after each station. By that point, the pressure had dropped enough to use the larger pipe.[43]

For the branch extending to El Dorado, engineers with Louisiana Standard quickly decided on a single eight-inch line.[44] Construction of the line, though laborious, proceeded in a straightforward fashion. While some crews started digging the pipeline trench, others distributed lengths of threaded pipe along the route. Still other crews followed behind, assembling the line by screwing together the pipe segments. A final crew buried the line, zigzagging it against alternative sides of the trench to provide some slack for contraction and expansion.[45]

Figure 3.1. In the years before welding and mobile pipe-threading equipment, crews laying crude oil pipelines threaded segments together manually. (Smithsonian 99-2240)

Crews constructing the line to El Dorado had the benefit of heavy machinery. When muscle power had been used to join threaded pipe, as it had in the construction of the original Tulsa–Baton Rouge line, a crew wielding long-handled pipe wrenches torqued each segment manually, attempting to get as tight a seal at the threads as possible (fig. 3.1). The longer and wider the pipe segment, the more difficult was the task, which was another factor weighing against the use of ten-inch and twelve-inch pipe. Railroad-mounted machines capable of screwing pipe segments together had been used since the 1890s, but those machines could only be used to construct lines along railroad tracks.[46] As heavy equipment became more mobile, additional routes became accessible to specialized digging and pipe-screwing machinery.[47]

The engineers who designed the El Dorado extension used threaded pipe, which had the disadvantage of being weak where threads had been cut into the pipe. Hence, engineers tended to specify a slightly thicker pipe than they would have otherwise used.[48] In addition, crude tended to seep through the threads and attack whatever coating had been used on the pipe. To join segments of pipe, some engineers had experimented with welding, a technology that had proven itself in wartime shipyards. However, no company had yet used this method for joining long runs of pipe and would not until the year after the El Dorado line went into operation.[49] Nonthreaded pipe could also be joined with rubber-packed couplings held together with bolts, but these couplings were expensive. In addition, crude tended to attack the rubber. Therefore, most companies limited their use of rubber-packed couplings to large-diameter gas pipelines, making it easier to connect the bulky pipe and eliminating potential problems with gas leaking out through poorly connected threads.[50]

Major river crossings caused the most problems for construction crews because each crossing had to be separately engineered. From the perspective of future maintenance, the most desirable solution was to run the pipe over a bridge, through a tunnel, or to suspend the pipe from cables. A more economical solution, though, was to dredge out a trench into which the pipe could laid. To lay the pipe, barges pulled previously assembled segments of pipeline into place and used heavy "river clamps" to weigh the line down. The weighted pipeline, surprisingly flexible over long lengths, settled into the trench as assembled.[51]

Crews tested each segment of a new pipeline for leaks only after it had been buried and the pumping station associated with that segment placed into operation. They then pumped the line full of water, exerting a pressure of about 750 pounds per square inch on the pipe. Next, with the line under pressure, crews inspected the entire route, looking for wet patches or water bubbling through dirt. Leaks were most common around threads, which early pipe-laying machines were notorious for crossing and damaging.[52] Wherever inspectors detected such leaks, crews exposed the line and clamped sleeves over the leaking joint.

Companies also had to watch for leaks after a line went into service. Operators at pumping stations could detect leaks only if they were large enough to cause a drop in the line's pressure.[53] If a leak were small or if a break occurred near the end of the line, where the pressure was already quite low, the drop in pressure would not be enough for anybody at a pumping station to notice. Indeed, after a pipeline went into operation, small leaks that could not be detected by station operators emerged as a troublesome problem. Most such leaks were caused by corrosion.

All pipeline companies fought a losing battle against corrosion, and small pools of oil were a common sight along most pipeline routes.[54] Even relatively small leaks, if unnoticed for a long period of time, could accumulate and damage water supplies. In some cases, leaking crude followed underground fissures and traveled some distance away from the line. In any case, farmers who experienced damage from pipeline leaks tended not to be forgiving. To reduce potential conflict, the right-of-way lease that some companies signed with property owners specified that if complaints of damage to "crops and fences" could not be "mutually agreed upon," the damages would be "ascertained and determined by three disinterested parties," one appointed by the landowner and one appointed by pipeline company, one appointed jointly.[55]

Pipeline companies also had other incentives to locate and repair leaks quickly. Although shipping contracts assumed that some fraction of the crude would be lost in transit, mainly to evaporation at pumping stations and tank farms, any oil lost still represented decreased profits. Furthermore, small leaks caused by corrosion eventually became larger leaks that could not be ignored. For these reasons, companies hired "line walkers" to inspect pipelines between pumping stations regularly. In the span of one or two days, these inspectors walked, drove, or rode horses along the entire segment for which they were responsible. Upon finding any signs of a leak, they telegraphed—or increasingly, by the time the El Dorado line was constructed, telephoned—for a repair crew.[56] Upon reaching and uncovering the leak, maintenance crews often made the necessary repairs by clamping a patch around the offending area with the line still full of crude. Crews knew that an empty line, full of vapor, could be the more dangerous and explosive alternative.

When Louisiana Standard constructed its El Dorado branch, knowledge of corrosion and corrosion protection had not advanced much beyond attempts to physically protect the line. Scientists described the corrosion process as a form of slow combustion that took place in the presence of oxygen, but these theories had not been translated into practical strategies for protecting pipe.[57] For example, in 1909, when Standard Oil constructed its first trunk line through Louisiana, crews coated the pipeline with red "ship-bottom" lead-based paint. In dry soils, they used a black tar paint. Experience, though, showed that neither coating offered sufficient protection. In 1913, barely four years after the original eight-inch trunk line had been laid, long segments of the pipeline had deteriorated. Crews replaced over forty miles of pipe and encased hundreds more in concrete. On subsequent lines, the company protected any pipe laid in acidic or wet soil by placing it in wooden forms and filling the forms with asphalt or concrete (fig. 3.2).[58] By 1921, companies also started protecting portions of new pipeline by first coating it with a bituminous mixture and then wrapping the coated pipe with asbestos felt. Every mile of pipeline required seventy-four double rolls of paper, thirty-three barrels of asphalt, and three barrels of liquid cement.[59]

Operators at pumping stations could detect large breaks and take corrective action fairly quickly. At worst, such breaks released no more oil than a failed storage tank, perhaps 50,000 to 100,000 barrels, but typically less.[60] When such breaks occurred, their effect on the surrounding area depended on the geography and land use. If the oil found its way to a lake or stream, the spill could be quite disruptive. However, if a break occurred in the middle of relatively flat terrain, the oil might bubble up to the surface, flow outward for several hundred feet, gradually evaporating and seeping into the soil. To those familiar with the sight of gushers and ponds of oil stored behind earthen dams, such a leak would appear serious but not disastrous.

When farmers complained about oil spills from pipelines or storage tanks, some pipeline engineers saw their complaints as little more than an excuse to collect damages. One pipeline engineer, Bert Hull of the Texas Pipeline Company, liked to repeat the story associated with a leak in a six-inch line that ran through a rice farm near the Sour Lake field in Texas. The leak, which occurred around 1911, was aggravated when the farmer flooded his field and caused the spilled oil to cover about ten acres of land.

Figure 3.2. Some sections of pipeline, such as this segment being constructed in 1919 by the Interstate Oil Pipe Line Company of Shreveport, were encased in concrete or asphalt to protect against corrosion. (American Heritage Center, University of Wyoming, P-448-pl)

According to Hull, the farmer received as much in damages as he would have for the best rice crop that could possibly have been raised.

Coincidentally, in the following year, the farmer's rice crop failed due to a fungus, except in those ten acres that had oil on it. Rice in those ten acres flourished. The farmer then started purchasing oil-soaked sediment that came from the bottom of storage tanks and, according to Hull, "hauled it away in tank wagons and spread it on his rice crop. And I want to tell you that in the following year, all of his rice was the most beautiful crop in Jefferson County, or Hardin County, or whatever county it was there. . . . There was just enough crude oil in there or something to kill the fungus in the rice."[61] To engineers like Hull, oil-fouled land was nothing to get excited about. Indeed, Hull's attitude did not differ significantly from that of one Pennsylvania state senator who, when debating a pipeline bill in 1883, dismissed concerns associated with leaks by asserting, "Why, oil is the best fertilizer in the world."[62]

Similarly, few people paid much attention to the effect of oil leaks and spills on groundwater unless someone complained. If a pipeline leak contaminated groundwater near a farmer's well and that farmer subsequently tasted a change in the water coming from that well, he or she typically sought redress.[63] Otherwise, the crude floated undetected on the top of the water table, contaminating soil and moving toward a distant point of discharge, perhaps toward another water well.

Field personnel also had to monitor storage tanks for corrosion and leaks. Corrosion typically attacked storage tanks near the roof, where moisture collected, or around rivets and bolts, where slightly different steels came into contact. Steel plates could also develop cracks leading to the complete failure of the tank. The risk of such failures could be reduced by the use of conservative designs, and engineers did take the consequences of such failures into account when designing tanks. For example, after reviewing the design of a tank installed in Providence, Rhode Island, an engineer with one oil company admitted that "we made a mistake in building these tanks a little lighter than perhaps was proper for work in the heart of the city" and recommended reinforcing those tanks.[64]

Efforts to Address Losses

In the early 1920s, volatile vapors escaping from storage tanks accounted for a significant fraction of the physical losses suffered by pipeline companies. Indeed, in 1921, one of the ten projects underway at the Bureau of Mines Petroleum Research Station in Bartlesville, Oklahoma, involved vapor losses. That study eventually influenced the design of storage tanks in all sectors of the industry.[65] The results also proved useful to insurance companies interested in reducing the risk of tank fires.

Heading the vapor loss investigation for the Bureau of Mines was J. H. Wiggins, an assistant petroleum engineer charged with systematically studying the losses due to evaporation that occurred as oil moved from field to refinery.[66] No problem could be better suited to the mission of the bureau. No one perceived the issue to be at all related to competitive strategy. In general, whatever competition happened in an oil field occurred before any oil reached storage tanks. The project also attracted attention because storage occupied an important position in the strategy of those who wished to stabilize the price of crude during periods of oversupply. If oil could be stored safely and inexpensively, then it could be kept until needed, without any pressure to dispose of it immediately and inexpensively.[67]

Wiggins first collected data on average-sized lease tanks that could hold from 250 barrels to 1,600 barrels, and he found oil field operators eager to cooperate. Most small producers did not have the resources to undertake such a study. Some were not even aware that a problem existed. In contrast to the storage tanks constructed by pipeline companies, lease tanks generally were small, constructed on site by "tankies" who moved from lease to lease (fig. 3.3). The size of the tank depended entirely on the needs of the operator, though batteries of several 500-barrel tanks constructed from wood or steel plate were common. Valuable gasoline—nobody knew exactly how much—continuously evaporated away from these poorly sealed tanks.

Wiggins put numbers to the losses suffered by a variety of tanks. For example, he discovered that oil stored for a month in a 250-barrel steel tank with openly vented roofs lost about 10 percent of its volume. Furthermore, the escaping vapors inevitably consisted of the lighter and more desirable hydrocarbons, reducing the value of stored crude by more than 10 percent.[68] After studying losses from these lease tanks, Wiggins went on to examine larger tanks used by pipeline companies and refineries. In one survey of a mid-continent field, Wiggins computed that the losses in moving crude from field to refinery were 6.2 percent of gross production.[69]

To fulfill its mission as a creator and disseminator of knowledge, the Bureau of Mines encouraged researchers such as Wiggins to publish data at various points in the process of an investigation. In addition to provid-

Figure 3.3. Many tanks in oil fields, such as these in the Soap Creek field of Wyoming in 1921, were poorly sealed and allowed significant amount of gasoline vapors to escape. (Herman C. Bretschneider Collection, box 7, American Heritage Center, University of Wyoming)

ing solid technical information, publications helped establish the bureau's technical authority and generated publicity that gave the organization a relatively high profile for a governmental agency. Therefore, Wiggins routinely published his results in reports and bulletins issued by the Bureau of Mines, as well as in various engineering bulletins and trade periodicals.[70]

Few investigations could be less controversial than the one undertaken by Wiggins, and companies supported the role that the Bartlesville station played in generating and disseminating information about how to avoid wasting valuable hydrocarbons. No one had any reason to question the assumptions, measuring tools, or estimates on which Wiggins relied. Even Standard Oil of New Jersey published the findings of its own vapor loss investigation, "in the hope that some benefit may be derived from them."[71] Everybody agreed that a problem existed and saw some advantage in solving that problem, especially if one could save money in the process or even earn money by helping others reduce their losses.[72]

The technical goal, of course, was to maintain a vapor seal in all storage tanks. Given that the volume of crude changes by several percent over the range of temperatures experienced in most oil fields, good pressure relief was also absolutely necessary. If the pressure inside the tank exceeded safe levels, one had to release and recover the associated vapor. Engineers also desired to minimize the amount of space that vapors could occupy. For lease tanks that had loose-fitting wooden roofs, or no roof at all, even the simple step of installing a gastight roof with some sort of pressure relief paid off. For larger tanks, Wiggins advocated the use of tanks with roofs that floated up and down to accommodate the level of oil, thereby reducing the vapor space to near zero. Indeed, tanks equipped with floating roofs came to be known as Wiggins tanks (fig. 3.4).

Not all studies having to do with storage tanks were as free of controversy as the study of vapor losses. In 1924, the spacing of storage tanks became an important issue when the National Fire Protection Association presented changes to a model ordinance that regulated the use, handling, and storage of flammable liquids. Although only a model ordinance, the document was something to take seriously, as many local governments would enact it word for word.[73]

Representatives from the petroleum industry opposed a section that required relatively large distances between storage tanks and adjoining property.[74] To meet the new requirements, many oil companies would have to purchase more land, which presented difficulties for refiners and retail companies operating in urban areas. Too generous a space requirement could prove expensive. For example, after a lightning strike caused an empty oil tank near Elvira Malaterta's home in Marcus Hook, Pennsylvania, to explode, she complained to the owners of the tank that she no longer felt safe during a storm. She asked the company to purchase her property. The company initially refused. However, Pennsylvania law required that tanks as large as the one in question be located over three hundred feet from all property lines, and the company ended up purchasing the property.[75]

Figure 3.4. To keep the vapor space in large storage tanks as small as possible, oil companies began using tanks with floating roofs, such as this Standard of Indiana tank photographed in Muskegon, Michigan, in 1929. (Library of Congress LC-US-Z62-55942)

As a result of the disagreement over the spacing requirement in the model ordinance, five organizations—the American Petroleum Institute (API), the Bureau of Explosives, the National Board of Fire Underwriters, the International Association of Fire Engineers, and the Associated Factory Mutual Fire Insurance Companies—formed a "conference board" to discuss the spacing requirements. The API immediately launched its own study. A team of engineers from several major oil companies compiled data for all reported tank fires in the ten-year period from 1915 to 1925. After contacting dozens of insurance companies, trade associations, and oil companies, they concluded that just over four hundred tank fires had occurred over the previous ten years, with most of the fires, regardless of their cause, associated with tanks that were not gastight. At refineries, 87 percent of tank-related fires were associated with containers that were not gastight. For pipeline stations, tank farms, and terminals, fires in non-gastight tanks accounted for over 90 percent of the total.[76]

In addition, previous to the API investigation, the president of Sun Oil, J. Howard Pew, had asked the president of the Franklin Institute, W. C. L. Eglin, to investigate whether lightning near a steel tank could induce sparks inside the tank.[77] Eglin indicated that any pipe extending into the tank could cause internal arcing if the pipes, perhaps due to rust, were insulated from the shell. However, the problem could be eliminated by securely grounding the pipe to the shell. In terms of risk from lightning, Eglin assured Pew that the real danger came not from steel tanks but from tanks with unsealed roofs.[78]

Based on such data, members of the API committee concluded that spacing was less important than other factors, including whether the tank was gastight, whether it was protected against lightning, whether it was protected by fire walls, the quality of tank construction, the level of tank maintenance, and the availability of fire-fighting equipment. They also decided that tanks holding crude oil required more space because crude always contained some water and could boil over. Tanks holding refined products, as would be the case in most urban areas, did not require as much separation space.

After estimating the "protection realized per unit of money expended," the API subcommittee recommended that no storage tanks be placed closer than one tank diameter to any property line and that tanks holding crude should be placed two diameters away. Both recommendations represented substantial reductions from the distances specified in the model ordinance. When representatives from the U.S. Bureau of Explosives supported the API-recommended distances, representatives of the National Board of Fire Underwriters and the Associated Factory Mutual Fire Insurance Companies agreed to compromise and reduce the spacing requirements in the model ordinance.[79]

Following the API report, a string of six disastrous tank and tanker fires over a six-day period stimulated further interest in how best to prevent tank fires. Combined, the fires consumed over sixty lives and eight million barrels of crude. In one case, bolts of lightning ignited three one-million-barrel reservoirs at a Union Oil tank farm in California. Eventually, the hot crude boiled over and ignited a row of steel tanks. Union Oil, which ended up losing six million barrels of crude oil, received about $9 million from insurance companies.[80] The Board of Fire Underwriters of the Pacific immediately doubled most premiums, tripling the rates in some cases. A few insurance companies stopped underwriting fire protection for tank farms altogether, leaving approximately one hundred million barrels of oil in California without insurance.[81] As insurance companies reevaluated the problem and assigned various levels of risk to different types of storage, unsealed tanks with wooden roofs emerged as the most important condition to avoid.[82] Spacing, though still an issue, ceased to be the central issue of concern to insurance companies.

By the mid-1920s, most large oil companies had taken some steps to reduce their vapor losses. In some cases, they invested in tanks with floating roofs. In other cases, they modified existing equipment. For example, at one tank farm, the Atlantic Pipe Line Company connected thirty-four 80,000-barrel gastight storage tanks to overflow tanks via breather tubes. At another, Shell reduced losses from two huge in-ground concrete reservoirs, each capable of holding 1,000,000 barrels, by sealing their wooden roofs with asbestos and keeping the vapor space above the oil filled with carbon dioxide. At other sites, operators eliminated vapor spaces in tanks by filling the bottom of the tank with as much water as necessary to keep the oil surface at the top of the tank.[83] By today's standards, relatively large vapor losses still occurred, but losses overall had been cut in half. Only in flush fields and in a few mature fields producing heavy crude did high-loss storage methods, including open pits, continue into the 1930s.[84]

By the early 1920s, then, the nationwide system for moving crude oil from field to refinery was relatively mature and consisted of an extensive network of gathering lines, storage tanks, and long-distance pipelines. In the process of moving crude, companies lost a significant amount of oil to vapor losses, storage tank fires, and pipeline leaks. Mechanical failures, such as those associated with the collapse of storage tanks, resulted in additional discharges and losses, as did the routine disposal of storage tank bottoms. Table 3.1 summarizes these discharges and losses, along with incentives and barriers associated with reducing them.

Table 3.1. Pollution- and Waste-Related Concerns in the Pipeline Sector, Ca. 1920

Concerns	Incentives to Change	Barriers to Change
Small leaks in pipelines	Repairing leaks when detected reduced the loss of oil, kept lines maintained, and avoided potential damage suits.	The difficulty of detecting small leaks and the lack of practical knowledge about how to prevent corrosion were the main obstacles.
Pipeline breaks	Companies had an incentive to strengthen pipe in areas where breaks were likely to occur (such as stream crossings).	Poor station-to-station and and station-to-crew communication often slowed down clean-up and repair efforts. No good technology for recovering spilled oil existed.
Vapor losses	Pipeline companies had a high incentive to study the problem and find ways to reduce losses.	Capital investment in existing tanks made quick adoption of new technology difficult. The need to pump oil in and out of storage tanks at pumping stations compounded the problem.
Fires at tank farms operated by pipeline companies	Insurance companies and self-insured companies sought ways to decrease the risk of tank fires and the amount of damage caused by fires.	Systematic investigations into the causes of tank fires were just beginning. Changes were slowed by the investment in existing facilities.
Failures of storage tanks	Engineers hoped to reduce tank failures by increasing the role of professional engineering organizations in setting design standards.	Engineering standards were slow to disseminate and not well enforced. As with fire prevention, determining how much to invest in preventing something that might not happen was difficult.
Leaks from tanks	Companies had an incentive to learn more about corrosion and how to prevent it.	No good technology for protecting steel tanks from corrosion existed.
Disposing of oil-soaked sediment from the bottom of tanks	Engineers desired to reclaim as much of the oil as possible from this sediment. Potential damage suits also provided an incentive to avoid the most sloppy practices.	In remote areas, burning or dumping was the most practical option.

Few people saw much need for state-enforced measures aimed at reducing leaks and spills from pipelines. Costs associated with the physical loss of a valuable product and the costs of damage and nuisance suits seemed to be incentive enough. Oil that entered a pipeline system became a documented asset, and any that escaped represented a direct financial loss to the pipeline company. In addition, pipeline companies had to maintain hundreds of miles of pipeline which crossed numerous streams and pieces of property. If ignored, leaks not only could result in significant losses but could also expose the company to damage and nuisance suits. Indeed, as engineers with the Bureau of Mines identified ways to prevent such losses, they assumed that companies would adopt new technology and practices simply because it paid to do so.

Of course, many technologies capable of reducing leaks and spills went unused because those technologies were not economical. For example, pipeline companies could have encased their entire pipeline system in concrete, but nobody chose to do so. Some companies surely would have used this option if the penalty for leaks and spills and the potential profit from transporting oil were high enough. However, in the absence of penalties or regulatory mechanisms, few companies were likely to make the necessary investments.

Some government-enforced regulatory mechanisms as well as some professional and industry-enforced regulations and standards did exist. For example, the design of storage tanks used in many urban areas were subject to fire and safety regulations, and engineers were expected to abide by those regulations. Engineers were also expected to meet the accepted standards of good engineering design set by their profession, and lawsuits could pivot around the question of whether those standards had been met. Insurance companies not only supported the development of professional standards and government-enforced fire and safety ordinances but could also impose their own design standards as a condition of coverage.

At the time, then, additional regulations based on controlling pollution or sustaining a certain level of environmental quality did not seem necessary. Certainly, engineers and technical managers did not ask whether the current mix of standards, regulations, and economic incentives were enough to maintain an acceptable level of environmental quality. If anything, engineers saw themselves protecting the public good by continually improving the efficiency of operations, decreasing the loss of a valuable material, and reducing the cost of transporting oil. And, in the early 1920s, as they turned their attention to the problem of corrosion, engineers in the pipeline sector of the petroleum industry hoped to reduce even further the frequency of pipeline leaks and the magnitude of system losses.

FOUR # Refineries, Pollution Concerns, and Technological Change

IN 1921, THE PETROLEUM that had taken millions of years to accumulate in formations far below the cotton fields of El Dorado, Arkansas, began reaching refineries throughout the eastern half of the United States. Some of these refineries were nothing more than a loosely organized system of pipes, stills, and tanks that stripped kerosene and gasoline from the crude, leaving behind a heavy oil that could be sold as fuel for industrial boilers. Other refineries—larger, more expensive, and more recognizable as forerunners of the huge, specialized plants of today—ran that heavy oil through additional processes, producing motor fuel and creating various types of grease, wax, lubricating oil, or asphalt-based coatings and binders.

Along with discharges of oily ballast from tankers, effluents from refineries became a subject of intense discussion in the hearings leading up to the Oil Pollution Act of 1924. Were laws necessary to prevent industrial plants, especially refineries, from discharging oily wastes into the nation's waterways? Advocates for strong regulations could not understand how one could penalize a ship for releasing oil into a harbor when a gas plant or refinery two hundred feet away did so with impunity. They pointed out that allowing land-based plants to discharge oily wastes into coastal waters would make enforcement even more difficult than it already was.

Refiners argued that it simply was not possible to reduce the concentration of oil in effluents to zero. They also knew that refining technology was changing quickly and that, over the last twenty years or so, engineers had significantly reduced the quantity of pollution-causing discharges released by refineries. To refiners, discharges from a well-operated facility, while large by today's standards, were about as close to zero as one could get. Technically, then, a prohibition on oily discharges would simply mean that all refineries, no matter how modern or how well operated, would always be in violation. Instead, they suggested that refineries be allowed to release quantities of oil not deleterious to health or fish.

However, advocates of strong pollution control legislation simply were not prepared to discuss specific levels of pollution-causing discharges. Rather than moving the discussion in the direction of measuring and monitoring discharges, they continued to speak in terms of a complete prohibition, resulting in an impasse that the representatives of industry used to their benefit. This failure to discuss effluent standards and methods of measuring and monitoring allowed arguments based on increasing efficiency to emerge as the dominant ethic in addressing pollution concerns related to effluents and emissions.

Transforming Crude into Gasoline

In 1915, approximately three hundred refineries in the United States produced a total of 4.8 million gallons of gasoline each day. Most facilities were old, relatively small, and capable of processing no more than one or two thousand barrels of crude per day (fig. 4.1). A handful of large refineries dominated the industry. Indeed, the largest fifteen plants accounted for almost half of all capacity (table 4.1). Over the next six years, the demand for gasoline approximately tripled, but the industry met this demand with far fewer facilities and much less crude than expected.[1]

Quite simply, as the demand for gasoline increased, engineers found ways to obtain more gasoline from a barrel of petroleum and to do so less expensively than before. Hence, when large firms built a new refinery or

Figure 4.1. Early refineries typically consisted of a few stills and storage tanks such as this 500-barrel-per-day refinery constructed by Standard Oil in Neodesha, Kansas, in 1897. (Charles Rothus Collection, American Heritage Center, University of Wyoming, neg. 176031)

renovated an existing one, they integrated the latest equipment and processes into their plants. Small refiners, unable to secure or risk the necessary capital for new equipment, continued to operate older and less sophisticated refineries. By 1920, the capacity of the largest fifteen facilities represented over half of the total capacity in the United States.

Refiners knew that the type of crude oil they started with made a difference. No two oil fields produced petroleum with the same color, viscosity, density, or odor—variations that reflected important differences in chemical content. Refiners, for example, placed a higher value on the light, paraffin-based crude from Pennsylvania than they did on the thick, sulfur-smelling, asphalt-based crude from parts of Texas. When comparing batches of crude coming from the same field, refiners relied on the density

Table 4.1. Largest Refineries in the United States, 1915

Company	Location	Capacity (barrels of crude/day)
Standard of California	El Segundo, California	65,000
Standard of Indiana	Whiting, Indiana	60,000
Gulf Oil	Port Arthur, Texas	55,000
Standard of New Jersey	Bayonne, New Jersey	45,000
Atlantic Refining Company	Philadelphia, Pennsylvania	35,000
Standard of New Jersey	Bayway, New Jersey	30,000
The Texas Company	Port Arthur, Texas	30,000
Union Oil	Oleum, California	25,000
Carter Oil	Norfolk, Oklahoma	25,000
Magnolia	Beaumont, Texas	25,000
Standard of Louisiana	Baton Rouge, Louisiana	20,000
Standard of California	Kern River, California	20,000
Standard of New Jersey	Jersey City, New Jersey	15,000
The Texas Company	Dallas, Texas	15,000
Cosden & Company	Tulsa, Oklahoma	15,000

Source: H. G. James, *Refining Industry of the United States* (Oil City, Pa.: Derrick Publishing, 1916), 7-11.

of the oil, measured in "degrees Baume," as a rough indicator of quality. The Baume scale, an inverted measure of a fluid's specific gravity, assigned a higher number to crude oil that contained more of the valuable lighter fractions. Fluids with a specific gravity of one, the density of water, were assigned a value of 10 degrees Baume. Fluids that were less dense were assigned a larger number, with pure gasoline having a value of around 90. If the operator of a producing lease held crude in storage for too long or, worse yet, allowed it to sit in open ponds, the Baume-scale "gravity" of the oil decreased as lighter and more desirable fractions evaporated.

Therefore, pipeline representatives, when purchasing oil from operators, measured the density of the oil using special hydrometers calibrated in degrees Baume. Depending on the measurement, they adjusted the price they were willing to pay for the oil. For example, oil extracted from the El Dorado field having a density below 33 degrees Baume probably had been sitting around in an open tank, allowing much of the gasoline in the crude to evaporate into the air. Hence, in the early months of El Dorado production, Louisiana Standard paid fifty cents for oil with a specific gravity less than 33 degrees Baume, sixty cents for oil between 33 and 35 degrees Baume, and seventy cents for any oil with higher readings.[2]

When comparing crude oil from different fields, refiners often talked about the "gasoline content" or the "kerosene content" of the crude. For example, El Dorado crude was approximately 9 percent gasoline. In comparison, some crudes from Kansas contained as much as 17 percent gasoline (table 4.2). In the nineteenth century, refiners had little reason to analyze their crude in much more detail. In processing crude, they relied upon experience, trial and error, and standardized tests. Indeed, the technology used by refiners changed little in the industry's first four decades.[3] Typically, companies hired chemists only to solve specific problems. In 1888, even Standard Oil had to hire an outside chemist, Herman Frasch, to develop a process for removing troublesome sulfur compounds from crude oil produced from fields centered around Lima, Ohio.

However, in the decade before World War I, as refiners attempted to transform as much of their crude oil as possible into motor fuel, they began to depend upon the skills of chemists. By the 1920s, all major refineries employed chemists who could analyze the content of a crude in detail and determine how to manufacture products from that crude meeting relatively strict specifications.[4] Even the most introductory descriptions of the refining process included some explanation of hydrocarbon chemistry, with authors illustrating the structure of the simplest hydrocarbons (fig. 4.2).[5]

The authors pointed out that these molecules were gases at atmospher-

Table 4.2. Content of Various Crudes (Average yields, no cracking)

Fraction	El Dorado crude	Kansas crudes	Oklahoma crudes
Gasoline	9%	17%	10%
Naphtha	22%	5%	6%
Kerosene	13%	22%	18%
Gas oil	12%	16%	15%
Remainder	44%	40%	51%

Sources: Jim G. Ferguson, *Minerals in Arkansas* (Little Rock: Ark.: State of Arkansas, 1922), 91; J. W. Coast, Jr., "Refining Petroleum," *Doherty News* (October 1919): 17; H. S. Bell, *American Petroleum Refining*, (New York: D. Van Nostrand Company, 1923), 6.

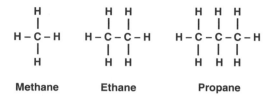

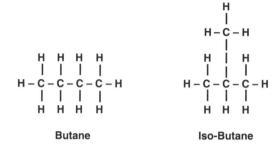

Methane   Ethane   Propane

Butane   Iso-Butane

Figure 4.2. By the 1920s, most books that provided an introduction to refining began by discussing the basic structure of the simplest hydrocarbons. (Illustration by author)

ic pressure and normal temperatures but that larger hydrocarbon molecules were liquids or, beyond a certain size, solids. Textbooks also showed diagrams and listed specific names for more complicated compounds, including many with branches, double bonds, and rings of carbon. However, the sheer number of possible combinations soon overwhelmed any attempt to be exhaustive, especially when diagrams of the larger molecules began to resemble sketches of chicken wire. For example, the molecule $C_{20}H_{42}$, which is a waxy substance in one configuration, has over 350,000 possible isomers.[6] Impurities such as sulfur and nitrogen compounds complicated matters even further. Hence, most of these introductory texts limited their discussion to general patterns, such as the paraffin series (straight chains of carbon with no double bonds); the olefin series (straight chains of carbon with double bonds); the naphthenes (cyclic structures with no double bonds); and the aromatic series (compounds containing a benzene ring, a cyclic structure containing double bonds).

Petroleum chemists saw crude oil as a soup of hydrocarbons, consisting of molecules ranging from dissolved gases to heavy solids that existed in countless tangled forms. Not only did the percentage of gasoline vary from crude to crude, but so did the chemical content of that gasoline. For example, the chemical content of gasoline distilled from El Dorado crude did not match the chemical content of gasoline distilled from Oklahoma crude. Although all gasoline consisted of molecules having the same approximate weight, typically consisting of molecules with five to seven carbon atoms, each cut still contained some unknown mixture of paraffins, olefins, naphthenes, and aromatics. The same could be said of other fractions, such as naphtha and kerosene. The type and quantity of impurities in each fraction also differed.[7]

In general, then, refiners saw their job as being, first, to separate this hydrocarbon soup into roughly homogeneous fractions and, second, to refine those fractions into various types of fuels, solvents, greases, lubricating oils, and coatings, with most products required to meet well-defined specifications. Given that the exact mix of molecules depended on the reservoir from which crude came, each new oil field presented different challenges.

From the beginning of the U.S. petroleum industry, refiners had separated their incoming hydrocarbon soup into various fractions by distillation. In the simplest batch process, obsolete by the 1920s, operators first charged a still with crude oil. Then, operators heated the crude enough to vaporize the lightest fractions, those with molecules containing five to seven carbons, but not enough to vaporize fractions heavier than that. The vapors then passed through water-cooled tubes that condensed the vapors back into a liquid.[8]

This first cut of liquid, which they called gasoline, distilled out as a very light, flammable mixture. The explosive quality of this fluid prevented its use either as an illuminant or as an industrial fuel, and therefore, in the early years of the industry, refineries had little use for this fraction. Most refiners disposed of their gasoline after distillation, often by running the fluid into a pit and burning it. With the success of gasoline-powered automobiles, however, refiners suddenly valued this fraction more than the kerosene fraction.[9]

After capturing the gasoline fraction, operators in charge of the simplest distillation process then raised the temperature of the crude, vaporizing the fraction consisting of molecules with between eight and eleven carbons. Some paint companies used this fraction, called naphtha, as a solvent, but they could not absorb all the naphtha that refineries had available. In the 1890s, Standard Oil had attempted to sell its excess naphtha as a fuel for specially designed stoves, but the effort met with little success. Among other things, insurance industries raised fire insurance premiums for homes and buildings that used the naphtha-burning stove.[10] Therefore, in the years before internal combustion engines, much of this fraction met the same fate as gasoline: disposal by dumping or burning. Naphtha, though, also proved useful as a component in motor fuel.

After removing the gasoline and naphtha, early refiners raised the temperature of the crude to vaporize the fraction sold as kerosene. Next, operators raised the temperature further, vaporizing a heavier fraction known as gas oil, which could be sold as an enriching agent for certain types of illuminating gas or as a fuel for diesel engines.[11] After the gas oil had been distilled out, the heavy oil that remained behind could be further processed to produce other products. Some crudes were more suitable for making wax, grease, and lubricating oil. Others were more suitable for tar, asphalt, road oil, and similar products. However, because the markets for many of these products were limited, much of the heavy oil ended up being sold as fuel for industrial boilers.

The temperature cutoffs defining each fraction were somewhat arbitrary, as were the number of fractions refiners decided to separate. In early refineries, operators had "look boxes" through which they could see the condensing fluid and often made judgments about where to direct the fluid based on changes in color and viscosity.[12] Even if a refiner desired to control the cutoff temperatures precisely, doing so would have been difficult. In addition to heating stills from the outside, many refineries also introduced steam inside the stills to agitate the crude and to prevent the bottom portion from overheating. Although operators monitored the temperature inside the still with pyrometers, nobody could guarantee that the entire stock was at the indicated temperature.[13] Moreover, the pressure in the still, which varied, also affected the temperature at which various fractions vaporized. Therefore, to meet particular product specifications, such as the flash point of a product, refiners typically blended various cuts having different properties as needed.

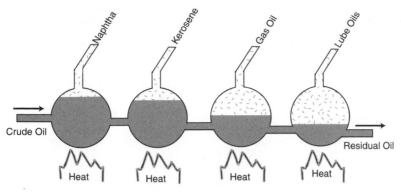

Figure 4.3. The first step toward continuous distillation employed different stills and heat sources to vaporize the different fractions of petroleum. (Illustration based on Harold F. Williamson, Ralph L. Andreano, Arnold R. Daum, and Gilbert C. Klose, *The American Petroleum Industry: The Age of Energy, 1899–1959* [Evanston, Ill.: Northwestern University Press, 1963], 124)

Increases in the demand for motor fuel sent refiners scrambling to pass more crude oil through their facilities and to get more motor fuel from each barrel of that crude. Their success in doing so shaped the technical agenda of that industry for the next half century. Two main innovations were responsible for the largest changes. One new process, thermal cracking, used heat and pressure to "crack" large hydrocarbons into smaller ones that could be used for motor fuel. The other innovation, continuous fractionation, allowed refiners to separate petroleum into various fractions more quickly and less expensively.[14]

Continuous fractionation allowed refiners to distill crude oil into its separate fractions twenty-four hours a day without operators having to recharge vats or sequence temperatures through the various cutoff points. An important step in this direction occurred in 1906, when Max Livingston, superintendent of Standard Oil's Point Breeze refinery in Philadelphia, installed a distillation process in which separate vessels were used for each fraction. The oil flowed continuously and sequentially from one vessel to the next, each designed to remove a specific range of hydrocarbons (fig. 4.3).[15]

Soon after, another engineer with Standard Oil, J. W. Van Dyke, introduced a related innovation: a distillation process that vaporized all the crude at the same time and eliminated the need to sequence crude through various temperatures. Hydrocarbon vapor containing a mix of all fractions passed through a series of condensing towers, each designed to condense different cuts from the vapor. However, Van Dyke designed the process as a batch process, requiring operators to recharge the system each time, which limited its usefulness.[16]

Over the next few years, various engineers modified Van Dyke's fractionation process to be continuous, which served as the forerunner of more sophisticated fractionation equipment known as "bubble towers." Refiners using bubble towers continuously vaporized all incoming crude. These vapors then passed into a tower consisting of various condensing plates at different levels, with similar cuts of hydrocarbons condensing at the same level. The lightest gases bubbled up through all the plates and through any liquid condensed on those plates before rising to the top of the tower. Bubble towers did not appear until the late 1920s and would not become standard equipment until the 1930s, but leading engineers knew what they were trying to achieve and had their eyes on that technology even before World War I.

Thermal cracking represented an even more significant innovation. Although continuous fractionation allowed refineries to distill crude oil faster, more consistently, and less expensively than before, the process did nothing to increase the amount of motor fuel one could obtain from a barrel of crude. Thermal cracking, on the other hand, did. By applying heat and pressure to larger hydrocarbons, refiners could break molecular bonds and transform heavier fractions of crude into gasoline.

Nineteenth-century refiners had been aware of cracking as a physical phenomenon. After vaporizing kerosene out of distillation vats, they discovered that they could produce more kerosene by overheating the heavy oil that remained. However, the kerosene produced by this method tended to have undesirable properties, including a foul odor that was difficult to eliminate. Furthermore, the process resulted in a large amount of petroleum coke (powdered carbon) that had to be shoveled out of the still, a laborious and time-consuming process.[17]

In the early twentieth century, as chemists became more involved with refinery operations, they became convinced that gasoline could also be produced from heavier fractions by subjecting those fractions to higher pressures and temperatures. Numerous questions, though, remained un-answered. Was there a particular cut of heavy oil that would produce the best results? What temperatures should be used and for how long? Would a more favorable reaction take place if the crude were in a liquid or vapor phase?[18] Those who attempted to answer these questions by experimenta-tion—such as William Burton, the head of refining for Indiana Standard (later Amoco) and his associate Robert Humphreys—proceeded at some risk to themselves and their technicians. The riveted vessels of the day were not designed to handle the desired pressures.[19]

In 1914, Indiana Standard put the first commercially successful thermal cracking process, the Burton process, into operation and proceeded to li-cense the technology to other oil companies. By then, stronger steels and a move from rivets to welding made practical the pressure tanks required for successful cracking. Thermal cracking soon established itself as an im-portant way to meet the demand for motor fuel without dramatically in-creasing the amount of crude oil required. By 1921, several different com-panies also reduced their own methods of cracking to practice and li-censed them as well.[20]

Thermal cracking prevented gluts of kerosene and gas oil from occur-ring. Without thermal cracking, refiners would have had to secure more crude oil to produce the desired amount of gasoline. In doing so, they would have also produced more of every other fraction.[21] With cracking, the gas oil fraction could be transformed into gasoline, and refiners did not need as much crude oil. Therefore, instead of becoming more avail-able as oil consumption increased, gas oil became less available. Even re-fineries that focused on the production of gas oil for sale to manufacturers of illuminating gas began licensing the new cracking technology to pro-duce more gasoline.[22]

If crude oil contained only pure hydrocarbons, no processes other than distillation, cracking, and blending would have been necessary. But all crude contains some impurities, including water, various salts, sulfur compounds, nitrogen compounds, and trace amounts of heavy metals. Removing these impurities was as important as separating the crude into fractions.

Except for the water and dirt that settled to the bottom of storage tanks, refiners removed impurities after the crude had been separated into different fractions. The extent to which refiners cleaned a particular fraction depended upon its end use and whether or not anybody would perceive the impurity as a nuisance. Gasoline and kerosene had to be stripped of impurities that interfered with their use as fuels. Lubricating oils, greases, and waxes also required highly specialized treatment processes. Products such as roofing material, sealants, and binders for road-building material required much less treatment.

Sulfur and sulfur compounds typically proved to be the most troublesome. "Sour" crudes—those high in sulfur compounds, with 1 to 5 percent of the crude by weight being sulfur—generally gave off dangerous gases and always produced foul odors upon combustion. "Sweet" crudes contained far less sulfur, often less than .5 percent, but motor fuel and lamp oil produced from that crude still had to be stripped of sulfur. The presence of sulfur also complicated the chemistry of crude. Each hydrocarbon molecule could include one or more sulfur atoms, significantly increasing the number of compounds possible. In addition, temperature and pressure caused some sulfur compounds to break down quickly, often into dangerous or malodorous compounds such as mercaptans, a family of molecules detectable by their highly nauseous smell.[23]

One sulfur compound, hydrogen sulfide (H_2S), is an extremely poisonous gas. If present in any crude exposed to the atmosphere, it wafts effortlessly out of solution, fatally threatening anyone who takes more than one good breath of the gas. Oil field hands working with high-sulfur crude feared the gas and could recount incidents in which escaping hydrogen sulfide gas had settled over low-lying areas at night, leaving fields of dead animals behind in the morning. Although oil field workers also could be fatally overcome by the gas, and some were, most knew enough not to take a second breath and to get away as fast as possible. Even then, some suffered symptoms such as temporary blindness.[24] Refiners removed any hydrogen sulfide present during the distillation process and piped the gas away to be burned.[25]

Other sulfur compounds were more difficult to remove but could not be left in refined products as they caused odors when burned and had a corrosive effect on engines and lamps. Refiners removed many of these compounds by washing distilled gasoline and kerosene with sulfuric acid, a practice that went back to the earliest days of refining.[26] Most refineries performed this "acid wash" by mixing the gasoline or kerosene with dilute sulfuric acid and stirring the mixture in an agitator. Typically they used between one and five pounds of acid for each barrel of product, but certain processes required more. One refinery used up to seventy-five pounds of acid per barrel.[27]

To stir the mixture, operators forced air into the bottom of the agitator and let it rise through the acid-laced crude. In the process, the acid dissolved many of the sulfur compounds present, which stayed with the acid as it settled to the bottom of the tank. After allowing the acid sludge to settle, operators piped the clean gasoline to another tank. There, they neutralized any acid that remained by washing the gasoline with caustic soda.

The acid wash did not remove all sulfur compounds. Some sulfur compounds required a "doctor" treatment, which involved adding a compound such as sodium plumbite (Na_2PbO_2) to the caustic wash. The lead, which could be recovered and reused, transformed the stubborn sulfur compounds into a less troublesome form. Another process, the Frasch process, had been developed in the 1890s. That process involved mixing about 1,500 barrels of the fraction to be "sweetened" with several thousands pounds of finely ground black copper oxide and then redistilling the mixture with the copper in suspension.[28]

In addition to removing sulfur, acid washes also removed some nitrogen compounds, gum-forming olefins, and tarry compounds. Typically, though, refiners worried less about nitrogen compounds than sulfur compounds. Nitrogen impurities did not produce unusual smells upon combustion, and nobody was concerned with the long-term effects of nitrogen oxides in the atmosphere. Refiners did not worry about the presence of heavy metals either. Most metals ended up in heavy asphalt products or stayed in the fraction sold as heavy fuel oil, eventually becoming part of the fly ash produced by oil-burning furnaces.

If achieving the right color and level of clearness were important, refiners also filtered the final product through fuller's earth, a special clay that contained minute, sponge-structured silica granules capable of adsorbing any tarry substances still present. Although a product's color did not affect its performance, consumers had come to expect a certain consistency in color. In 1922, engineers at one refinery discovered the importance of consistency when they decided to filter their gasoline. The farmers who purchased most of that refinery's gasoline had become used to the previous bronze color and did not like the change.[29]

In general, then, removing impurities was more of an art than a well-defined procedure. Most refiners learned by experience what treatment processes worked well. For example, when washing products with acid, refiners could affect the type of compounds dissolved or decomposed by adjusting the quantity of acid, the concentration of the acid, the contact time, and the temperature. Too little contact time might result in too few sulfur compounds being dissolved. Too much contact time could result in desirable compounds being dissolved or altered. The particular choices used by a particular refiner depended on the type of crude being refined, the requirements of the final product, the costs of sulfuric acid, and, to some extent, tradition and whim.

Process Wastes, Disposal Practices, and Pollution

By the 1920s, engineers and technical managers clearly recognized the ability of refinery wastes to pollute streams and waterways (fig. 4.4). Indeed, in a 1922 handbook for petroleum engineers, the author of the refining section instructed engineers responsible for determining the site of new refineries to consider "whether drainage effluent may later form the basis of a pollution suit." That consideration, though, fell seventh in a list of locational factors, well behind access to transportation routes, access to crude supplies, and the availability of water for condensers and boilers. The same author also advised engineers to make themselves aware of local smoke control laws.[30]

The author of another treatise on petroleum technology noted that the public's response to pollution was not all that predictable. He pointed out that "many streams already foul, after further pollution, however slight, are remembered in the public mind as water ways of previous extreme purity."[31] Therefore, he advised engineers not to be complacent even though many bays, rivers, and streams were already polluted, mostly due to raw sewage from towns and cities. In practice, though, unless a town or city

Figure 4.4. By the early 1920s, many refineries, such as this refinery in Ponca City, shown in 1925, were large enough for waste disposal to be taken more seriously. (Continental Oil Company Collection, American Heritage Center, University of Wyoming)

used the receiving body of water as a source of public drinking water, most people expressed little concern with contaminants such as ammonia, phenols, and sulfides. Even when a town or city drew water from the stream receiving a refinery's wastewater, no sense of urgency existed as long as the wastewater did not create foul tastes (fig. 4.5).[32]

Although textbooks warned engineers to comply with local laws, authors rarely gave specific examples containing numbers and detailed descriptions. The result was that nobody really knew what types and levels of discharges were appropriate. In practice, most refineries simply attempted to keep the concentration of their wastes below the point at which people identified their facilities as the source of particularly noxious waste products. Therefore, despite the listing of pollution—or, to be

Figure 4.5. Not everybody perceived discharges from refineries as a concern, such as the boys swimming in front of this Ponca City refinery around 1915. (Continental Oil Company Collection, American Heritage Center, University of Wyoming)

more precise, nuisance and damage suits—as a concern, younger engineers received mixed messages as to how far they had to go in preventing pollution.

This ambivalence toward pollution control carried over into actual practices. At one refinery in Oklahoma, wooden planks prevented emulsified oil in a holding pond from flowing off company property. An engineer who joined the refinery in the early 1920s later noted, "it was my job to see that whenever we had a really heavy rainstorm, somehow or other some of the planks of the dam would become dislodged thus permitting us to lose a quantity of this unusable and unwanted material."[33] This engineer, making the comment fifty years later to illustrate the "primitiveness of oil refineries of the nineteen twenties," also noted that he was "quite sure that there were farmers planting their corn crop each spring between the refinery and the river with no other expectation than to claim damages for the loss of their crop by reason of refinery spillage."[34]

However, refinery engineers generally did not see themselves as polluters. They realized that refinery wastes had decreased significantly over time. They could point to developments that allowed companies to make greater amounts of gasoline with fewer vapor losses, leaks, and spills. Hence, most refinery engineers perceived themselves as actively working to reduce pollution-causing waste.

Indeed, some believed that, in light of the dramatic drop in refinery wastes associated with unwanted fractions, concern over pollution seemed trivial. At a government hearing in the mid-1920s, G. S. Davison, the president of Gulf Oil, recalled the days when the Allegheny River was "coated with the floating tar seeping or discharged from a neighborhood petroleum refinery." Those days, he asserted, were over because there was no longer "the smallest fraction of the crude oil, whether solid, liquid, or gaseous, which cannot be made to serve a useful purpose." As evidence of the change, he pointed to the return of "the mosquito plague" in the area around Gulf Oil's Port Arthur refinery because the mosquito-killing wastes that had previously covered the water had disappeared. Only incidental losses, which he estimated as being between 2.6 and 8.0 percent of throughput, remained.[35]

But what did refineries do with their acid sludge, caustic soda, oily water, emulsified oil, unusable residue, spent chemicals, and used filtering clays? How did refineries dispose of their waste products and did any technical personnel perceive there to be a problem? Precise answers depend on the specific refinery, with variations in size, location, and processes being significant. Table 4.3 summarizes the general pollution- and waste-related concerns faced by refineries in the early 1920s.

Refineries known as skimming plants, which distilled gasoline, kerosene, and gas oil from the incoming crude and then sold everything that remained as industrial fuel, typically generated less waste than refineries that transformed heavy oil into waxes, greases, lubricating oil, and asphalt-based products. The type of crude processed by a refinery also mattered. A crude containing a high concentration of sulfur and other impurities resulted in greater losses and more spent acid sludge than crudes

Table 4.3. Pollution- and Waste-Related Concerns in the Refining Sector, Ca. 1920

Concerns	Incentives to Change	Barriers to Change
Vapors from refinery storage tanks	Cost of lost material encouraged efforts to research problem and find ways to prevent vapors from escaping.	Capital investment in existing tanks made quick adoption of new technology difficult.
Smoke and other by-products of combustion	Smoke and obnoxious odors could be identified as a nuisance and trigger law suits.	The technology to reduce emissions of sulfur dioxide, whether from burning fuel or flaring gases, was limited.
Flaring of gases	Flared gases could potentially be used as fuel or product.	The technology to separate hydrogen sulfide from other gases was undeveloped, and flaring was often the more economical option.
Acid sludge	Reclaiming sulfuric acid saved money and burning recovered hydrocarbons as fuel eased a disposal problem.	Many refineries found it more economical to continue dumping the sludge.
Oily water being discharged into rivers and bays	Pollution suits and local laws encouraged refineries to skim away any oil that floated on the surface of the water.	Emulsified oil was difficult to detect and remove. Also, storm bursts could overwhelm a refinery's capacity for separating oil from wastewater.
Leaks from tanks, pipes, and valves	Preventing leaks reduced the loss of product and the chance of fire.	Detecting and preventing underground leaks was difficult. Also, altering the physical plant to direct surface leaks towards sewers was expensive.
Sludge and sediment from tanks and oil-water separators	Little incentive to change disposal practices.	Few alternatives to existing dumping practices were available.
Spent caustic	Reclaiming caustic reduced the quantity of material once needed to replace.	Spent caustic was seen as a useful way to control the pH of wastewater.
Discharges due to fire or explosion	The loss of equipment and production time provided a high incentive to prevent fires and explosion. Insurance companies also provided some incentives.	Accidents were seen as difficult to prevent, and changing the safety-related culture of refinery personnel was difficult.
Presence of various chemicals in waste water	Chemicals that caused a foul taste in drinking water usually resulted in pressure to take action.	Single sewer systems made capturing and treating such wastes difficult.

containing few impurities. Although refiners sought to transform every ounce of crude they received into a marketable product, roughly 4 percent of the crude that entered the average refinery never made it out the door as a product.[36] And that amount added up. At a refinery capable of processing 11,000 barrels of crude each day—a large refinery at the time, but still six or seven times smaller than the largest in operation—a loss of that magnitude translated into approximately 175,000 barrels of oil each year.

In general, to compute losses, engineers compared the volume of crude entering the refinery with the volume of products leaving. However, keeping precise records was difficult because the volume of petroleum and petroleum products changed with temperature. Thermal cracking further complicated matters because that process, in breaking large molecules into smaller ones, produced a fraction less dense than the feedstock and, in the absence of any loss, generated a net increase in volume. Keeping track of weight rather than volume would have helped, but refiners found measuring transactions by volume—usually by changes in the level of oil in a storage tank—more convenient. Refineries also used some portion of their throughput as fuel, and not all refineries recorded that use too carefully.

Although nobody kept too precise an account of how much oil was lost from what source and although those amounts varied from refinery to refinery, engineers had a general idea of what happened to material that never made it out as products. All of it exited in one of several ways: through the refinery's sewer, by physical removal of sludge or spent material that contained the oil, by seepage into the ground, or by emissions to the atmosphere. For example, refineries using unsealed storage tanks lost almost 2 percent of all petroleum and petroleum products as vapors, with much of that loss occurring during transfer operations when fluid entering a tank could push out a tankful of vapors.[37] Gases flared during the refining process also represented a significant loss of hydrocarbons. For example, light hydrocarbons created during cracking, mainly butane and propane mixed with impurities such as hydrogen sulfide, could not be stored or marketed economically and were often flared.[38]

When washed with sulfuric acid, relatively large amounts of a distilled fraction, from 4 to 20 percent by volume, also ended up as part of the sludge generated during that cleaning processes.[39] Many refineries disposed of this sludge without any recovery effort. Oil-soaked sediment removed from the bottom of storage tanks and oil disposed of with spent filtering material added to these losses. Steam that came into contact with oil also carried away some hydrocarbons, as did leaks in valves, hoses, condensers, and pipes. For the most part, reducing such losses did not require any revolutionary new technology but incremental changes associated with designing better sewers and storage tanks, installing processes for the recovery of spent material, and finding ways to use gases as fuels. However, the effort and costs associated with making such incremental changes at older refineries were significant.

All water used at a refinery, as much as 500,000 gallons for every thousand barrels of crude processed, eventually left the refinery as steam or through one or more sewers. In the report on oil pollution that the Bureau of Mines prepared during the hearings leading up to the Oil Pollution Act of 1924, engineers estimated that approximately 5 percent of the oil flowing through a refinery also found its way to these sewers. Although skimmers captured approximately 80 percent of this oil, the remainder flowed out with the wastewater.[40]

Refineries used most of their water to condense hydrocarbon vapors back into liquids. Ideally, this water ran through condenser tubes and never came into contact with any oil. However, leaky tubes and condenser fittings, often plagued with corrosion problems, always allowed some oil, under pressure, to enter the water. Refineries also turned large quantities of water into steam, some of which came into direct contact with hydrocarbon vapor and carried absorbed oil and gases into the sewer after condensing. Water and steam used to clean tanks and equipment also exited through a refinery's sewer system, as did oily water that operators drew off from the bottom of settling tanks. Oil leaking from myriad valves, pipes, tanks, and vessels also made its way to the plant's sewer system or, in the case of many older refineries, seeped into the bare dirt. Such leaks added up, with approximately one-half to one percent of a refinery's throughput following this route. Heavy rains eventually washed any surface oil on the refinery's grounds into the sewer system as well.[41]

According to one handbook on refinery operations written in 1923, the sewer system at a refinery should catch all wastewater and spilled oil, with all oil skimmed away before being discharged. The recovery of this oil was important, the author noted, "not only from a financial standpoint, but because if allowed to flow into streams or other bodies of water, contamination would follow."[42] Despite efforts at well-managed refineries to keep oil out of the final effluent by skimming, a significant amount of emulsified oil remained dispersed throughout the water, and could not be easily separated from the wastewater. Any sludge, spent clay, or dirt that made it into a refinery sewer only increased the ease with which emulsification occurred, and nobody measured the amount of oil actually escaping past a refinery's oil-water separator in the form of emulsions.

Refineries that had only one sewer system for both wastewater and storm water faced another problem. Most oil-water separators consisted of a settling pool with some method for skimming oil from the surface. Under most conditions, the settling pool could handle the wastewater flowing out of the refinery. However, during a downpour, runoff from the refinery grounds and from the land surrounding the refinery poured into the refinery sewer, flooding the settling pool and carrying large quantities of oil into the receiving body of water.[43] Such incidents sometimes led to complaints. For example, after reading an article about the hearings on oil pollution, R. J. Hansen, who fished regularly off the shore of New Jersey, wrote a letter to Commerce Secretary Herbert Hoover. Hansen complained that one facility "let hundreds of gallons of oil run into the basin" after "every heavy rain."[44] In addition to taking advantage of storms to get rid of troublesome wastes, the owners of this facility—in this case, a producer of manufactured gas from coal—probably operated a poorly designed sewer system.

Sludges and spent chemicals also posed a disposal problem. In the nineteenth century, most refineries disposed of their sludges and spent chemicals by dumping them in a convenient spot near the refinery. By the 1920s, operators at the largest refineries, especially those in populated areas, could no longer dispose of their wastes in so casual a manner. The sheer quantity of wastes being generated forced them to put more thought into disposal methods, which resulted in some attempts to recover and recycle spent chemicals and filtering material. Smaller refiners in unpopulated areas continued to pay much less attention to their disposal practices.

Refiners certainly wanted to transform as much as possible of the crude oil they received into marketable products. At the very least, the heaviest fractions could be sold as fuel oil. In one case, engineers at a refinery operated by Sun Oil stored a waste by-product for years, believing it had some value. They started storing the material after developing a new process for cleaning lubricants produced from a particular type of Texas crude. In that process, naphthenic soaps and sulfonates sank to the bottom of the treating tanks, which operators drew off as waste. The company stored this material in tanks until the mid-1920s, filling several 50,000 barrel tanks before engineers developed a use for this material.[45]

Figuring out what to do with their acid sludge represented a larger challenge to refiners. Washing cuts of petroleum with sulfuric acid removed 4 percent or more of the material being treated, resulting in sludge consisting of sulfuric acid, tarry compounds, and sulfur-containing organic matter. An average-sized refinery, one that processed about 11,000 barrels of crude per day, generated between six and thirty tons of this acid sludge daily. Most refineries disposed of their acid sludge by some combination of dumping, flushing, and if the oil content were high enough, burning.[46]

Alternatives to disposal did exist. In 1891, one Standard Oil refinery, which had previously dumped its acid sludge into the Atlantic Ocean, began selling that sludge to the Rasin Fertilizer Company for use as feedstock in the manufacture of fertilizer. By 1911, some Standard Oil refineries started manufacturing their own sulfuric acid and, in the process, reclaimed as much of their own acid sludge as possible.[47]

Most refineries, though, were too small to justify the recovery of waste acid. Even if a refinery had the luxury of being located near a fertilizer manufacturer willing to purchase their acid sludge, storage and transportation costs still could make disposal the more economical option. For example, in 1924, the American Agricultural Chemical Company offered one refinery $3.50 per ton for any waste liquid removed from its sludge. Each day, this refinery processed about 15,000 barrels of crude and generated about ten tons of acid sludge from their "light oil" agitators. Although the company routinely "cooked" the sludge to remove some of the liquid before disposal, engineers determined that capturing, storing, and transporting the liquid would not be worth the effort. The company also produced a batch of "heavy sludge" but was hesitant to "cook" off any liquid because that made it too difficult to haul the heavy sludge away in trucks.[48] Such refiners also had less of an incentive to recover more of their spent caustic. Mixing the caustic waste with waste acid kept the pH of their effluent within a reasonable range. Given that scientists had identified a low pH as being particularly detrimental to the aquatic life, mixing the two wastes together appeared to make good sense.

Refiners also had to dispose of the spent fuller's earth they used as a filtering material. After the spongelike granules of that material became clogged with tars and coloring matter, some refiners attempted to regenerate its filtering capacity by burning off the impurities that had been captured. However, fuller's earth regenerated in this way proved less effective than the original clay. Hence, most refiners ended up dumping tons of spent clay, along with any absorbed hydrocarbons, in convenient areas near their plants.[49]

Refineries also emitted large quantities of wastes into the air. In the early 1920s, most facilities looked more like small manufacturing plants than the jungle of exposed reaction vessels and pipes that we now associate with refineries. Companies still housed their equipment inside brick buildings. Nobody, though, would mistake a refinery for a manufacturer of shoes or some similar product. Various types of visible emissions poured out of all refineries: including plumes of steam and smoke from boilers and flares. Vapors associated with the loss of volatile hydrocarbons represented another type of emission, as did the corrosive fumes from acid sludge that came into contact with water.

In general, refineries made much less progress in reducing their emissions into the air than they did in reducing the quantity of pollution-causing contaminants in their waste water. Thermal cracking actually increased the amount of material that individual refineries released to the atmosphere. In breaking down heavy molecules into lighter ones suitable for gasoline, the cracking process also created light gases too volatile for gasoline. Initially, many refiners flared those gases to the atmosphere. Still, thermal cracking reduced the absolute size of the waste stream. After all, cracking allowed refiners to supply the levels of gasoline demanded by U.S. consumers with much less crude oil than otherwise would have been necessary.

Unlike concerns associated with water pollution, which were being debated at the national level, concerns associated with emissions were typically local. Civic leaders sometimes pressured industrial plants to burn their fuel more efficiently so as to eliminate any dark smoke being generated. They would also pressure facilities to take action when odors, such as those associated with emissions of sulfur dioxide, were particularly obnoxious. In response to such pressure, individual refineries could take limited action. For example, they could install higher stacks and stop burning their acid sludge. Debate, though, typically remained at the local level.[50]

Refineries and the Oil Pollution Hearings

In late 1923 and early 1924, as the hearings leading up to the Oil Pollution Act of 1924 came to a close, refiners found themselves fighting to prevent their oily discharges from being prohibited by federal legislation. As it happened, the increases in efficiency secured by refineries over the previous twenty years provided a base for arguing against the need for strong pollution control legislation. Refiners noted that they had come a long way in reducing the discharge of pollution-causing waste on their own. Why not let them just keep going down that path?

Initially, it appeared as if representatives from the refining industry and the National Coast Anti-Pollution League had reached an agreement. Soon after the Bureau of Mines released its report on oil pollution, Sedley Hopkins Phinney, the secretary of the Anti-Pollution League, informed Commerce Secretary Hoover that "a spirit of mutual trust" had emerged between his group and representatives from the American Petroleum Institute. At a meeting with the Anti-Pollution League, J. H. Hayes, the attorney for Jersey Standard, introduced a resolution that stated:

. . . inasmuch as the elimination and control of pollution in navigable waters is a problem that will require the continued, united, and cooperative efforts of all parties interested or affected thereby, it is recommended that a permanent advisory board be organized to meet after March 1, 1924, to be composed of representatives from the Federal government, States and Municipalities bordering on navigable or coastal waters, business men, affected industries, and members of the Women's Committee, for the purpose of cooperating in the elimination of pollution, promoting and assisting in educational campaigns, and for advising and educating in the development of methods and devices for the elimination and control of pollution and to ascertain the facts relative to such pollution as they may exist at any time, and to make recommendations for the elimination of any such pollution to the League.[51]

A second part of the resolution called for a federal investigation to develop "the means to eliminate or neutralize the elements which may be found to cause pollution" and to do so "in the interest of public economy and efficiency." Phinney also indicated that the API agreed to "support a bill similar to last year's Frelinghuysen Bill to be somewhat amended by a committee on which they will have adequate representation."[52]

When it came time to discuss specifics, however, any sense of cooperation quickly dissolved. According to Hayes, the point on which the groups differed revolved around whether oily discharges from land-based sources should be included in the proposed legislation. Their meetings ended abruptly when attendees could not agree on an answer to this question.[53]

So many pollution control bills were then introduced, each with subtle differences, that few could keep the bills straight (table 4.4). Even those who introduced the bills tended to stumble when asked to articulate the differences between their bills and others. Most participants, though, knew the main issues. First, and most controversial, was the question of whether the legislation should prohibit the discharge of oil both from ships and land-based plants or from ships alone. Second, should the legislation affect only discharges in coastal waters or discharges in inland rivers and lakes as well? Third, who should enforce the legislation? Bills varied on other grounds as well—such who was liable for fines, whether oil in sewage should be prohibited, and whether other forms of industrial pollution should be prohibited—but debates over those differences remained in the background.

At congressional hearings held to discuss this last wave of bills, representatives from the oil industry steadfastly defended their position. At one

Table 4.4. Major Oil Pollution Bills, 68th Congress, 1st Session

Bill	Source Being Regulated	Area of Enforcement	Responsibility for Enforcement
HR 51	Ships only	Coastal waters only	War Dept.
HR 612	Ships only	Coastal waters only	Commerce Dept.
S 1388	Ships only	All navigable waters	War Dept.
S 936 HR 3319	Ships and land-based plants	Coastal waters only	War Dept.
S 42 HR 203	Ships and land-based plants	All navigable waters	War Dept.

point, the API's director of research, Van H. Manning, pointed to statistics collected by the port supervisor of New York, who already had the right to control discharges of oil into the harbor. Of the 125 violations that had been issued by the port supervisor, most applied to discharges from floating craft. Only two incidents involved refineries.[54] Manning also cited data as to the amount of waste oil being discharged by Jersey Standard's Bayway plant. According to Manning, that refinery discharged twenty million gallons of water each day and, in that water, only two barrels of oil escaped. Based on these numbers, the concentration of oil in that plant's effluent was about .0004 percent. Though certainly low (the concentration was estimated, not actually measured) the exact figure did not really matter. In twenty million gallons of water, even significant quantities of oil seem negligible when expressed as a concentration.

Another speaker from the API, Francis McIlheny, an attorney for Sun Oil, was more to the point. He repeated Manning's numbers and pointed out that "it is absolutely impossible to get out every atom of oil. A chemical analysis at one of our outlets will show oil. It may not be visible to the eye or injurious to the rivers, but as long as our refineries are there, there is a certain amount of oil going into the water." If land-based plants were included in the legislation, he emphasized, "we will be guilty of a technical offense."[55]

The discussion moved toward the issue of using some concentration of oil—McIlheny suggested 1 percent—to distinguish between acceptable and unacceptable levels of oil. The senator questioning McIlheny noted that if only .0004 of 1 percent of the effluent consisted of oil, the cutoff could be much smaller than 1 percent. However, the attorney for Jersey Standard, J. H. Hayes, then moved that discussion in another direction by suggesting language more flexible than a specific concentration. He recommended the phrase "not deleterious to health or fish." However, both Hayes and McIlheny also reiterated their original position: refineries should not be included in any legislation because they would divert attention from the real problem. The legislation, McIlheny argued, should focus on the "elephants," not the "mice." In this analogy, refineries fell under the category of "mice." Ships were the "elephants."[56]

Sanitary and marine inspectors from New York did not understand the logic of these two spokespeople for the API. They pointed out that even if refineries were as well behaved as Manning asserted, other facilities were not. Specifically, facilities manufacturing gas from coal routinely discharged tarry material into the waters around New York and needed to be controlled. Although only two refineries had been cited for violations, twenty gas plants had been cited (table 4.5). Furthermore, the sanitary and marine inspectors pointed out, not all refineries were as well behaved as Manning suggested. Even if a refinery did have a sufficiently large oil-water separator, and not all did, no separator could handle storm drainage. In some newer refineries, storm water entered a different sewer than oily wastes and spills, but those refineries were the exception.[57]

Although proponents of strong pollution control legislation made solid arguments in favor of including land-based plants in any legislation, they failed to address tough questions associated with measuring, monitoring, and enforcement. Indeed, most speakers continued to talk in terms of prohibiting oily discharges, not regulating those discharges. Serious discussion of how much oil could be discharged without being a problem never took place.

Table 4.5. Pollution Violations Issued By the Supervisor of New York Harbor, 1919–22.

Source	No. of Violations
Oil-burning steamships	82
Gas plants	20
Oil tankers	6
Piers	4
Ship repair yards	3
Tugboats	2
Refineries	2
Coal tar plants	1
Sewer department	1

Source: U.S. Bureau of Mines, "Pollution By Oil of the Coast Waters of the United States," preliminary report, Sept. 1923, 33.

As to the question of which government agency should actually enforce the legislation, few involved in the hearings had strong feelings. The original Rivers and Harbors Act of 1899 gave that responsibility to the secretary of war, mainly because the Army Corps of Engineers had responsibility for enforcing laws associated with navigation. Hence, almost all of the bills kept the secretary of war responsible for enforcement. However, Commerce Secretary Hoover, as overseer of fisheries and steamship practices, wanted the Commerce Department to be responsible for enforcement and lobbied for that change on the grounds that his steamship inspectors could take on the job at a trivial cost.[58] The bill that Hoover supported was the only bill that specified the commerce secretary as the enforcing agent.

In the final vote, Hoover's position—that the legislation should focus only on the discharges of oil from ships because that effort represented a wise first step—carried weight. Most legislators saw him as a practical engineer, an effective administrator, and an outspoken advocate of conservation. These qualities suggested that his judgment could be trusted on an issue having to do with resource use, technology, and economics. Hence, when Hoover articulated support for the most lenient version of the bill, arguing that it represented a practical and positive step in the right direction, many legislators sided with him.

Not unexpectedly, those seeking stronger legislation strenuously sought to change Hoover's mind. A few, hoping to pressure Hoover, accused him of being an agent of Standard Oil. This strategy accomplished nothing other than to drive a wedge between them and Hoover.[59] In the end, the weak legislation that Hoover supported passed. However, in the process of reconciling the different versions of the bill passed by the Senate and House, the responsibility of enforcement stayed with the War Department. The strain between Hoover and those who supported a stronger bill undermined any lobbying effort to place enforcement in the Department of Commerce.

In the end, the leap from oil slicks to the specifics of regulating industrial effluents proved too great. The final legislation simply addressed the immediate problem of oil-burning ships and oil tankers discharging oily residue into harbors and coastal waters. Although a provision of the act called for further investigations, refineries and other industrial plants generally were left to chart their own course and to move at their own pace in the area of pollution control. In reporting on passage of the bill to his employer, attorney Francis McIlheny noted that the legislation contained "the ideas of the Committee of the Petroleum Institute" and represented a bill "of which the oil industry cannot, I think, properly complain."[60]

What to Do with Tankers?

IN THE CONGRESSIONAL HEARINGS leading up to the Oil Pollution Act of 1924, participants recognized that oily discharges from tankers and oil-burning steamships represented the major source of scum floating in the nation's harbors. The problem was straightforward. Much of the residue that stuck to the sides of fuel and cargo tanks ended up being flushed into the sea. Given that the number of oil-burning steamships and oil tankers had increased significantly over the last decade, the practice of discharging oily water into harbors had become a serious problem.

The way to solve this problem, however, was not as clear. Should the government simply prohibit ships from discharging oily water into the harbor? If so, how would that prohibition be enforced? Technically, any power that U.S. authorities had over foreign ships extended only three miles out. Yet oily wastes discharged ten and twenty miles away could easily reach shore. Or should the government provide a way for ships to dispose of their oily wastes? But then who would pay for this service? If ships were expected to pay for the service, what would prevent them from discharging their oily water just outside the three-mile limit?

Other issues also complicated matters. In general, people knew that any oil pollution was undesirable and, therefore, made no attempt to quantify what concentration of oil represented a problem. For example, in its report on oil pollution, the Bureau of Mines described sources of pollution as being "minor" or "significant" and described harbors as being in "good" condition or "bad" condition, but rarely attempted to be more precise. At one of the early congressional hearings, someone asserted that two tablespoons of oil could cover a square mile of water, and this became a "fact" repeated, in one form or another, over and over again.[1] Depending on how one interpreted this piece of information, the situation could look either harmless, with a relatively small amount of oil responsible for what seemed to be a big problem, or hopeless, as preventing the discharge of such small amounts might be impossible.

In any case, ships discharged far more than a "tablespoon" of oil. Oil-burning steamships and cargo tankers, including those distributing heavy fuel oil up and down the Atlantic coast, routinely flushed thousands of gallons of oily water and tarry residue into the sea when they emptied their ballast water or cleaned their tanks. Many people in positions of authority hoped that the entire problem could be addressed with new technology. They reasoned that if shipowners could recover the oily residue left in their tanks and put it to good use, they might have enough of an incentive to change their practices without the pressure of government-enforced regulation. In that case, there would be no need for anybody to be more precise about what concentration of oil constituted "pollution." The problem would just go away. No such technology, however, emerged.

With the Oil Pollution Act of 1924, Congress avoided, or rather delayed, addressing many of the fundamental questions associated with oily discharges from tankers. The act simply prohibited ships from discharging any oily water within the three-mile limit. The act specified no concentration thresholds, required no monitoring or reporting, and included no provisions for preventing discharges further from shore. In practice, most oceangoing ships began flushing their tanks well beyond the three-mile limit. This solution reduced the most obvious symptoms—large amounts of oily scum in harbors and along beaches—but delayed efforts to seriously address the concern.

Experience with Tankers

In 1921, the year in which Congress held its first hearing on oil pollution, most ocean-going tankers in the United States carried oil from the Gulf of Mexico to the Atlantic seaboard. Even oil extracted from interior fields such as the newly developed field in El Dorado, Arkansas, reached the East Coast via this tanker route, with pipelines first carrying the oil to loading stations on the Gulf. For long distances, tankers were simply more economical than pipelines. They also allowed companies to reach their own refineries without having to use another firm's pipeline.

The companies that pioneered this tanker route initiated it around the turn of the century when the famous Spindletop field near Beaumont, Texas, flooded the local market with heavy crude. That crude contained little kerosene, then the main product obtained from petroleum, and failed to command the attention of Standard Oil. Instead, independent companies such as the Texas Company (later Texaco), Guffey Petroleum (later Gulf Oil), and Sun Oil began marketing the oil for other purposes, primarily as an enriching agent for certain types of illuminating gas and as fuel oil for industrial boilers and steamships. To move that heavy crude to areas where some demand for gas oil and fuel oil existed, these companies bypassed Standard Oil's pipeline system and shipped it by tanker to cities on the East Coast.[2]

The use of bulk tankers by these relatively small companies did not represent a major technological innovation. By the time Spindletop started producing oil, between seventy and eighty ships already carried petroleum in bulk.[3] The Royal Dutch Oil Company and the Shell Transport and Trading Company used many of these tankers to carry crude oil from the Baku region of Russia to ports in Europe. The Anglo-American Oil Company, the British affiliate of Standard Oil, also imported a significant amount of crude by tanker from the United States. These global companies also shipped some refined products in bulk, mainly kerosene to Asian markets.[4]

Indeed, oil companies had been transporting oil by oceangoing ships for almost a half century. Initially, they transported the oil in wooden barrels that had to be loaded and unloaded individually. To avoid this effort, oil merchants soon installed bulk tanks on their ships, which they filled using pumps. Crews quickly discovered that any spillage in the gaps between these tanks and the ship's hull produced lingering vapors that could explode unexpectedly. Hence, the captains of such ships prohibited open flames and put off making repairs that required hot rivets until all vapors could be eliminated. The added tanks also made inspecting the hull's interior surface difficult. Naval architects eventually addressed these problems by making the hull an integral part of the oil tank, thereby eliminating the most dangerous and hard-to-reach gaps.[5]

The earliest ships with integral tanks faced problems with oil shifting and sloshing during rough seas, which compromised a ship's stability and placed significant stresses on structural members. To limit the oil's movement, shipbuilders then installed longitudinal bulkheads that prevented any oil from crossing the centerline of the ship. They also constructed tanks with expansion trunks—that is, a narrow upper section—which further limited how far the oil could shift (fig. 5.1). A prototype of the modern tanker, one that incorporated features such as integral tanks and expansion trunks, first sailed in 1886 as the *Gluckauf*. British-built and German-owned, this tanker could carry approximately 20,000 barrels of crude. Interestingly, the ship was designed with dedicated ballast tanks, which meant that seawater used for ballast did not come into contact with any oil and remained free of oil. This practice of reserving space for dedicated ballast tanks, however, soon died out.[6]

The ships put into service by the Texas Company, Guffey Petroleum, and Sun Oil approximately fifteen years after the *Gluckauf* did not require any major innovations in tanker design. Indeed, the Texas Company and the Sun Oil Company simply retrofitted existing cargo ships. In the case of Sun Oil, the company already had contracted with a shipyard to convert a Great Lakes cargo ship into a tanker by adding oil-tight tanks, longitudinal bulkheads, pumps, and all the necessary piping. Originally, the company had planned to transport crude oil from Mexico to its gas oil refinery on the Delaware River.[7] The Texas Company, which ended up selling much of its Beaumont crude as industrial fuel to utilities, also converted several cargo ships for use as tankers.[8] The Guffey Petroleum Company, financed by the Mellons of Pittsburgh, initially made a deal with the British-based Shell Transport and Trading Company to ship its crude to

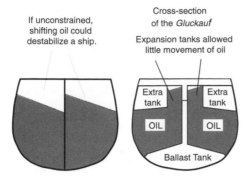

If unconstrained, shifting oil could destabilize a ship.

Cross-section of the *Gluckauf*

Expansion tanks allowed little movement of oil

Extra tank

Extra tank

OIL

OIL

Ballast Tank

Figure 5.1. The *Gluckauf*, which sailed in 1866, was the first ship to incorporate the major features of a dedicated oil tanker, including tanks integrated into the hull and expansion tanks that prevented cargo from shifting. (Illustration based on Harold F. Williamson and Arnold R. Daum, *The American Petroleum Industry: The Age of Illumination, 1859–1899* [Evanston, Ill.: Northwestern University Press, 1959], 643)

buyers in London. Eventually, though, Guffey Petroleum purchased its own tanker.[9]

These tankers represented a large investment to these small but growing companies. Therefore, when production from Spindletop dropped off in 1904, the companies that operated these ships quickly sought new sources of oil. All three companies eventually built pipelines to carry oil from new fields in Oklahoma to terminals on the Gulf. In securing a steady supply of oil, they established the tanker run from the Gulf of Mexico to the East Coast as a significant part of the system for transporting crude in the United States.

Over the next two decades, tanker designs became increasingly standardized. In the years just before World War I, the Mexican Eagle Oil Company, which had a contract to supply the British Navy with fuel, ordered nineteen dedicated tankers from British shipyards.[10] These tankers, ten capable of carrying about 16,000 tons (110,000 barrels) of oil and nine capable of carrying about 9,000 tons (60,000 barrels), incorporated much of the industry's accumulated experience with tankers into their design.[11] Many ships constructed during the war followed the same general design, and by 1921, almost 250 dedicated oil tankers having an average capacity of about 9,000 tons were in operation.[12]

Tanker operations also became increasingly standardized. By the early 1920s, most companies trained crews specifically for tanker service, emphasizing how to use the maze of valves, pumps, and hoses needed to load, unload, and clean the ship's labyrinth of oil-carrying compartments. Guides and manuals left it up to the captain to decide what to do with any oily water that had to be discharged, but one author, Captain Herbert John White, noted that "the officer who has the tanks cleaned and ready to receive cargo in the quickest time after the ship's arrival at port is the one who is more highly thought of by his owners."[13]

Tankers and Pollution-Causing Discharges

As summarized in table 5.1, numerous pollution-related concerns were associated with the operation of oil tankers. For example, to have all tanks cleaned before reaching port often meant flushing tanks with seawater and discharging the oily water into the sea.[14] Few refineries had facilities capable of receiving and treating oily slop and oily ballast water.[15] Captains ordered crews to perform a more thorough cleaning when repairs were scheduled or when changing from a "dirty" cargo of heavy crude to "clean" refined products such as lubricating oil or gasoline. In some cases, such as when a ship had just carried a heavy creosote oil, crews first sprayed down tanks with a "wash oil" that dissolved whatever residue had been left in a compartment, a procedure usually executed when the ship was at a refinery.[16]

The reason crews cleaned tanks so thoroughly before repairs was to prevent an explosion. Although shipbuilders and oil companies had settled on a relatively stable design for tankers by the 1920s, captains of these tankers and the insurance syndicates that assumed the risk of accident and fire realized that empty tankers were vulnerable to explosion. When full, a tanker lacked the dangerous mixture of oxygen and hydrocarbon vapor needed to support an explosion. However, after a tanker unloaded its cargo, the vapors that remained proved dangerous. If the ratio of air to hydrocarbons reached a ratio of about four to one, the slightest spark could set off an explosion strong enough to tear the hull apart.

Any residue that settled to the bottom of a cargo tank, along with any oil sticking to the inside walls, constantly created new vapors, preventing a crew from simply venting the tank. The most thorough way to rid a tank of these dangerous vapors involved heating the oil by injecting steam into the tanks, hosing down the walls of the tank with sea water, and wiping down the inside surfaces. Given that about one-half of one percent of the original cargo remained behind after unloading, the residue in a tanker

Table 5.1. Pollution- and Waste-Related Concerns Associated with Oil Tankers, Ca. 1920

Concerns	Incentives to Change	Barriers to Change
Discharge of oily ballast water and slop from cleaning tanks	Public complaints about oil slicks and oil-coated beaches led to the threat of federal legislation.	No good solution available to tanker captains other than to flush tanks as far away from shore as possible. Enforcement of regulations was difficult.
Discharges due to accidents, collision, and grounding	Insurance companies and their certification organizations pressured shipowners to develop safe practices.	The law of the sea limited liability of the shipowner to the value of the ship and its cargo after the accident.
Disposing of oil before repairs at drydocks	Good fuel or product could be pumped out and stored in tanks.	Amounts not worth the effort to store often ended up being discharged.
Hose drainage and spills during transfer operations	If harbors were not already covered with oil slicks, the source of such spills was easy to identify.	The amounts spilled did not represent a large economic loss to oil companies. Most docks were not equipped to capture any spills that occurred.
Operator-caused accidents during loading and unloading	Ship owners had a significant incentive to train officers charged with setting valves.	Small accidents often went undetected.
Intentional discharges of fuel or cargo during storms	Insurance companies pressured captains to operate safely.	Until load lines were implemented, as they were in England, there was no easy way to prevent such practices.
Leaky hulls	Oil trails from such leaks advertised the ship's poor condition.	Riveted hulls that leaked were difficult to repair.

capable of carrying 9,000 tons (about 60,000 barrels) of oil could easily exceed 45 tons, all of which had to be removed for a full cleaning of all tanks.

Empty tankers leaving refineries on the East Coast sometimes loaded refinery wastes to dump on the return trip. For example, after unloading a cargo of crude oil at a refinery along New York's East River, one captain would, through "mutual gentlemanly understanding," take on whatever slop the refinery needed to get rid of. He noted that, in the years before 1910, much of this slop consisted of "pure, unadulterated gasoline which had no value except in limited quantities." After clearing beaches along the East Coast on his trip back to Texas, the captain gave orders to discharge the refinery waste directly into the sea.[17]

Tankers sometimes discharged crude oil as well. To transport as much oil as possible, captains sometimes carried more oil than their ship could carry safely in bad weather. If they encountered a storm, they simply released cargo until the ship gained enough buoyancy to weather the storm. Insurance syndicates, led by Lloyd's of London, eventually established rules that prohibited ships from carrying more weight than they could safely carry in all conditions. To help enforce this rule, they refused to certify a ship unless it had load lines—including a Plimsoll mark, which indicated the ship's maximum safe load—painted on its hull. Ships were not insured if they left port with their Plimsoll marks below the water surface.[18]

Tankers also lost oil from other sources. Fuel leaks in the engine room, for example, drained into the ship's bilge and mixed with other wastewater. In most cases, crews simply pumped this bilge water overboard. Oil left in hoses, small spills during loading and unloading, and drips from leaky valves and connections represented other types of oily discharge from tankers and oil-burning ships. Large quantities of oil also could be spilled when a crew member turned a valve at the wrong time or forgot to close a valve at the right time. Given the maze of valves and pipes on tankers, this type of mistake happened more frequently than shipowners would have liked.[19]

When tankers were being loaded or unloaded, some oil typically reached the water because most docks had no way to capture any oil that drained from hoses after they were disconnected (fig. 5.2). That oil simply ran off the dock and into the water. Some crews addressed this problem by pumping a little water through the hoses at the end. Therefore, when they stopped pumping and broke the connection, the hoses contained only water.[20] But without any method to catch spilled oil at the dock, some oil was bound to reach the water.

Leaky ships were also a problem. Before World War I, shipbuilders con-

Figure 5.2. Many docks at which tankers loaded and unloaded, such as this Los Angeles Harbor dock shown in 1919, were unequipped to capture spills that occurred. (Library of Congress LC-US-Z62-124117)

structed hulls by riveting steel plates together. If hulls were exposed to sea-water alone, oxidation tended to seal gaps between plates. Oil, though, tended to work its way between riveted plates.[21] Leaks often occurred above the waterline, where the pressure of water did little to prevent the oil from escaping. Naval architects generally specified higher standards of riveting when designing ships specifically to hold oil, but problems still occurred when cargo ships were converted to tankers. Similar problems arose when shipowners converted coal-burning boilers to oil-burning boilers and stored fuel in compartments along a riveted hull. Although shipyards turned to welding during World War I, ships with riveted plates continued to operate for many years after.[22]

Shipyards also faced challenges disposing of oily residue from the ships they repaired. When repairs had been scheduled well in advance, captains could clean their tanks thoroughly in open water and pump all wastes into the sea. However, damaged ships sometimes arrived at shipyards before any tanks had been cleaned. Even if the yard had facilities for collecting oil and slop, workers were not always careful in preventing spills. More frequently, though, yards were simply not equipped to handle large amounts of oily slop and oily water. Investigators collecting data for the Bureau of Mines report on oil pollution noted that crews at one repair yard had to remove 1,200 barrels of oil and sludge from a ship before they could replace seventeen rivets. The effort required a substantial amount of manual labor and, in this case, the crew accomplished the task without polluting the harbor. However, not all crews had investigators from the Bureau of Mines watching over their shoulders, and the water around most shipyards testified to this fact.[23]

The amount of oil being discharged into coastal waters also increased in the years after World War I because more steamships—regardless of their cargo—began burning oil in place of coal. Indeed, inexpensive oil from Spindletop played a large role in encouraging this use of petroleum. Initially, some ships burned straight Spindletop crude, no fractions removed, as fuel.[24] Over time, ships stopped using straight crude for fuel oil. Not only did the gasoline in crude became more valuable but insurance companies also started prohibited ships from burning fuel with a flash point above a certain limit. In practice, though, this requirement simply meant that shipowners purchased the residual after refineries stripped away the valuable lighter fractions. If anything, the change increased the percentage of tarry residue that remained behind in fuel tanks. The British Navy also set new standards for fuel oil, but its goal was to protect crews from the noxious sulfur gases given off by most Mexican oils, not to reduce the oily discharges its ships released.[25]

Oil, of course, could also be spilled due to collisions and grounding. Unlike intentional discharges of oily water from tankers, which port officials had come to see as a problem by the early 1920s, accidental releases of oil due to collisions or grounding received little attention. Most people regarded such accidents as beyond the control of shipowners. In their 1923 report on oil pollution, investigators with the Bureau of Mines distinguished between accidents that occurred despite all precautions and accidents in which the responsible parties failed to take proper precautions, but such distinctions were difficult to make.

The American Steamship Association (ASA), which assisted the Bureau of Mines and the American Petroleum Institute in collecting data about the sources of oil pollution, certainly did not shy away from admitting that tanker accidents occurred. Indeed, in 1924, with congressional hearings on oil pollution still underway, the ASA published a brochure that portrayed tankers as difficult to control. The ASA distributed its brochure as part of a campaign opposing the construction of a bridge across the Mississippi approximately two miles above the corporate limits of New Orleans. Oceangoing oil tankers routinely used this portion of the river. According to the ASA, strong currents near the bridge rendered steering difficult and forced ships to travel at high speeds before they could "be gotten under control." Furthermore, a "mistake of a pilot in giving an order" could result in a tanker going "off her proper course and across and under the influence of the current, which no human effort could break in time."[26] If those difficulties were not problem enough, the brochure made it clear that "machinery and steering gear will break down, despite all diligence to make them sound." The authors suggested that if a tanker lost power upstream from the proposed bridge, the bridge was a sure hit. The ASA emphasized that if such an event occurred, huge amounts of oil would spill into the river and cause a fire that "would immediately become known in all shipping and insurance circles."[27]

Shipping representatives also noted that marine insurance associations already regulated shipping practices. These insurance associations, which had a vested interest in preventing collisions and groundings, defined the precautions that shipowners had to take and enforced the implementation of those precautions. At the same time, these insurance associations had little interest in secondary damages caused by accidents. After all, in the case of an accident, the law limited the liability of a shipowner to the value of the ship and its cargo.

For example, three months after the ASA released its brochure on the dangers of tankers, the captain of the tanker *Llewellyn Howand* drifted off course and hit rocks off the shore of Newport, Rhode Island. To protect navigation and to prevent property damage, the Army Corps of Engineers set the ship and its oil on fire. An insurance broker who happened to be visiting Newport noted that "there was considerable talk in Newport of endeavoring to recover damages from the owners of the ship, but if they must depend, which seems to be case, on the value remaining on the wreck for recovery, the chance would seem very small."[28]

Technological Change and the Oil Pollution Act of 1924

To eliminate intentional discharges of oily slop and oily water from ships, two general possibilities existed. A relatively complicated system of shipboard oil-water separating devices, barges, and land-based facilities could be used to concentrate, gather, and dispose of the oily waste. Or tankers and oil-burning steamships could discharge their oily water further from shore.

Initially, port officials in New York focused on the use of scavenger barges. In 1921, soon after the specter of oil pollution gained wide attention in New York City, the commissioner of docks and ferries, Murray Hulbert, organized a committee consisting of representatives from shipping companies, insurance associations, and the petroleum industry. Among other things, this committee, which referred to itself as the Oil Pollution Committee of the Port of New York, suggested pumping the waste residue of oil-burning ships, usually an emulsified mixture containing approximately 20 percent oil and 80 percent water, into barges. In theory, the barges would then sell the waste oil to a land-based plant that could separate and reclaim the oil for some productive use. Hulbert's group eventually announced that "two or three oil-salvaging companies had begun to operate and that five or six more will soon enter the field."[29] Later that year, when Hulbert appeared before the U.S. House of Representatives to describe conditions in New York, he expressed confidence that the entire problem could be better solved through cooperation, not legislation.[30]

As it turned out, Hulbert and the others proved too optimistic. The plan to use privately owned barges to salvage oily residue discharged by oil-burning steamships never worked. The operators of these barges quickly discovered that the slop they collected had no value. Refineries would only take it for a price, forcing the barges to charge as much as fifty cents to one dollar per barrel for their services.[31] That fee encouraged more ships to routinely clean their tanks by emptying and filling them before reaching the harbor.[32]

One legislator then suggested that all ships be forced to pay for the scavenger service and that no ship be allowed into port until that fee was paid. Others complained that such an approach would be difficult to enforce fairly. Would ships be charged according to the amount of oily slop they transferred to the scavenger barge? If so, what would prevent ships from flushing their tanks in open water so as to reduce their fee? Furthermore, several speakers pointed out that imposing fees unilaterally on foreign ships would have consequences for U.S. ships calling on foreign ports. For example, some feared that Britain might start requiring U.S. ships to respect British laws regarding load lines.[33] They argued that if the use of barges were necessary, ports should provide the services free of charge. Then all ships would have fewer reasons to discharge their oily water and residue outside the three-mile limit.

Another complication emerged. Some barge owners also tried to avoid the costs of disposal. Witnesses at congressional hearings testified that some scavenger barges ended up dumping loads of oily emulsions just outside the harbor from which they collected it. Some barges even pumped their loads into the empty compartments of outgoing tankers, which then discharged the oily waste in open water.[34] In practice, then, discharges from the scavenger barges would also have to be regulated if the strategy were to be successful.

When it became clear that barges alone would not solve the problem, many people expressed hope that a practical oil-water separator could be developed. Even with barges, some need for a practical shipboard oil-water separator existed. While barges might be able to collect relatively concentrated wastes from the fuel tanks of oil-burning steamships, they could not handle the large volumes of ballast water discharged from cargo tankers. Given that the ratio of oil to water in tanker ballast was low—typically around one part in a hundred—tankers would need some way of concentrating their oily wastes before transferring that waste to barges.

Various manufacturers and entrepreneurial engineers claimed that they could produce a shipboard device that would pay for itself.[35] One supporter of this solution compared it with legislation that required air brakes on trains. Railroad owners, he noted, initially resisted this legislation but after gaining some experience with air brakes could not be forced to remove the equipment.[36] It was on this note that Congress put its hearings on hold and issued a resolution calling for an international maritime conference on oil pollution.

In general, people with their eyes on the proposed international maritime conference supported legislation that focused only on ship-based sources of oily discharges. First, any legislation that focused on ships alone could be passed with little resistance. Everybody, even the American Steamship Association, agreed that ships were a problem. Refiners, however, promised to fight legislation that included land-based plants. Second, at the international conference, they wanted to hold up a model piece of legislation that all nations could adopt in their efforts to stop oily discharges from ships. From this perspective, any legislation that mentioned land-based sources would simply complicate matters.[37]

However, like the promoters of scavenger barges, those who promoted oil-water separators also proved far too optimistic. No practical shipboard device capable of separating clean oil from large volumes of oil-water mixtures existed. Indeed, the report on oil pollution released by the U.S. Bureau of Mines killed the hope that some technological silver bullet existed.[38] Although many companies employed oil-water separators in land-based applications, those devices were not constrained by space or weight. Nor were those devices expected to work on the types of emulsions found in cargo tanks. The entrepreneurs who marketed shipboard oil separators deliberately blurred these points. They performed tests that verified performance, but those tests took place under controlled conditions involving light oils and little emulsification.[39]

Because no solution that paid for itself existed, the Interdepartmental Committee on the Pollution of Navigable Waters postponed any maritime conference until such a solution could be found.[40] In the meantime, Congress, through the Oil Pollution Act of 1924, prohibited ships from discharging any oily ballast or oily slop within three miles of shore. Great Britain had passed a similar law, the Oil in Navigable Waters Act, in the previous year.[41] In practice, then, ships were free to flush their tanks outside the three-mile limit until an enforceable international agreement prohibiting discharges in international waters could be reached.

In 1926, as a first step in securing an international treaty prohibiting oily discharges from ships, the Coolidge administration convened a preliminary international conference. Representatives from nine nations of Western Europe, Canada, and Japan attended. By that time, however, the pressure for action in the United States had diminished. Most ships already discharged their oily residue far from shore, and conditions in harbors had improved. The most obvious problems—the fire hazard represented by oil-soaked docks and damage to bathing beaches due to oily residue washing ashore—had been addressed. Consequently, active lobbying by the fire insurance industry had ceased and public interest had dwindled. Indeed, at the conference, representatives from Great Britain, not the United States, emerged as the strongest supporters for comprehensive action.

In the discussions that took place, attendees did address an issue that Congress had avoided. They quantified what they meant by "oily water." They concluded that although there was "no hard and fast line dividing oily mixtures which are harmful from those that are practically innocuous," a mixture "containing more than .05 of one per cent of crude, fuel oil, or diesel oil should be regarded as constituting a nuisance." Essentially, then, they decided to classify any mixture containing less oil than 500 parts per million as being clean enough to discharge anywhere.[42]

Attendees had more difficulty reaching agreement on how far from shore the prohibition on oily discharges should extend. Opinions varied from something less than 50 miles to a complete prohibition throughout all international waters. In the end, those attending the conference accepted 50 nautical miles as the norm, with provisions to accept prohibitions of up to 150 miles from certain coasts if special conditions justified that distance. Some attendees still desired to require all oil-burning ships to carry equipment capable of separating oil from the emulsions that formed when crews cleaned cargo tanks or discharged oily ballast, but the lack of practical equipment for accomplishing this task prevented any consensus in this area.

In the end, no international treaty emerged. Instead, shipowners agreed informally not to discharge any oily water within fifty miles from shore. Even if attendees had reached an official agreement, enforcing its provisions would have been difficult. Captains who chose to discharge oily water within the voluntarily imposed limit of fifty miles—or even within the three-mile limit specified in the Oil Pollution Act of 1924—risked little chance of being discovered, especially if the discharge occurred at night. Furthermore, no easy way existed to measure the oil content of water being pumped overboard, making the 500 ppm threshold quite flexible. [43]

The absence of any treaty prohibiting the discharge of oil in international waters undermined any hope that the Oil Pollution Act of 1924 might encourage oil companies to design cleaner tankers. Still, some changes did occur. Oil companies did pay more attention to oil spilled during loading and unloading operations. And as steamships gave way to diesel-powered ships, which required a more refined fuel, problems caused by flushing and cleaning fuel tanks became much less of a problem. However, oil companies did little to reduce the amount of oil discharged into the sea by their tankers. Captains simply flushed their dirty cargo tanks further from shore. With the number of tankers steadily increasing, this solution merely delayed the day of reckoning.

PART II

Fighting Pollution under an Efficiency Ethic

Validating a Guiding Ethic

AFTER PASSING THE OIL POLLUTION ACT of 1924, legislators had a choice to make, a choice that no one actually articulated or even saw the need to articulate. They had to choose between two general approaches. In regulating uses of the shared environment, should they first establish clear objectives and then take action to reach those objectives? Or should they simply attempt to optimize the economic benefits extracted from a set of resources? As different as these two approaches are, few people in the early 1920s saw any explicit need to choose one approach over the other.

Most people in positions of power took for granted that they should attempt to optimize the benefits extracted from a set of resources. Few people saw any need to ask how those "benefits" should be defined, measured, distributed, and otherwise quantified. Furthermore, most people believed that this approach would, in the long term, also address concerns associated with industrial pollution. After all, it seemed reasonable to expect that firms would continually identify and eliminate sources of waste and inefficiency, and they assumed that this would eventually eliminate pollution-related nuisances. Hence, reaching consensus on environmental objectives and putting regulations in place to achieve those objectives did not seem necessary.

The rhetorical power of arguments rooted in efforts to eliminate waste and increase efficiency demonstrated itself many times in the mid-1920s, and each success further validated the power of this approach in the eyes of those who put their faith in it. In petroleum-related decisions, efficiency-based arguments carried the day in several key debates: whether tetraethyl lead should be allowed as a gasoline additive; how to proceed in efforts to reduce refinery pollution; and how to reduce the waste of oil caused by poor extraction methods. As a result, this guiding ethic influenced the general strategy for addressing pollution concerns over the next several decades. Until that ethic failed, those who desired to establish environmental objectives and to regulate industrial activity so as to achieve those objectives found themselves fighting a losing battle.

The Debate over Leaded Gasoline

The debate over leaded gasoline took place soon after Congress passed the Oil Pollution Act of 1924. Good reasons to market tetraethyl lead as a gasoline additive certainly existed. Adding a few drops of the compound to every gallon of gasoline solved two major problems, one related to the performance of engines and another related to the conservation of petroleum. At the same time, researchers with the U.S. Public Health Service had just completed a study on the toxicity of lead and knew that it accumulated in the body. As it turned out, health-related cautions proved impotent in the face of the tangible benefits promised by leaded gasoline.

What were those benefits? By the 1920s, automobile manufacturers knew that they could increase the performance of an engine by increasing its compression ratio. However, as manufacturers lengthened an engine's piston stroke, and hence the amount by which the piston compressed its fuel, some components of gasoline, such as heptane (C_7H_{16}), burned in such a way that produced a violent pressure disturbance and a loud pinging sound. Other components of gasoline burned more smoothly and resisted this behavior, with iso-octane (C_8H_{18}) being among the best. To prevent engines from knocking and pinging, refiners had to use more iso-octane, a component of petroleum already in demand as an aviation fuel. Therefore, in 1922, governmental and industry leaders concerned about the nation's oil supply welcomed the announcement by Charles Kettering, the head of the General Motors Research Corporation, and Thomas Midgley, Jr., a research chemist, that a few drops of tetraethyl lead in every gallon of gasoline eliminated engine knock.[1] Using leaded gasoline would allow auto companies to manufacture higher performance engines and allow refiners to save iso-octane for use in other applications and to get more gasoline out of every barrel of crude oil.

Officials with the Public Health Service, however, questioned the wisdom of putting tetraethyl lead into gasoline.[2] After all, researchers such as Alice Hamilton had recently established the toxicity of lead-based compounds. What would happen when millions of automobiles exhausted small amounts of lead into the atmosphere of crowded cities? Midgley and Kettering, who made their first public sale of leaded gasoline in early 1923, paid little attention to those concerns. Soon after, General Motors and Standard Oil formed a joint venture, the Ethyl Corporation, to market the additive on a larger scale.[3]

Supporters of tetraethyl lead pointed to it as an example of scientific conservation—a science-based technological development that allowed more efficient use of available natural resources. They saw the compound as innocent until proven guilty, arguing that abandoning such a compound, an apparent "gift of God," was unthinkable.[4] Midgley knew, as did all chemists familiar with tetraethyl lead, that the compound was dangerous. It could attack the central nervous system after being absorbed through the skin. Indeed, Midgley had first-hand knowledge of its toxicity. In the course of his research, several assistants working with the compound died from overexposure.[5] But, as Kettering and Midgley pointed out, nobody necessarily had to come into contact with the additive in its concentrated form.

To allay concern that exhaust from engines burning leaded gasoline would be dangerous, the Ethyl Corporation contracted with the Bureau of Mines to study the health effects of leaded exhaust. Researchers with the bureau had recently determined the ventilation requirements for New York City's Holland Tunnel and so had some experience addressing such questions. The Ethyl Corporation also initiated its own investigation, contracting with Robert Kehoe of the University of Cincinnati Medical School to study the implications of using the additive.[6]

Soon after the Ethyl Corporation took these steps, the issue exploded. On October 25, 1924, an accident at a pilot plant for manufacturing tetraethyl lead exposed forty-five workers to vapors containing the additive. Thirty-five became violently ill. Five workers, in a state of delirium, had to be placed in straitjackets. They died within the week, which newspapers reported under spectacular headlines. Doctors diagnosed them as having acute lead poisoning. Several months later, accidents at a tetraethyl lead plant operated by Du Pont killed four more workers. The Ethyl Corporation decided to suspend production on May 1, 1925, until Kehoe and the Bureau of Mines could complete their work.[7]

Kehoe suggested that the accidents proved nothing other than the need for a safer manufacturing process.[8] Workers at many industrial plants worked with dangerous compounds. The real unknown was the long-term health effects of leaded exhaust from automobiles. How toxic was that exhaust? What would happen in urban areas with tens of thousands of cars burning leaded gasoline? R. R. Sayers, the chief surgeon of the Bureau of Mines and a surgeon with the U.S. Public Health Service, addressed those questions by exposing rabbits, guinea pigs, monkeys, and dogs to various doses and concentrations of tetraethyl lead vapors and leaded engine exhaust. Although Sayers publicly announced that he found no evidence supporting a ban on leaded gasoline, he structured his final report as a presentation of data in which he never actually states whether the use of tetraethyl lead in gasoline was safe or unsafe. Only through the use of phrases such as "no symptoms were noted" and "appear normal and healthy" does he imply that no reason to condemn the additive existed.[9]

Critics of tetraethyl lead discounted the reports from Kehoe and the Bureau of Mines on the grounds that their work was funded by the Ethyl Corporation. As a result, the U.S. Surgeon General held a general conference on the matter. Following that conference, the Public Health Service published a report supporting the Bureau of Mines and indicated that "there are at present no good grounds for prohibiting the use of ethyl gasoline."[10] As a matter of safety, though, the Ethyl Corporation abandoned its former practice of having gasoline station operators add the compound on site. Instead, they had refineries mix the additive in bulk and dyed the final product red to distinguish it from regular gasoline. With those measures in place, public concern over the use of leaded gasoline died away.[11]

The debate over tetraethyl lead raised a fundamental question closely related to the issue of industrial pollution. Should pollution-causing substances discharged into the shared environment be prohibited until one could show they were safe? Or should firms be allowed to make such releases, or, as in this case, products that resulted in such releases, until the health danger was proven? In either case, how does one measure a threat to the general public and where does one draw the line between safe and unsafe?[12] The supporters of tetraethyl lead believed that firms should not be constrained when no definite health threat could be demonstrated, and public sentiment generally supported their view. Most people sought the removal of products only when those products triggered acute illnesses with readily identifiable symptoms. Products that contributed to chronic illnesses, with symptoms identifiable only in the long term, generated much less concern.[13] As a result, the opponents of leaded gasoline, who were concerned about the ability of lead to accumulate in the bloodstream over long periods of time, found themselves fighting an uphill battle.[14]

As in the debate leading up to the Oil Pollution Act of 1924, reformers seeking to regulate industrial activity so as to maintain a certain level of environmental quality failed to convince industrial leaders, government officials, and others in positions of power that action was necessary. The idea that tetraethyl lead represented a way to maximize the nation's use of its oil supplies proved too attractive, and those who saw the efficient use of resources as key to the progress of industrial society saw no need to question their approach.[15] If anything, the end result of both debates simply reinforced the faith that most engineers and technical managers placed in the philosophy of utilitarian conservation.

Neither the debate over leaded gasoline nor the earlier oil pollution debates had much effect on the petroleum industry. Tetraethyl lead had emerged from the laboratories of General Motors, and its discovery as an antiknock additive was outside the refining industry's general interest in manipulating hydrocarbons through cracking and other processes. Some companies, most noticeably Sun Oil, even refused to use the compound for another twenty years. Sun, intent on never paying tribute to Standard Oil, avoided the use of tetraethyl lead through the continued use of high octane blends and advertised its gasoline as having "no poisonous anti-knock agents." This strategy suggests that the stigma of leaded gasoline was attached to the additive, not the petroleum industry as a whole.[16] As for the debate over oil pollution in coastal waters, the directors of the American Petroleum Institute (API) framed the issue as one involving shipping practices, not the discharge of industrial waste by the petroleum industry. Spokespeople for the API steadfastly maintained that refineries, which had significantly reduced their pollution-causing discharges over the last two decades, would continue to show improvements as they became even more efficient.

That the oil pollution debate did not have a great impact on the thinking of industry executives was apparent when the API's newly formed education committee met to discuss how to improve the industry's public image. In the oil pollution hearings, the API had emerged as a powerful and capable organizational actor, and the creation of an education committee represented one of several moves aimed at consolidating the API's role within the petroleum industry. The committee explicitly expressed its goal as being the "elimination of distrust and suspicion within the industry and in the mind of the general public."[17]

In listing the sources of distrust and suspicion, the committee made no mention of any issue related to pollution. Instead, members of the committee—all prominent figures within the petroleum industry—focused on issues such as "fake stock promotions" and "frequent price fluctuations," making it clear that the public's perception of oil companies as being run by crooked industrialists concerned them most. Also on their minds was the poor publicity that the Teapot Dome scandal had created for the industry. Another concern involved the perception that larger companies, mainly the spawn of Standard Oil, took advantage of small oil producers.[18]

Despite the relatively well-defined goals of API's education committee, the members had no real idea about how to reach these goals. To assist them, members of the committee invited public relations experts from other industries to make suggestions. In the style of a congressional committee gathering information in a hearing, a process with which some in attendance were all too familiar, the guest speakers first gave a brief statement and then addressed questions posed by committee members. One publicity agent pointed out that the petroleum industry had to educate the public and inform them that the difference between profit and loss on every gallon of gasoline was a fraction of a cent. Furthermore, he suggested, the oil industry had to show there were literally hundreds of companies competing for business, not a few companies acting in a monopolistic fashion.[19]

Bruce Barton, who developed advertising campaigns for automobiles, suggested the committee take a slightly different angle. He noted that his mother grew up in a village, never leaving it for thirty years until she got access to a car. Forget about "the ballot freeing women," he told them, "the thing that has freed women is gasoline! You advertise it as something that is sticky and smelly. I say it is the water of life."[20] This mention of oil as being sticky and smelly was the closest anyone came to talking about pollution.

The API, Industrial Standards, and Pollution Control

In the same year that Congress passed the Oil Pollution Act, the API issued its first official set of industrial standards—that is, standard specifications of products, materials, and processes for easy reference by engineers and purchasing agents. The first API standards established two classes of lap-welded pipe and three classes of seamless pipe, allowing engineers and purchasing agents to specify state-of-the-art products with a minimum of research, negotiation, and expense.[21] Engineers who referenced these specifications could be more confident that the pipe they ordered would not burst under pressure or stretch like taffy as crews lowered a mile or more of the pipe into an oil well.

Indirectly, this move into developing standards reinforced the notion that efforts to increase efficiency and reduce waste would continually reduce pollution-causing discharges. First, engineers framed their standard-setting efforts in terms of reducing waste and increasing efficiency. Second, the general dissemination of standardized equipment and practices helped to prevent accidents, leaks, and spills. With standardization, oil well equipment, pipeline systems, and refinery processes gradually became more robust and less prone to accidents, failures, fire, explosions, spills, and leaks. Finally, the committees established to create industry-wide standards eventually began developing and disseminating standards related to waste disposal practices.

By the time the API entered the standard-setting business, professional engineering societies had already established a process for creating voluntary industrial standards by consensus. Advocates of this process argued that standard specifications not only simplified business transactions and reduced problems associated with incompatible equipment but also eliminated the wasted time spent with recreating specifications over and over again. The directors of the API simply grafted this process into their organizational framework. In doing so, they brought the professional activity of oil company engineers under their control and reinforced the API's position as a trade organization capable of speaking for the entire industry.

The consensus process for setting voluntary industrial standards emerged in the second half of the nineteenth century. At the most basic level, the lack of precise weights and measures represented a barrier to the dissemination of new manufacturing techniques. For example, in the years before the Civil War, improvements in machine tools allowed government arsenals to produce muskets, rifles, and pistols with interchangeable components. However, the lack of precise national standards prevented those manufacturers and arsenals from communicating information involving dimensions. Instead, each armory or manufacturer maintained a system of master gauges, and machinists were expected to produce pieces within an acceptable tolerance of the relevant gauges. Until adequate national standards were established and disseminated, no precise dimensions could be reliably communicated without the physical exchange of gauges.[22]

New technologies and more sophisticated commercial transactions also gave rise to new kinds of standards. For example, the development of telephone networks, electric lighting, and electric traction systems were accompanied by more precise units of electrical measurement. Not only were standard definitions of the ohm and other units of electrical measure required to help scientists and engineers communicate with one another, but these units were also required for commercial transactions. Specifying "good quality copper wire" was no longer enough.[23] Chemical companies, oil companies, steel manufacturers, and railroads all faced a similar need for more sophisticated ways of verifying that the goods they purchased and, sometimes, the goods they sold met precise engineering specifications.

Charles Dudley—the co-founder of the American Society for Testing and Materials, an organization dedicated to the creation of material and process specifications—began his career writing purchasing specifications for the Pennsylvania Railroad. When Dudley, a university-trained chemist, joined the railroad in the 1870s, the company ran a highly decentralized and poorly organized purchasing effort. Purchasing agents had no systematic way of verifying the quality of the fuel, lubricating oils, coatings, steel, and other material they purchased. Dudley's specifications allowed departments throughout the company to use the same criteria for evaluating and controlling the quality of this material. At one point, Dudley promoted a particular specification for steel, but several manufacturers opposed it, and Dudley found himself in the middle of a controversy over the proper role of standard specifications. This experience led Dudley to establish the American Society for Testing and Materials, stressing that standard specifications should be created by consensus, with all those affected by a standard given a chance to participate in shaping it.[24]

Engineering societies also had a significant interest in creating industrial standards. When the American Society of Mechanical Engineers (ASME) held its first annual meeting in 1880, the issue of standardization stimulated a contentious debate between academically trained engineers and their shop-trained employers, many of whom held prominent positions in the new society. The academically trained engineers believed in stamping industrial practices with the mark of professional authority, while their employers worried that the ASME was entangling itself in commercial matters. Although the governing board of the ASME officially decided not to endorse any standards, testing procedures developed by ASME committees eventually came to be known as standards. As the number and influence of academically trained engineers increased, the importance of engineering societies as an expression of professionalism also increased. Each new society desired to define its own engineering standards in order to identify its members as professionals and separate them from shop-trained technicians.[25]

Congress also expressed a new interest in standards. In 1900, the editors of *Scientific American* argued that if American manufacturers were to achieve their full potential, the nation required "a standardizing bureau such as is provided for manufacturers in other countries."[26] Many in Congress agreed. In an expression of this sentiment, legislators asked the National Academy of Sciences to study Germany's Physikalische-Technische Reichsanstalt and England's National Physical Laboratory. Soon after, in 1901, Congress chartered the National Bureau of Standards to establish standards related to fundamental units of measurement.[27]

In addition to defining units of measurement, engineers with the Bureau of Standards also began creating standard specifications for products purchased by the federal government, and they backed their specifications with a program of testing and research. For example, when various departments of the federal government experienced problems with faulty light bulbs, researchers from the Bureau of Standards created specifications for light bulbs that other departments could use when making their purchases.[28] In general, such efforts generated little controversy. However, when the bureau suggested that its specification for federally purchased concrete be made into a national standard and that the government test and certify all commercial cement, companies protested vigorously.[29]

Close cooperation between industry and government during World War I reinvigorated the interest of engineering societies in creating standards. In 1918, five major engineering societies and the National Bureau of Standards established the American Engineering Standards Committee (AESC) to serve as a clearinghouse for all standards of national interest. Members of this committee also included representatives from the American Society for Testing and Materials and other standard-setting organizations, insurance associations, trade groups, and technically grounded government agencies.[30] Herbert Hoover, who saw standardization and uniform specifications as key components in his well-publicized program to eliminate waste in industry, made a special point of securing the cooperation of the AESC.[31]

The philosophy of the AESC certainly complemented Hoover's own philosophy. Paul G. Agnew, a physicist with the Bureau of Standards who became the head of the AESC, argued that standard specifications rationalized commercial transactions. He emphasized that "in the flow of products from farm, forest, mine, and sea, through processing and fabrication plants, and through wholesale and retail markets to the ultimate consumer, most difficulties are met at transition points—points at which the product passes from department to department within a company, or is sold by one company to another or to an individual. The main function of standards is to facilitate the flow of products through these transition points."[32]

Although the AESC has survived in various guises, with the American National Standards Institute (ANSI) being a direct descendant, the technocrat's paradise that Agnew envisioned never emerged. The AESC, in declaring itself to be the clearinghouse of all national standards, failed to gain the full cooperation of key organizations. The American Society for Testing and Materials, for example, resisted becoming too closely associated with the work of the AESC. Still, no other body professed to serve the same function, and the AESC continued to promote itself as a central player in the process of sponsoring, creating, and disseminating industrial standards of national interest.[33]

Therefore, in 1924, when the API sanctioned its first set of standards, engineers active in professional societies already saw creating industrial standards as a significant component of their professional identity. Indeed, engineering societies such as the American Institute for Mining and Metallurgical Engineers, the American Association of Petroleum Geologists, and the American Society of Mechanical Engineers were already setting standards affecting the oil industry. However, the directors of the API, in embracing the process of setting industrial standards, had no intention of yielding any control to engineering societies. J. Edgar Pew, the newly appointed head of API's oil field standardization committee, explicitly avoided any entanglement with the National Bureau of Standards, the AESC, or any organization that would reduce the say his committee had over any oil field standards that were created.

Integrating the new technical committees into the API required organi-

zational changes. Up to that point, the API had been a management organization that took on technical problems as needed. So, to integrate technical committees into the organization on a more permanent basis, the directors of the API established technical committees and charged them
with tasks that were of general interest to the entire industry. After putting
this structure in place, Pew and the other API directors quietly began
shifting technical work sponsored by engineering organizations to API-
sponsored committees. Most of the engineers on the various committees
worked for companies managed by Pew and his associates, so shifting
the work to fall under API sponsorship was neither difficult nor contentious.[34]

Under this arrangement, the API provided engineers working for oil
companies with an outlet for professional development and gained control of any standard-setting activity that affected oil companies. In general, these new committees began sponsoring technical activity that promised to simplify transactions in the petroleum industry. For example, in
1928, the API Division of Production sponsored an effort to create standard procedures for measuring, sampling, and testing crude oil. At the
time, key procedures—such as those for determining the amount of water
and sediment in a batch of crude, the amount deducted for evaporation
during transportation, and correction factors due to differences in temperature—varied from region to region and from company to company.
In some cases, procedures defined by agencies in two different states differed.[35]

The heads of large oil companies desired to avoid technical activity that
might cause friction between companies. Hence, in 1928, when W. R.
Boyd, the vice president of the API, was asked to develop a technical
group to discuss refining technology, he hesitated because he thought
such a group would be "uncontrollable." J. Edgar Pew agreed and advised
Boyd against the effort, indicating that few companies would support
their "technical men" meeting for that purpose. He also pointed out that
every company thought "its technology was the best in the world" and did
not want to let out any secrets.[36]

But the directors of the API actively encouraged standardization efforts
that facilitated transactions and encouraged safe practices.[37] Indeed, by

1933, the API's Division of Refining supported twelve technical committees. In addition to developing standard specifications for components such as pumps, valves, and motors, these committees also recommended procedures for preventing accidents and fire, disposing of wastes, and preventing corrosion. The API also supported committees in its transportation and distribution divisions, each pursuing technical activities of general interest to all oil companies.[38]

Hosting technical committees gave the API more control over standards that might have broader consequences for the industry. The narrow interest in many standards set by API committees justified narrow participation in those committees. However, not all standard-setting efforts fell into this category. Product specifications, safety practices, and standards for waste-disposal practices certainly were of interest to other groups. The API had good incentive to be vigilant in the content of such standards. A single reference to a voluntary standard in a local ordinance could make that standard something less than voluntary.

Indeed, the ground that separated a "standard" from a "regulation" could be traversed fairly quickly. As head of the American Engineering Standards Committee, Paul Agnew explicitly held up the consensus process as a mechanism for creating national standards on which state regulations could be based. For example, according to Agnew, "one type of electric motor, in order to be legally safe in the State of Pennsylvania" had to be made "legally unsafe in the State of Wisconsin, and vice versa." To address such regulatory barriers to trade, he saw product safety, fire-prevention ordinances, and construction codes as all falling under the AESC umbrella.[39]

Product specifications that imposed constraints on the industry were nothing new to refiners. In the early days of the kerosene-based petroleum industry, when few product specifications had to be met, refiners often included some quantity of naphtha and some amount of gas oil in the kerosene they sold as lamp oil. They did so, of course, to get as much salable lamp oil out of their crude as possible. However, if they included too much naphtha, the kerosene became too explosive for safe use. If they included too much gas oil, the kerosene smoked heavily.[40] Because few enforceable specifications existed, early consumers found that the quality of lamp oil varied considerably from manufacturer to manufacturer and, in some cases, from batch to batch produced by the same company. One manufacturer of lamp oil, hoping to reassure customers, advertised its product with the slogan, "It will not explode."[41]

As accidents involving explosive kerosene climbed, insurance companies began lobbying for government-enforced regulations. Specifically, they wanted oil companies to manufacture lamp oil meeting well-defined specifications, especially in regard to a product's flash point. Some states already had explicit requirements, but those requirements, and the methods by which the flash point was measured, varied from state to state. Refiners who wished to market their products in different states were forced to meet the requirements of each state.[42] In their efforts to develop standards, insurance companies received tacit support from Standard Oil. From the start, part of Rockefeller's strategy had been to produce a "standard" product that consumers could trust. Given that refineries outside the Standard fold could gain sales by producing a lower quality, less expensive kerosene, Rockefeller and his associates generally did not oppose regulations that forced other manufacturers to offer at least the same quality and consistency as Standard Oil. If nothing else, price wars with refineries that had to meet a given standard of quality were easier to win.

In the years after World War I, refiners also had to meet gasoline-related specifications, and they desired to retain as much control over those specifications as possible. Hence, when several states passed legislation that required all gasoline sold in those states to meet certain product specifications, leaders in the petroleum industry resisted. They argued—not without cause, especially since they were trying to develop a fuel that resisted knocking—that the detailed specifications developed by the federal government suppressed innovation.[43]

Insurance companies were also interested in standards, such as standard practices associated with fire safety, wiring codes, and building codes. Furthermore, they generally had leverage in getting companies to adopt those standards. Underwriters could increase premiums or refuse to provide insurance to companies that did not adapt the practices recommended by insurance associations. In the refining sector, the regulatory role of these associations increased when, in the early 1920s, some oil companies joined mutual insurance associations as an alternative to purchasing traditional fire insurance. Advocates of these insurance pools argued that they helped to reduce the indiscriminate writing of policies involving widely differing risks, which sometimes forced "honest and careful policyholders" to pay the losses of the dishonest and careless. The premiums paid by members of the association covered only damages suffered by other refinery owners in the same insurance pool.[44]

Policing of the risks at each refinery in the insurance pool also increased, leading to interest in standard fire and safety procedures. In 1926, to verify that all members of its association were avoiding unnecessary risks, the National Petroleum Mutual Fire Insurance Association surveyed conditions in twenty East Coast refineries. Inspectors noted wide variations in the seriousness with which refiners approached fire prevention.[45] Although the organization possessed no direct authority over any of the refineries, such surveys encouraged the insurance industry to develop recommended practices that could be referenced in insurance contracts. The API committee on fire prevention eventually took on this task of developing recommended fire and safety procedures.

By the early 1930s, the API had established the basic institutional structure necessary for coordinating technical work deemed to be of general interest to the petroleum industry. Through a system of technical committees and subcommittees, engineers employed by petroleum companies participated in efforts to make the industry safer, more efficient, and to some extent, less polluting. The extent to which the API succeeded in promoting and controlling its standard-setting activities is evident in the eventual use of the word "API" by oil workers to be synonymous with the word "standard" or "standard practice." By the 1940s, oil field workers who said that a product was "strictly API" meant that it was a standard component and safe to use. Or, if somebody made a mistake, a fellow oil worker might say, "That's API for him."[46]

The API's move into waste disposal procedures came in the late 1920s, partly in response to a report required by the Oil Pollution Act of 1924. This report, issued in 1926 by General H. Taylor, the chief of engineers for the Army Corps of Engineers, noted that conditions in coastal waters had generally improved since passage of the Oil Pollution Act of 1924.[47] The authors of the report emphasized a decision-making process based on optimizing economic benefits, saying that "what constitutes the highest and most economic use of a waterway aside from its use as a public highway of commerce involves a study of local conditions and the proper adjustment of the various conflicting interests of the locality. The growth and industrial development of a locality may depend on the use of its streams for the economic disposition of its domestic and industrial wastes; and should the pollution of streams in such a locality be prohibited, the future development of the section might be seriously impaired."[48]

On the other hand, the Corps of Engineers also indicated that some pollution from oil still occurred, and this pollution interfered with the "natural process of purification." They observed that due to a lack of the oxygen needed for the natural decomposition of sewage, domestic wastes underwent the "more offensive process of putrefaction." In addition, the authors noted that oily discharges attacked fish and shellfish by destroying the organisms on which they fed and by killing their larvae. The authors then concluded that refineries could easily reduce the oil they discharged by installing "efficient recovery devices" with no "undue or excessive burden." Hence, in the end, the Corps of Engineers report recommended that industrial plants, including refineries, be prohibited from discharging oil into the nation's waterways.[49]

The recommendation demanded a response from the American Petroleum Institute, especially after a new pollution control bill was introduced in the House of Representatives. It called for a nationwide prohibition against the discharge of industrial wastes from "any manufacturing plant or any shop or any establishment whatever."[50] In the absence of any response by the petroleum industry, Congress was likely to adopt the recommendations made by the Corps of Engineers. Therefore, at their annual meeting in 1926, the directors of the API created a committee to study the sources of oil pollution. They did not disagree with reports of past conditions but expressed a wish "to learn *the conditions as they exist today*" so that they could take the appropriate measures, with "the theory being that evils of pollution could be remedied more effectively by cooperation from within the industries than by legislation from without."[51]

Advocates of stronger legislation could not understand why Congress did not pass the amendment simply on the strength of the recommendation by the Army Corps of Engineers. Indeed, a representative from Pennsylvania who had attended hearings leading up to the Oil Pollution Act of 1924 noted that he had been "through so many of these hearings that I haven't the faintest suspicion that these [representatives of the petroleum industry] can say anything which I haven't heard before." The API's new investigation undermined that objection by promising new information.

The API proceeded by charging its technical committees with collecting information about waste disposal practices in the industry. By 1928, after 81 engineers had visited 274 oil refineries, 254 marine terminals, and 177 oil fields, the API compiled the results and circulated a report to its members. The report, a curious mix of serious recommendations and smooth generalities, concluded that oil refineries were "taking every practical precaution to prevent the escape of oil from their plants" and that, of the refineries surveyed, wastewater from 181 plants contained practically no oil. At the other 93 refineries, committee members recommended corrective measures. The report also pointed out sources of oil pollution not under the control of petroleum companies, including oily wastes from other industries, used crankcase oil discharged into sewers, and underwater seepage in areas such as the Santa Barbara Channel and Santa Monica Bay.[52]

The report also noted ways in which oil companies could reduce pollution and suggested that the API hire seven qualified engineers to help individual oil companies with their pollution control efforts. One director of the API reacted to this recommendation by saying he would not support employing "a large number of men to police the industry" because this would "create friction with many oil companies" and "savor of paternalism on the part of the Institute." Yet, even he described the survey as "a really remarkable report" and approved the formation of an antipollution committee responsible for creating pollution control procedures.[53]

The API effort, backed by some behind-the-scenes maneuvering, proved useful in deflecting legislative action.[54] In 1930, after the API had released its report and when Congress considered another amendment to regulate land-based plants, representatives from the API pointed to their report as evidence supporting their position. They argued that oil companies were not only making substantial progress in reducing discharges of oil but also had a plan to reduce those discharges further.[55]

In the end, all bills that proposed to amend the Oil Pollution Act of 1924 failed, and the promise of self-regulation took the place of legislative action. This promise came in the form of API technical committees creating recommended procedures for the disposal of refinery wastes. The API's influence over the technical committees that created these standards ensured that all recommended practices would be as stringent—or as lax—as the industry deemed appropriate. Furthermore, individual companies presumably used recommended procedures only if those procedures made operations simpler and safer without significant net costs, if any. As a form of self-regulation for protecting workers, the general public, and environmental quality, these standards and recommended procedures were effective only to the extent that individual companies reduced costs associated with material loss, accidents, fires, insurance premiums, and nuisance and damage suits. However, in the 1930s, there were plenty of costs to reduce and, consequently, room for real change. From the perspective of engineers, participating in the development of these pollution control standards was an important component of their professional identity and consistent with their general goal of using standards to increase efficiency and reduce waste.

Efficiency and the Conservation of Oil

Another issue involving use of the shared environment by the petroleum industry also emerged in the mid-1920s: the conservation of oil by regulating the rule of capture. In the debates that followed, which lasted about a decade, arguments rooted in an efficiency ethic again trumped all others. This time, however, industry leaders found themselves on the losing side.

Starting in the years after World War I, self-described "conservationists" pointed to laws governing the ownership and extraction of oil as a major obstacle to rational oil production. According to leading geologists and petroleum engineers, poor production practices encouraged by the unregulated rule of capture resulted in operators extracting only 10 to 15 percent of the oil available in most fields. As much as 85 to 90 percent of the oil remained underground, stripped of viscosity-reducing natural gas and trapped behind pockets of water-wet sand and unable to flow toward producing wells.[56]

Petroleum geologists and, increasingly, petroleum engineers pointed out that oil field operators could increase the productivity of the entire field by cooperating. Furthermore, they could lift larger amounts of oil with much less capital. By placing one well every forty acres or so, producers could drain the same amount of oil from the reservoir as they could with dozens or even hundreds of wells in that same area. In addition, by extracting oil at a controlled rate, rather than in one mad rush, producers could prevent the loss of natural gas, which played an important role in keeping the viscosity of the oil low and in keeping the pressure of the reservoir high. Cooperation would also reduce the erratic infiltration of salt water into the producing area. Finally, managed production would also avoid large drops in the price of oil due to temporary overproduction.

The unregulated rule of capture represented the main barrier to cooperation. As long as one operator in a field decided to extract oil as rapidly as possible, which the law allowed, then all other operators had no real choice but to follow suit. Otherwise, uncooperative neighbors would end up capturing the largest share of oil. Even if the rush for oil damaged production and consumed far more capital than necessary, oil operators inevitably chose to go after the oil while they had a chance. Proponents of oil conservation advocated changing the laws governing oil extraction so as to encourage companies to operate entire pools as a single unit.

Initially, several voices on the periphery of the petroleum industry led the effort to regulate production. Henry L. Doherty, the eccentric and innovative founder of Cities Service Oil and Gas Company, stood out as the most vocal. Mark Requa, a protégé of Herbert Hoover who served as the government's petroleum administrator during World War I, also pushed for changes, as did George Otis Smith, the head of the U.S. Geological Survey.[57] All three tried to get the Coolidge administration to take action.

In arguing against unregulated production, Doherty took the most technical approach, focusing on the role of natural gas in facilitating production. Many oil producers who encountered natural gas simply flared that gas to the atmosphere. Some operators flared off millions of cubic feet of gas just to get twenty to thirty barrels of oil per day. Natural gas, Doherty argued, played a crucial role in the production of oil. That gas— whether dissolved in the oil or existing in separate pockets—provided some of the force necessary to propel the oil through the formation that held the oil, a fact demonstrated by J. O. Lewis of the Bureau of Mines.[58] Doherty also asserted, and later demonstrated, that at pressures and temperatures found deep underground, dissolved natural gas dramatically lowered the viscosity of oil. Oil that initially flowed like water could flow like molasses when stripped of its gaseous component. Therefore, wasting this gas not only represented the loss of a potentially valuable material (Doherty sold gas through his utilities) but also reduced the amount of oil that could be recovered from the field.[59] To prevent this waste, Doherty argued that the unregulated rule of capture should be modified.

Mark Requa took a more philosophical approach in arguing against unregulated production—usually too philosophical for the oil company executives he addressed. In one speech, he divided the growth of industry into three epochs: the period of pioneering, the period of rapid development, and the period of maturity. In Requa's pioneering period, natural resources were abundant, but operators had little capital or transportation facilities at their disposal. Pioneers in the oil industry created the capital they needed through the hasty extraction of natural resources, and with that capital, constructed the transportation systems and refineries necessary for the rapid development of the industry. "Now," Requa argued, "we have passed two of the great periods of our growth . . . wherein it becomes necessary to scrutinize not only our resources, but also our methods of utilization, to the end that the maximum use may be obtained." He then called for conservation through cooperation, for leaders in the petroleum industry to meet and decide how best to extract oil from domestic fields.[60]

George Otis Smith of the U.S. Geological Survey looked at the issue from the perspective of national security. When Smith first started calling for United States involvement in the development of foreign oil fields, the approximate annual production of crude in the world's main oil-producing regions stood as follows:[61]

| United States | 443 million barrels |
| Mexico | 163 million barrels |

Russia	30 million barrels
East Indies	16 million barrels
Eastern Europe	13 million barrels
South America	9 million barrels
Persia	6 million barrels

Smith realized that foreign fields would become increasingly important over time and wanted large companies to establish channels of supply from foreign fields as soon as possible. Otherwise, foreign-owned oil companies—mainly British firms—would secure those channels.[62]

Though not directly involved in the Teapot Dome scandal, Smith, as an early advocate of setting aside oil reserves for the navy, had been scalded by the affair. This experience only reinforced his support for new laws governing oil extraction. In the Teapot Dome scheme, the secretary of the interior, Albert Fall, surreptitiously leased portions of Teapot Dome—an oil field in Wyoming that had been set aside as a naval reserve—to Harry Sinclair and Edward Doheny, both prominent figures in the oil industry. In other naval reserves, where private firms were already extracting oil, such leases made sense by ensuring that the navy received some oil before private firms depleted the entire field. In the case of Teapot Dome, though, geology was thought to protect the oil from being drained away by wells on private land. Hence, opening the protected portion of the field to private firms made little sense, especially if they only paid a small royalty fee for the right to produce. Leasing the field in secret and giving preferential treatment to a couple of cronies only made matters worse. The resulting investigation and cover-up put the laws governing oil extraction in the headlines.[63]

For officials such as Smith, the question was how to encourage foreign exploration by slowing down domestic production, so that sufficient supplies of domestic oil would be available well into the future. He saw the "feverish haste" to produce oil as an artifact of World War I, when "the urge of national self-preservation" demanded it. He applauded the industry's accomplishments but bemoaned its inability to slow down.[64] Now that they had the chance, he urged companies to develop fields throughout the world, which would allow them to reduce the destructive pace at which they developed domestic fields.[65]

Not everybody saw uncontrolled production and the resulting glut-induced low prices as a bad thing. Any oil company that concentrated its investment in the refining sector certainly enjoyed the surplus of oil encouraged by the rule of capture, as did motorists who purchased inexpensive gasoline. Small producers with relatively small leases also heard calls to regulate production with suspicious ears. Would such regulations limit the amount of oil they could extract? Antitrust watchdogs also looked askance at production controls. Given that the federal government had just initiated an antitrust suit against forty-eight oil companies for trying to monopolize the interstate trade of gasoline, any plans that encouraged companies to "coordinate" production raised more than a few eyebrows.[66]

Foreign oil imported by a few large companies complicated the matter further. Although some producers supported production controls in the hope of eliminating price-reducing gluts, they certainly did not want to see foreign imports, such as those from Mexico or Venezuela, rise. To small producers, imported oil simply made it more difficult to stabilize prices at a profitable level. Before supporting any controls on their production, they wanted to see a tariff placed on foreign oil.[67]

To sort out the tangle of issues and to recommend a policy, Doherty, Requa, and Smith urged members of the Coolidge administration to organize an investigatory board. Finally, in late 1924, after the oil pollution hearings had come to an end and after the initial shock of the Teapot Dome scandal had faded, Coolidge formed a Federal Oil Conservation Board (FOCB), which consisted of high-ranking officials in his administration. Specifically, the board included the secretary of the navy, Curtis Wilbur; the secretary of the interior, Hubert Work; the secretary of war, Dwight Davis; and the secretary of commerce, Herbert Hoover. Coolidge charged the board with identifying ways to conserve the nation's petroleum supply by eliminating wasteful practices.[68]

A "Committee of Eleven" formed by the American Petroleum Institute quickly responded to the creation of the FOCB by assuring President Coolidge that there was "no cause for apprehension as to the adequacy of the petroleum supply for our national defense and security, and that under the customs and practices of the industry today there is no appreciable or avoidable waste."[69] They backed their assurances with a report that pointed to six billion barrels of easily extractable reserves, twenty-six billion barrels that were potentially recoverable through yet-to-be-developed methods of extraction, and huge but unknown amounts to be discovered in new fields and with deeper drilling. The authors of the API report asserted that, if the price of oil rose high enough, motor fuel could be extracted profitably even from coal and oil shale. Therefore, according to the official position of the API, "the play of competition and the law of supply and demand" were all that was needed to secure the nation's oil.[70]

Henry Doherty, a director of the API and the institute's lone voice in favor of reform, dismissed the report. He noted that "the mere mention of the report of the committee of eleven before any meeting of petroleum scientific men is always sure to provoke a snicker that nothing will suppress and no matter how many of the bosses are present either."[71] Here then was an issue that divided engineers from their employers. Most petroleum engineers saw efforts to optimize the amount of oil extracted from a field, a task requiring their skills and expertise, as more sensible than protecting the narrow interests of specific companies.

In its first official action, the FOCB assembled a staff to create and send surveys to leaders in the industry, including surveys on "waste in production," "waste in storage and pipe lines," and "waste in refining."[72] Some of the survey questions were highly technical, asking respondents to quantify material losses suffered at various points in the process of extracting, transporting, and refining oil. Other questions, such as those asking about techniques and practices, were more open-ended. Although most oil company executives agreed with J. E. Pew of Sun Oil, who perceived the surveys as the result of "clerks" attempting to run the businesses of others, most—including Pew—responded.[73]

On questions related to transportation and refining, respondents generally agreed with one another. Most even agreed that the extraction of oil would be more efficient if operators cooperated in conserving natural gas and increased the spacing between wells. However, ideas as to how this might be accomplished varied widely, with many respondents strenuously opposing any suggestion that federal legislation might be desirable. Even industry leaders hesitantly receptive to reform could not reach any consensus as to what specific policy the federal government should adopt.[74]

After analyzing these surveys, the FOCB invited representatives from the oil industry to a set of hearings held in early 1926. The hearings, well attended by the heads of major oil companies, opened with W. S. Farish (president of the API and vice president of Jersey Standard), A. L. Beaty (president of the Texas Company), and W. J. Teagle (president of Jersey Standard) making brief statements. Each reiterated the API's official position and pointed out their "mutuality of interest" in eliminating waste. All three also encouraged more study.[75]

Then, Henry Doherty spoke, to the irritation of his colleagues. The rule of capture, he asserted, was just a rule, and rules could be altered. Furthermore, he noted that because the problem affected a national market, it required legislation at the national level. No single state, however interested in conservation, could rely on state-enforced laws alone.[76] Indeed, in 1915, when the Oklahoma legislature had given state authorities the power to limit production to market demand, the statute proved ineffective and unenforceable. Producers sold to a national market, and few were willing to accept the convenient fiction that state authorities in Oklahoma could somehow serve as arbiters of that market.[77]

In any case, most oil producers present resisted any change, state or federal, to the rules governing oil extraction. Doherty's early business experience operating utilities which were already heavily regulated undoubtedly made him more comfortable with the idea of government-enforced regulations. E. W. Marland of Marland Oil, in contrast, vehemently opposed any effort to restrict his choices. A major competitor of Doherty's in Oklahoma and a future governor of that state, Marland argued that the more he drilled, the more oil he recovered. If he drilled too many holes on a single lease, Marland admitted, he might not make any money, but that was his concern, not the government's. Hence, Marland objected to any legislation that would prevent operators from drilling as many wells as they wanted, even if it meant forcing neighboring operators to drill more wells than they wanted. The Constitution of the United States, Marland asserted, did not allow such legislation.[78]

As the hearing progressed, each new speaker revealed another piece in the complex tangle of interests that prevented easy consensus on how to proceed. Not surprisingly, most people were concerned that changes would render their business strategies and investments obsolete. For example, producers who owned relatively small leases scattered throughout various fields tended not to support production limits. They realized that limits would undermine the advantages of this checkerboard strategy.[79] Refiners who sold significant quantities of petroleum as fuel oil expressed concern over a sentence in the FOCB mission statement that suggested limiting petroleum to uses where coal could not be employed. Let the market take care of that, J. F. Pew of Sun Oil argued. He asserted that if scarce oil caused prices to rise, liquid hydrocarbons from shale and coal could be produced at a profit.[80] Charles Kettering, who pursued the discovery of tetraethyl lead as the head of the General Motors Research Corporation, also expressed little concern over the potential depletion of oil. He suggested that automobiles would become more efficient as prices rose. According to Kettering, the average car went only fifteen miles on a gallon of gasoline, but one gallon contained enough energy to propel a car over 440 miles, giving the automobile industry lots of room for improvement.[81]

L. V. Nicholas, representing independent producers, spoke in favor of production controls. The problem, as he saw it, was that large oil companies could afford losses in the production sector. Hence, these large companies took advantage of small producers by encouraging overproduction that resulted in cheap oil. As evidence, he quoted the executive of one integrated company as saying "we do not particularly care in what branch of our operations a profit is made or in which a loss is sustained, so long as the consolidated balance sheet shows a proper and satisfactory balance."[82] A prominent professor in the emerging field of petroleum engineering, Lester C. Uren, also recognized the power that vertically integrated companies held over small producers. Uren suggested that small producers were "misfits" in the petroleum industry, unable to reach the scale of operation necessary to attain competitive levels of efficiency.[83]

Advocates of production controls changed few minds at the first FOCB hearings. However, in the summer of 1926, just several months after industry leaders gathered to be heard, opinions began to shift. The discovery of a large oil field near Seminole City, Oklahoma, unleashed a glut of production, large enough to affect the price of crude throughout the mid-continent. This discovery and a series of even larger discoveries eventually encouraged the directors of the API to reverse their position and support production controls.

With the discovery of the Seminole City field, Oklahoma's conservation statute was put to the test once again. And once again, officials with the Oklahoma Corporation Commission, the agency empowered to enforce the statute, experienced little success.[84] Anything short of full cooperation from producers, which the commission never had, undermined whatever voluntary action a few producers were willing to take. Oil poured into the market. The discovery of more oil in 1928, this time in Oklahoma City, complicated matters further. Part of the field lay within the city limits where property was already divided into small lots. Derricks sprang up everywhere—in backyards, playgrounds, and parking lots. A few rigs even stood within sight of the Capitol, mocking those responsible for enforcing the state's conservation statute.[85]

If symbols are important—and they are—the blowout of "Wild Mary" on the outskirts of Oklahoma City in early 1930 qualifies as a significant symbol, serving both as a metaphor of the oil gushing out of Oklahoma and as a portent of things to come. Such blowouts, or gushers as they were called in the period before World War I, had once been fairly common.[86] By the early 1920s, though, blowout preventers—devices that slammed shut and stopped the escape of fluid when crews lost control of a well— had been patented and put to use in fields having high reservoir pressures. However, not all drilling contractors chose to use such devices, and most large fields still experienced blowouts at one time or another.[87]

In the case of Wild Mary, problems started after the crew drilling the well reached oil-bearing sand early on the morning of March 26, 1930. As members of the crew removed the drill pipe, they failed to keep the hole filled with drilling mud. A powerful rush of oil and gas shot into the air (fig. 6.1). Ten days later, after numerous attempts to attach a new master-gate to the surface casing, natural gas, oil, and sand still spewed from the hole, covering land around the well with a film of crude. By that time, Wild Mary had become a national news story, with newspapers and radio stations throughout the United States tracking the progress of those attempting to stop the flow of oil. As 20,000 barrels of oil and 200,000,000 cubic feet of natural gas shot out of the hole each day, listeners across the United States followed with interest. Crews eventually gained control of Wild Mary by cutting new threads into the surface casing, which had remained in place only because crews had cemented it there, a procedure required by Oklahoma's conservation statutes. The company that owned the well then had to clean up the mess, something companies did not always do when blowouts occurred in more remote areas. In this case, the company paid for houses in the area to be repainted and for thousands of oil-soaked acres to be plowed under and replanted.[88]

Wild Mary, officially known as Indian Territory Illuminating Company's "No. 1 Mary Sudik," could not claim to be either the most spectacular or the last blowout the industry would ever see. Indeed, even as oil companies developed the Oklahoma City field, engineers in Romania were attempting to control a well that would burn for two years and crater over one thousand acres of land.[89] In the United States, automatic blowout preventers would become more common and blowouts less frequent, but drillers would continue to lose control of their wells. Texas alone would see over one hundred blowouts in the 1930s.[90]

Wild Mary—spewing oil on the outskirts of an urban center in which legislators were wondering how to control the flow of oil—served as a convenient metaphor for the state of the industry. Indeed, the glut of oil flowing out of Oklahoma played a significant role in convincing the directors of the API to reconsider their stand against production controls. By 1930, in a gradual reversal of position, the API came to support state legislation aimed at encouraging the orderly development of fields.[91]

If anybody in the oil industry remained unconvinced about the wisdom of putting limits on the rule of capture, the discovery of the East Texas field in 1930 brought them to their senses. This field, forty-two miles long and four to eight miles wide, dwarfed any previous field in the United States. Within two years of the discovery well, over 3,600 oil wells had been drilled in the field, most capable of producing significant amounts of oil. East Texas, with help from a national economic depression, threatened the stability of the entire industry. In 1929, the average price per barrel of crude in U.S. markets was $1.29. By 1931, the average price had dropped to 65 cents.[92]

Figure 6.1. In 1930 in the Oklahoma City field, the blowout "Wild Mary" symbolized the industry's inability to control the glut of oil being poured onto the market. (Library of Congress LC-U5-Z62-5590)

Oil producers in East Texas faced an even greater challenge than producers elsewhere. At its lowest, the price of crude oil in East Texas dropped to ten cents per barrel. Nobody could make any money unless they owned refineries capable of transforming that crude into gasoline. Hence, after constructing makeshift, inefficient, and highly polluting refineries, entrepreneurs began selling gasoline at cutthroat prices. Although this route to market short-circuited the far more efficient transportation and refining system operated by large companies, East Texas crude was so inexpensive that small refineries did not have to be particularly efficient. The larger companies, attempting to maintain pre-1930 gasoline prices, could not compete. Established refiners who previously enjoyed inexpensive oil now saw low prices as a major problem.[93]

In Texas, state legislators ordered the Texas Railroad Commission, at the time a relatively obscure agency with authority over oil conservation statutes, to impose production limits. Producers immediately challenged those limits by raising questions about their purpose. Were those limits meant to eliminate "economic" waste by matching production to meet market demand or to eliminate "physical" waste associated with poor production practices?[94] In late 1931 as oil continued to flow unimpeded, the governors of both Oklahoma and Texas imposed martial law in strategic oil fields, temporarily shutting down production in those fields. After troops succeeded in establishing order, the Railroad Commission, its authority reinforced by a conservation statute known as the Market Demand Act, limited production from most oil wells in the East Texas field to twenty barrels per day.[95]

To some extent, production controls only aggravated the situation. Many oil field operators began to produce "hot oil," that is, oil in excess of their "allowable" production. The incentive to do so was high and those who sold hot oil initially looked upon production controls as being no different than the concurrent and openly violated prohibition against alcoholic beverages.[96] Hence, many resorted to hidden pipes, fake valves, and secretive nighttime operations to keep the oil flowing.[97] In addition, independent marketers who purchased gasoline refined from illegally produced crude tended not to pay gasoline taxes, enabling them to undercut further retailers who paid taxes. Some kept their costs even lower by distributing their gasoline directly from tank trucks.[98]

The heads of the major petroleum companies sought to fight the suppliers of cheap gasoline on several fronts. First, the API created a "code of ethics" that discouraged established oil companies from engaging in cutthroat competition or supporting marketers of cheap gasoline in any way, and held meetings to enforce that code.[99] The institute also put pressure on retailers selling inexpensive gasoline from tank trucks by initiating a

Figure 6.2. The Conroe field in Texas, viewed here in the 1940s, was one of the first to be developed in an orderly fashion. (Smithsonian 99-2237)

campaign to have states enforce tax laws more rigorously. In addition, they supported efforts to declare the sale of gasoline directly from tank trucks illegal on the grounds that such operations represented a fire hazard. Finally, the API worked with government legislators and administrators at both the federal and state level to create the legal and administrative mechanisms necessary to set and enforce production limits.[100]

The mechanisms finally put into place included: the Connally "Hot Oil" Act, which allowed the federal government to limit the interstate shipment of hot oil; an Interstate Oil Compact that provided participating states with a way to coordinate how much crude each state could produce; estimates of the demand for crude performed by the U.S. Bureau of Mines so that reasonable state quotas could be set; and state conservation bureaucracies capable of allocating and enforcing production allowables. By 1935, with these mechanisms in place, the era of unregulated production was over (fig. 6.2).[101]

Industry leaders who had aggressively opposed controls a decade earlier now spoke forcefully in favor of those controls. For example, E. W. Marland, who had defended his constitutional right to drill wells before the Federal Oil Conservation Board, now helped to establish the Interstate Oil Compact Commission, the body that coordinated the production levels of major oil-producing states.[102] Throughout, the advocates of production controls had consistently articulated their goals in terms of increasing efficiency and eliminating waste. Hence, an efficiency-based ethic had once again demonstrated its power to shape policy and guide decisions affecting use of the shared environment.

Environmental Objectives and Utilitarian Conservation

In his article "Letting the Grandchildren Do It: Environmental Planning During the Ascent of Oil as a Major Energy Source," Joseph Pratt rightly characterizes the response of participants in the leaded gasoline and oil pollution debates as a missed opportunity in environmental planning. Indeed, the notion of environmental planning did not even fit into the intellectual framework of most participants. Environmental planning suggests that people must first reach consensus on environmental objectives and then put policies in place to achieve those objectives. In the 1920s, few people saw any need to explicitly define such objectives.

Rather, in most decisions affecting use of the shared environment, participants saw the environment as a set of resources and sought to optimize the value extracted from those resources. Toward this end, technical arguments based on optimizing efficiency and reducing waste appeared sufficient. Asking questions about environmental quality, or however they would have referred to that notion in the 1920s, seemed unnecessary to most people. Those who expressed what might be called an environmental ethic—the argument that certain aspects of the environment should be protected regardless of how it affected economic activity—were in a minority and their views carried little weight. Even many advocates of strong pollution control legislation argued in terms of utilitarian conservation, justifying regulatory mechanisms by associating losses with a failure to put resources to their highest economic use.[103]

In practice, this efficiency-minded utilitarian approach to environmental decision making assumed that everybody agreed on the objective: to optimize the economic benefits extracted from a set of resources. Theoretically, the decision-making process then became an optimization problem to be solved by experts who collected data, studied the various options, and determined which course of action maximized the benefit extracted from those resources. This assumption, of course, breaks down when different groups value the same resources in different ways, as was the case in debates over the oil pollution of navigable waters and the use of leaded gasoline.

In debates over use of the shared environment, efficiency-based arguments served as the ticket of entry and the currency of choice. The alternatives that various groups championed often fell in line with vested interests, but arguments in support of those alternatives still had to be grounded in efficiency-based arguments. Unless a group could justify entering the decision-making process on the basis that it could show how to improve efficiency and increase the economic value extracted from a set of resources, those controlling the process felt justified in dismissing any challenge to their decisions.

In debates over regulating waste disposal practices at refineries and prohibiting the use of leaded gasoline, efficiency-based arguments proved to be a powerful ally to groups resisting government-enforced actions. Refiners, for example, promised that standard-setting efforts and more efficient technology would eventually make refinery pollution a moot issue. In the debate over leaded gasoline, the Ethyl Corporation emphasized how tetraethyl lead helped the nation use its oil resources more efficiently, and despite concerns with the long-term health effects of leaded exhaust, the company was given a green light to proceed. However, in the debate over production controls, arguments rooted in the logic of utilitarian efficiency trumped industry's initial resistance to regulatory mechanisms, further validating the power and sufficiency of an efficiency-based decision-making ethic in the eyes of those who put their faith in it.

SEVEN

Success and Failure in the Oil Fields

A PROVISION OF THE OIL POLLUTION ACT of 1924 required the secretary of war, as overseer of the Army Corps of Engineers, to determine what polluting substances were being deposited into the navigable waters of the United States and the extent to which they endangered navigation, commerce, and fisheries. The law charged the secretary to report back to Congress within two years. The report eventually issued by the chief of engineers, General H. Taylor, contained few surprises. Anybody who had read the previous investigation of oil pollution performed by the U.S. Bureau of Mines would have been on familiar ground.[1]

However, the Corps of Engineers mentioned one concern that had not been discussed in the oil pollution hearings: salt water released from oil fields.[2] The American Petroleum Institute (API), in a report released two years later after its own investigation into pollution concerns, admitted that "until recent years, very little care has been exercised to prevent oil waste water from being carried from producing areas into natural drainage."[3] The way in which oil producers eventually addressed problems caused by the disposal of oil field brines further validated the notion that

Table 7.1. Disposal Practices at the Richland Field in Texas, 1925

Company and Lease	No. of Wells	Summary of Comments by Inspector
Atlantic Co., Swink lease	7	"Waste oil and b.s. [bottom sediment] disposed of in pits; no waste oil or b.s. leaving lease."
Humphreys Co., Webb lease	9	"No waste oil or b.s. leaving lease and no b.s. in storage."
Gulf Prod. Co., Elkins lease	4	"No waste oil or b.s. leaving lease. No b.s. in storage."
Gutman and Appleman, Swink lease	3	"Has pit with 50 bbls of b.s. and waste oil in it; they pick up this oil and treat it back once or twice or week. No b.s. or water leaving lease."
J. K. Hughes, Swink and Brown lease	5	"No salt water going across Simms or Allison leases as they have lines to Pin Oak Creek. B.s. is pumped into a pit and burned there."
Rowan, Davis lease	2	"They have installed a treating plant and are now cleaning out their b.s. instead of letting it flow to Pin Oak Creek as before."
Simms, Allison Lease	8	"They are piping all their salt water into Pin Oak Creek; none going across Allison farm. They have no b.s. in storage or leaving the lease."
Sun, Brown lease	16	"No b.s. or waste oil leaving lease as they have three small pick-up pits and are very careful about their waste oil. No b.s. in storage."
W. B. Tucker, Davis lease	9	"They have a large pick-up pit next to Creek but they keep their b.s. and waste oil picked up and none leaving lease."
C. A. Tucker, Davis lease	4	"We have been assured by management that proper bleeders will be put in to care for their b.s. and salt water."
Barkley and Meadows, Davis lease	6	"They have a small pit and are supposed to burn their b.s. and waste oil but most of it goes into a branch of the Pin Oak Creek. They are pumping their salt water into three small reservoirs but these overflow and the salt water goes into Pin Oak Creek."
Republic Prod. Co., Webb lease	7	"They are taking care of their salt water in a large reservoir. No salt water or b.s. leaving the lease."
Derby Oil, Brown lease	1	"Taking care of their salt water in a small reservoir that is not large enough to hold the water. Whatever oil gets into Pin Oak Creek is picked up by private pick-up pits."

Source: G. E. Martin to Green Prescott (Pure Oil), November 5, 1925, box 38, series 7, Sun Company Collection, Hagley Museum and Library.

self-regulation rooted in the ideals of utilitarian conservation could serve as a guiding ethic in addressing pollution concerns.

What to do with the Salt Water?

By the time the Corps of Engineers identified the disposal of oil field brines as a pollution problem, many oil field operators were already making some effort to prevent their waste salt water, tank bottom deposits, drilling mud, and spilled oil from reaching nearby streams. A few even trucked away the sludge that came from their tanks. For those who took action, damage and nuisance suits provided much of the incentive. In 1925, one survey of a Texas oil field resulted in a summary of how each lease handled its wastes and reflected this concern with housekeeping (table 7.1). The purpose of this particular survey was to identify oil field operators who put other operators at risk by fouling nearby streams and damaging relations with farmers, ranchers, and town officials.

In most oil fields, the disposal of salt water emerged as the main pollution-related problem. Not many disposal options existed, and problems only grew worse with time. As producers extracted more oil, the salt water surrounding that oil moved toward the well, forcing operators to lift an ever-increasing percentage of water and making pollution-related complaints even more likely.

The ideal solution, of course, would have been to extract and sell the chloride and other minerals present in the waste brine. After all, people had been using brine for this purpose long before Edwin Drake drilled his oil well. Indeed, Drake had used the drilling techniques of local salt producers. Herbert H. Dow, who founded Dow Chemical in 1895, even had used a brine well as the source of his bromine.[4] However, Dow had drilled a well specifically for extracting a brine high in bromine. No oil was involved. In 1926, though, one company did begin processing waste brine from a nearby oil field and, from each barrel of brine, recovered about fifty pounds of sodium chloride (table salt), ten pounds of calcium chloride, three pounds of magnesium chloride, and a small amount of iodine. In all, the plant produced seventy tons of salt from the 3,000 barrels of brine processed each day.[5]

Still, however much oil producers would have liked to sell their waste brine to salt and chemical companies, doing so usually was not a practical option. Companies extracting salt from brine generally found it more efficient to drill and operate their own brine wells. Not only was transporting brine difficult, different formations produced brines with different concentrations of desirable minerals. Extracting salt from the ocean or mining dry salt from mines often proved even less expensive. For oil producers, other less satisfying solutions had to be explored.

In an effort to help oil producers, the U.S. Bureau of Mines began collecting whatever information they could about how producers actually disposed of their salt water. By 1929, they were publicizing what they had learned. As it turned out, oil producers in California had the most direct solution. They simply piped their waste brine to the Pacific Ocean. By 1930, producers in California could, if necessary, pipe over 200,000 barrels of waste brine to the ocean each day.[6] In Kansas, of course, that option was not available. There, most producers created "evaporation ponds," shallow pits surrounded by a brim of dirt. In parts of Texas, producers stored their salt water when streams were dry. They released this water only when rain turned those dry channels into fast-moving streams. The bureau's engineers also discovered that some oil producers had no disposal problem. In one Texas field, ranchers even asked oil field operators to run their waste brine into a ditch from which cattle could drink. Apparently, water produced in that field had a salt concentration low enough for the cattle to tolerate.[7]

Often, oil producers did not realize they had a disposal problem until someone complained. For example, oil companies operating in the Darst Creek field of Texas ran about 12,000 barrels of salt water directly into the Guadalupe River each day. Producers in this field discovered they had a disposal problem only after farmers using the Guadalupe to irrigate their pecan groves complained about damage to their trees.[8] When told of the problem, J. E. Pew, the head of production for Sun Oil, instructed his employees "to take care of this water at any expense rather than take this chance [of damaging the pecan trees], particularly if we should be getting any tort of this kind where we are associated with some small producer in doing it, as we would be the victim."[9]

Eventually, producers working the Darst Creek oil field formed a Salt

Water Disposal Company that impounded the brine behind two large reservoirs, each capable of holding over a year's worth of wastewater. During rainy periods when streams and rivers reached their highest flow, operators released the salt water into two different rivers, the Guadalupe and the San Marcos. Because the San Marcos was five miles away, the Darst Creek Salt Water Disposal Company leased a right-of-way to the river along which landowners waived any right to claim surface and subsurface drainage damage. The disposal company recovered its costs by billing well owners in the field for a prorated amount of the total cost.[10]

Given their lack of options, producers accepted the costs associated with keeping salt water out of streams as a way to prevent damage suits. Ranchers and farmers simply assumed that a certain level of water quality should be maintained and that all economic activity should be forced to respect those standards. Producers generally accepted that constraint as legitimate. In part, oil field operators accepted this environmental constraint because technology existed to monitor the salt content of water and to correlate those measurements with identifiable effects.

Indeed, when disputes arose, inspectors with agencies such as the Kansas Board of Health and the Texas Game, Fish, and Oyster Commission monitored the salt content of streams and knew how much salt could be tolerated by fish, crops, livestock, and humans. Nobody suggested that streams had to be absolutely pure; they recognized that all streams contained dissolved solids from other sources as well. However, they could identify fairly defensible standards of quality, such as the concentration of chloride safe for human consumption (around 250 parts per million, the level recommended by the U.S. Public Health Service); the concentration that most species of fish could tolerate (up to 5,000 ppm); or the concentration that various types of crops, farm animals, or livestock could tolerate (2,000 to 13,000 ppm). For example, in 1934, Texas law prohibited anybody from raising the total chloride content of streams above 2,000 ppm, which officials with the Game, Fish and Oyster Commission saw as sufficiently low for most fish in those streams.[11]

New concerns related to brine contamination emerged in the early 1930s, with the most significant problems occurring in Kansas and Texas. Experience showed that neither evaporation ponds nor the controlled release of salt water into streams were acceptable long-term solutions. In both cases, producers had to find new ways to dispose of their waste salt water. Health officials in Kansas discovered that the "evaporation ponds" used by oil producers actually complicated the disposal problem. Any evaporation that did occur simply created a more concentrated brine. Furthermore, most ponds did not have enough surface area to evaporate the amount of wastewater being produced. Most of the salt water simply leaked out through the porous soil, eventually reaching aquifers that lay below. By the early 1930s, after the Bureau of Mines completed its initial survey of disposal practices, these evaporation ponds had contaminated shallow aquifers throughout the oil-producing regions of Kansas, forcing many farmers to relocate their freshwater wells and generating a flurry of litigation against oil field operators. An ongoing drought further aggravated the problem.[12]

Getting rid of waste salt water was also a problem in the newly discovered East Texas oil field in 1930. The sheer size of this field magnified problems with pollution beyond those associated with smaller fields. Streams in the area simply could not handle the amount of salt being lifted by wells in that area. Production controls, recently established by the Texas Railroad Commission, did not help much either. Even though the Texas Railroad Commission allowed producers in the field to extract only around twenty barrels per day from each of their wells, producers could lift as much salt water as needed to meet their allowable quota of oil. In marginal areas of the field, operators lifted over ten barrels of water for every barrel of oil.[13]

To keep law suits, damage claims, and fines to a minimum, major oil companies in the field formed the East Texas Anti-Pollution Committee and charged the committee with finding ways to "keep the streams of East Texas free from oil, b.s. [bottom sediment], drilling mud, and waste water."[14] The committee accomplished relatively little. Two years after its formation, by which time fifteen thousand oil wells had been drilled in the field, many operators still stored their wastewater and oily emulsions in open pits, allowing storm water to sweep the waste into nearby streams (fig. 7.1).

Representatives from the Texas attorney general's office and the state

Figure 7.1. Waste brine damaged many streams running through the East Texas field. (Smithsonian 99-2360)

Game, Fish, and Oyster Commission expressed concern and asked the East Texas Anti-Pollution Committee to take stronger action. In response to such concerns, the head of Humble Oil's production department (an affiliate of Jersey Standard) advocated the distribution of pamphlets that explained the sources of oil field pollution, the indirect costs of that pollution, and methods of prevention. He noted that Humble had created such a pamphlet and credited it with cutting his company's liability from damage claims and law suits arising from pollution in East Texas from $4,200 in 1933 to $1,900 in 1934.[15]

Initially, the API's Committee on the Disposal of Oil Field Wastes proposed publishing the pamphlet but decided against the idea in fear "that an API code, becoming public property, would furnish ammunition to damage claim lawyers." A better alternative, members of that committee

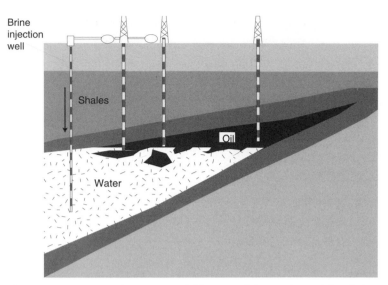

Figure 7.2. Engineers in the East Texas field suggested that producers inject their waste salt water back underground, as illustrated in this east-west cross section of the uniquely structured field. (Illustration by author)

believed, was to have each company issue a pollution control pamphlet in its own name.[16] Companies such as Cities Service and Phillips Petroleum then printed their own versions; the one by Phillips Petroleum described waste and pollution as "going hand in hand" and suggested that "cleanliness costs nothing and earns large dividends."[17] Pamphlets, however, could not stop the flood of salt water.

Another committee, the East Texas Engineering Committee, addressed the brine disposal problem in a more comprehensive manner. Among other things, this committee concluded that hydrostatic pressure accounted for approximately 96 percent of the field's lifting energy and that every barrel of fluid lifted, oil or water, reduced the reservoir pressure. In some parts of the field, they noted, operators were lifting 300 barrels of water to obtain their twenty barrels of oil. To prevent the field's hydrostatic pressure from dropping too quickly, the engineering committee recommended that companies stop producing from any well in which the water-to-oil ratio exceeded a certain value. In that way, no operator would be allowed to waste the field's natural lifting energy.[18]

Members of the engineering committee also proposed three options for disposing of the increasing quantities of salt water coming from wells: gather and store the water until the Sabine River reached flood stage and then release it; pipe it to the Gulf; or inject the salt water back into the formation from which it came. Of the three options, most members of the committee advocated injecting water back into the formation as the most sensible (fig. 7.2). Not only would injection get rid of the brine and solve a disposal problem, but it would also help maintain reservoir pressure and increase the efficiency with which producers extracted the available oil.[19]

Injection as the Disposal Solution

Around the same time that engineers in East Texas began recommending the use of brine injection wells, health officials in Kansas also began thinking about using such wells to replace so-called evaporation ponds. When Kansas officials approached the Bureau of Mines for advice, they discovered that the bureau had been following developments associated with gas, air, and water injection for some time. They learned that some oil companies had already been experimenting with fluid injection as a way to maintain reservoir pressure in productive fields and to rejuvenate depleted fields. However, in the period before production controls, when individual producers extracted as much oil as they could in competition with their neighbors, coordinating effective injection programs proved difficult.

John Carll, a prominent geologist with the Pennsylvania Geological Survey, had first observed the effect of injected water on oil fields in 1880. He noted that if water flooded one oil well, production in nearby wells increased before being overtaken by water.[20] Hence, around the turn of the century, by which time production in Pennsylvania had dwindled to a trickle, many operators began experimenting with water flooding—using fresh water, not brine—as a way to rejuvenate depleted fields. However, they did so secretly and on a small scale. Manufacturers of high-quality lubricating oils paid top dollar for Pennsylvania crude, and some operators feared that injecting water into the oil sands would kill whatever production remained. Gradually, though, operators demonstrated that they could increase production by using injected water to sweep oil toward nearby wells. In 1921, Pennsylvania legislators passed laws explicitly legalizing water flooding.[21]

Producers in Oklahoma, California, Texas, and Kansas showed more interest in injecting fluids to maintain reservoir pressure. For example, in 1926, the Cities Service Company implemented a large-scale air injection program for pressure maintenance on leases it owned in Kansas. After engineers with the company discovered that natural gas lowered the viscosity of oil, they started injecting natural gas instead of air. But in the years before states placed limits on the rule of capture, pressure maintenance efforts such as the one attempted by Cities Service were rare. Why would one company pump their natural gas back into a formation if operators on neighboring leases were lifting oil and flaring gas from that same formation with abandon?[22]

By the mid-1930s, production limits and prorationing schemes were in place that encouraged cooperation by eliminating the mad race to extract oil. In addition, new tools for modeling underground formations made efforts to control the dynamics of a field more feasible.[23] By the time oil field operators in Texas and Kansas considered solving their brine disposal problem by injecting the brine back into water-saturated portions of the oil-producing formation, it was both legally and technically feasible to do so.

Therefore, in late 1933, when the East Texas Engineering Committee proposed to test what would happen if they injected waste brine back underground, fifteen of the twenty major oil companies in the field agreed to help fund the effort. As it happened, the committee failed to reach an agreement with the Texas Railroad Commission on where to inject the water or how to compensate leaseholders who would be adversely affected. Three years passed before the committee injected any water into the formation.[24] Meanwhile, officials in Kansas moved forward with their plan to encourage the use of injection wells.[25]

To engineers with the Bureau of Mines, brine injection technology was the silver bullet they had been looking for. Not only did that technology promise to solve the brine disposal problem, it also promised to make oil fields more productive. Oil producers, they accurately predicted, would rush to adopt the technology if its feasibility and value could be demonstrated.

Not surprisingly, the first Kansas oil field operators to inject waste salt water back into the ground ran into problems. Subsurface formations that initially accepted large quantities of waste brine soon refused to accept any more, even at relatively high pumping pressures. Engineers from the bureau's Bartlesville Petroleum Research Station visited the sites and catalogued numerous reasons why injection wells clogged up. In some cases, algae, which flourished after the deep water came into contact with oxygen, quickly plugged the pores of the receiving formation. Scale associated with the corrosion of metal pipes—accelerated by salt, sulfur, and other compounds in the brine—could also serve as the plugging agent, as could chemical reactions between incompatible brines. Engineers with the Bureau of Mines soon developed a fairly straightforward array of diagnostics, corrective chemicals, and procedures to condition the brine before injection.[26]

With these correctives, the use of injection wells in Kansas spread rapidly. By 1936, the Kansas State Corporation Commission had licensed 499 wells throughout the state. Of these, producers used 325 wells in conjunction with secondary recovery efforts.[27] Engineers with the Bureau of Mines estimated the initial cost of the larger injection systems to be about $11 per barrel per day throughput, meaning that a system designed to inject 5,000 barrels per day would cost about $55,000.[28] Operating and maintenance costs aside, the per unit cost of such systems, therefore, was slightly less than that associated with the storage dam method employed in the Darst Creek field, which amounted to about $16 per barrel per day throughput.

Oil producers in East Texas adopted the new technology more slowly than producers in Kansas. In the same period that Kansas operators put almost 500 injection wells into service, oil companies in East Texas failed to make much progress at all. By the end of 1937, oil producers operating in the East Texas field had drilled over 20,000 oil wells into the massive reservoir and were lifting over 100,000 barrels of brine per day. Furthermore, the quantity of salt water being lifted doubled from year to year. Most operators had no choice but to run this water into disposal pits akin to the old evaporation ponds in Kansas.[29]

As conditions in East Texas deteriorated, representatives from the larger oil companies revived discussions about organizing a salt water disposal corporation.[30] Although engineers experimented with a few injection wells over the next several years, companies still could not agree on a coordinated plan of action. By 1939, prominent engineers advocated a massive program of strategic injection to take care of both the problem of disposal and the problem with the reservoir's dropping pressure. Although the amount of salt water being lifted had climbed to 200,000 barrels per day, companies were injecting only 6,100 barrels per day into test wells.[31]

Two incidents eventually encouraged the Texas Railroad Commission to work more closely with oil companies in organizing a salt water disposal corporation. First, to protect the Neches-Angelina watershed from salt contamination, Texas officials brought an injunction against all operators in one portion of the field.[32] Second, federal officials, responsible for securing a steady supply of petroleum during World War II, expressed concerned over the field's rapidly dropping pressure. They argued that too large a drop in reservoir pressure would cause natural gas to come out of solution, making the oil more viscous and less fluid. In a series of well-attended public meetings, E. L. DeGolyer, a respected petroleum geologist and the wartime assistant to the federal oil coordinator, made it clear that if operators and the Railroad Commission could not cooperate in implementing an effective injection plan, the federal government would take action in the interest of national security.[33] Within two months, twelve oil companies had organized the East Texas Salt Water Disposal Corporation. This nonprofit corporation quickly constructed a system of gathering pipelines, water conditioning plants, pumps, and injections wells. By the end of the war, the East Texas Salt Water Disposal Corporation was injecting over 250,000 barrels of brine each day, an amount that doubled over the next two years.[34]

Throughout the 1940s and into the early 1950s, the Bureau of Mines took a leading role in disseminating information about the use of injection systems to dispose of waste salt water, to maintain field pressure, and to aid in the secondary recovery of oil from older fields. Engineers at the bureau's Bartlesville station even organized tours of waterflood projects in Oklahoma.[35] By the early 1950s, the use of injection wells had become routine and were judged to be the best way to solve brine disposal problems (fig. 7.3).

Figure 7.3. By the late 1940s, many oil producers collected their waste brine for treatment and reinjection. (Library of Congress LC-US-Z62-124118)

The success of using injection wells for brine disposal, pressure mainte-nance, and secondary recovery operations reinforced the conservationist assumption that optimizing the use of resources overlapped with efforts to address pollution concerns. For example, officials with the Arkansas Board of Health credited the introduction of secondary recovery opera-tions in their state with saving many Arkansas streams and rivers from salt pollution. In 1937, when Arkansas first passed laws prohibiting the discharge of salt water into streams, a grandfather clause exempted fields already in production from control. At that time, as much as 85 percent of the fluid being lifted in mature fields such as El Dorado was salt water, and all of that water still could be discharged into streams and rivers with no legal repercussions. After oil producers in mature fields started using injection systems to facilitate the secondary recovery of oil, they began pumping their waste brine into injection wells instead of letting it run into streams.[36]

Reinforcing Trends

In many ways, the success of organizations such as the East Texas Salt Water Disposal Corporation represented the finest hour of those who saw efforts to increase the efficiency of industrial operations as a long-term solution to pollution-related concerns. Indeed, some authors writing about the oil industry have pointed to salt water disposal corporations as evidence that those in the industry were "among the earliest environmentalists."[37] However, applying that description to those responsible for brine disposal and waterflood projects misses the mark. The engineers responsible for such projects did not necessarily see environmental quality as a socially defined environmental constraint on industrial activity. Rather, they assumed that environmental quality was the inevitable result of increasing efficiency and reducing waste. That assumption would eventually fail.

Failure, though, was the farthest thing from the minds of petroleum engineers in the late 1930s. Technological innovations associated with locating and managing oil reservoirs, efforts to conserve oil and stabilize prices by controlling production, and the need to find better methods of brine disposal all reinforced one another and pointed to nothing but success. The desire of trained petroleum engineers to use their newly acquired skills also played a significant role. In short, new techniques for acquiring geophysical knowledge made managing reservoirs as a geological unit more possible just as economic and legal reasons to do so also emerged.

In the United States, systematic development of the science and technology necessary to acquire and apply geophysical knowledge took off in the 1920s, with the initial focus on exploration tools. At the beginning of the decade, geophysical prospecting was still in its infancy. For example, in 1920, Norman H. Ricker, a newly minted Ph.D. from the physics department at Rice University, sent proposals to oil companies describing ways in which he could help them locate salt domes through the use of geophysical measurements. No oil company showed any interest in his ideas. Eventually, he took a job with Humble Oil investigating how best to recover spent fuller's earth generated at Humble's refinery in Baytown, Texas.[38]

By 1924, however, conditions had changed. Several European firms had begun marketing geophysical prospecting instruments for locating salt domes. Among them were two German firms, one that employed a torsion balance to locate gradients in the earth's gravitation field and another that measured the reflection of seismic waves. Wallace E. Pratt, the chief geologist for Humble Oil, expressed interested in the latter instrument but worried about entrusting sensitive information to German-speaking crews employed by the geophysical contractor. Therefore, Pratt decided to develop geophysical capabilities within Humble and made inquiries at Rice University as to who might be knowledgeable about such methods. He was directed to Ricker, already one of Humble's employees. By 1925, Ricker was supporting seismic crews operating on the Gulf Coast. Other oil companies soon followed Humble's lead.[39]

By the early 1930s, most oil companies of any size owned some geophysical equipment or contracted with companies that operated such equipment. In 1929, Empire Oil, associated with Cities Service (later CITGO), owned four torsion balances and eleven magnetometers, but these instruments did not lead to any oil discoveries. The company then sold its equipment and began contracting with Geophysical Service Incorporated and the Seismograph Service Company to supply the necessary crews and equipment. Soon after, however, Cities Service again purchased its own seismograph equipment and used it not to find new oil fields but to eliminate leases in unfavorable locations.[40]

Tools associated with gathering data about a producing oil field also arrived on the scene. One of the most significant developments was electric well logging, a technique in which crews measured the resistivity and spontaneous potential of formations by running electrodes along the wall of the well hole. Introduced in the United States by Conrad and Marcel

Schlumberger in 1929, this tool allowed petroleum engineers to record changes in these variables over the length of the hole on a strip of chart paper. These logs, which required interpretation, served as a visual representation of the formations that lay below. By correlating numerous electric logs from the same field and combining these data with information obtained from drilling and core samples, engineers could create reasonable three-dimensional models of an oil field at a relatively low cost.[41]

Around the same time that firms began making use of electric logging, engineers at the Bureau of Mines Petroleum Station in Bartlesville, Oklahoma, developed several instruments that allowed engineers to measure important quantities such as temperature, pressure, viscosity, water content, and the gas content of oil at the bottom of a well. Together these instruments allowed engineers to learn more about the dynamics of an oil-bearing formation than was previously possible.[42]

The move toward regulated production dramatically increased the demand for engineers who could acquire and interpret geophysical data. With regulated production came unitization, in which multiple leases were operated as a unit, and prorationing, in which regulatory agencies specified how much oil a producer could extract from a well. Different states used different prorationing schemes and unitization rules, but all were based on efforts to maximize the efficiency with which producers recovered oil.[43] Regulators and companies alike suddenly had to make decisions based on the amount of oil a formation contained and the movement of fluids in that formation. To be successful in providing the information others wanted, petroleum engineers had to learn something about the legal system governing oil extraction and the specifics of regulatory rules. Toward this end, schools such as Texas A&M at College Station, the University of Texas at Austin, and the University of Oklahoma in Norman began socializing engineers into an industry that had made the shift from unregulated to managed production. By 1936, the education of petroleum engineers, for all practical purposes, had become standardized.[44]

Advances in drilling technology, which gave rise to a problem that had not existed when courts originally upheld the rule of capture, also provided some incentive to regulate production. In the mid–nineteenth century, courts acknowledged their ignorance about what happened underground.

They likened oil to a wild animal that crossed property boundaries of its own will. Hence, they applied the same law that applied to wild animals, the rule of capture, to oil. If a person caught the animal—in this case, oil—on their property, they owned it. Furthermore, with impact drilling, the only technology available in the early oil fields of Pennsylvania, holes could not deviate too far from the vertical, which meant that wells generally stayed inside boundaries defined at the surface. However, impact drilling had since given way to rotary drilling.

With rotary drilling, keeping a hole straight was not as easy as one might think. Although one segment of drill pipe might appear strong enough to resist bending and twisting, a string of pipe about a mile long resembled a strand of wet spaghetti.[45] Drillers who placed too much weight on the bit caused the hole to deviate off the vertical, especially at points where a hard formation sloped away at an angle. Although most people in the industry realized that no hole drilled with rotary equipment was perfectly straight, few realized how far off the vertical most wells actually were until the 1920s. In one case, drilling crews separated by a third of a mile deviated enough off the horizontal to establish a hydraulic connection. Drilling mud from one well started flowing freely into the other well.[46] In 1925, J. N. Pew of Sun Oil wrote to Elmer Sperry and asked whether Sperry could design an instrument to help keep holes straight. Pew noted that increasing depths magnified the problem. "Ten years ago," he wrote, "1,500 or 2,000 feet were considered deep. Today it is not unusual to drill 6,000 feet."[47]

Subsequent measurements indicated that many wells went down only about 70 to 80 percent of the distance drilled.[48] Therefore, a well that used 6,000 feet of pipe might only be 4,500 feet deep. Furthermore, such holes did not necessarily deviate smoothly from the vertical but usually spiraled around, opening into a wider spiral with increasing depth. Given the flexibility of long lengths of pipe, drillers began visualizing the pipestring as a wire hanging from the crownblock, which they weighted down by attaching a few segments of especially heavy pipe, called drill collars, just above the bit. These heavy collars allowed drillers to keep the pipestring in tension. Collars also promoted straight holes because they had little clearance with the sides of the hole being drilled.[49]

Not all oil companies wanted to drill straight wells. For example, producers operating from leases along the coast of Huntington Beach, California, where oil was known to be just offshore, experimented with ways to get at that oil from shore. As engineers developed better instruments for monitoring the pressure on the bit, drillers discovered that they could control the deflection of a hole away from the vertical with a whipstock, a special attachment that could force a drill bit to deviate off the vertical in the desired direction. By 1935, companies operating in Huntington Beach had reduced directional drilling to a practical technique.[50]

Although directional drilling might be useful along the shore, it could cause problems elsewhere, especially in fields divided into small leases. If drillers could intentionally slant a well in any direction, they could also reach oil reservoirs not directly below the lease on which they were drilling. That ability also gave producers more reason to be suspicious of neighboring wells. In short, by the 1930s, drilling technology was also at odds with the unregulated rule of capture.[51]

By the late 1930s and early 1940s, petroleum engineers no longer saw oil as a wild animal. Oil fields had been tamed. Efforts to achieve the maximum efficient recovery of oil, which historian Edward Constant has referred to as "the cult of MER," had come to be embedded in the rules governing the industry, in the social organization of producers, and in the training of petroleum engineers.[52] In the process, the industry had also addressed a serious pollution concern by using a pollution-causing waste—oil field brine—to increase production. This success dramatically reinforced the assumption that efforts to increase the efficiency of industrial operations would continue to address pollution concerns until pollution ceased to be a nuisance.

A Crack in the Ethic

Although brine injection systems solved the most pressing brine disposal problems, they did not eliminate all concerns. If located near a poorly plugged abandoned well, an injection well could force brine up the abandoned hole, allowing it to move from one stratum to another, sometimes contaminating groundwater in the process (fig. 7.4). Problems with brine flowing between strata first surfaced when leaky abandoned wells

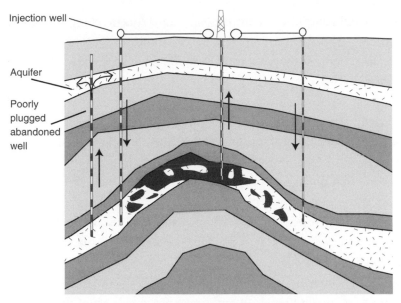

Figure 7.4. Poorly plugged abandoned wells or corroded casing in an injection well could provide injected brine with a path to freshwater aquifers. (Illustration by author)

suddenly came to life as artesian salt springs. For example, in the late 1930s, one inspector with the Kansas Board of Health discovered an abandoned well releasing a steady stream of brine directly into the Kansas River.[53]

In response to the concern, the Kansas Board of Health initiated a joint study with the Bureau of Mines to determine the most efficient way to locate and plug leaky abandoned wells. However, engineers with the Bureau of Mines had no simple solution. All oil-producing states already required oil field operators to cement abandoned wells above and below fluid-bearing formations, but not all operators followed the rules. Also, in older fields, many wells had been abandoned before regulatory agencies established adequate plugging procedures and enforcement mechanisms. In addition, injection wells with corroded casing could also provide a path to other formations, including freshwater aquifers.

As in the past, nuisance and damage suits provided some incentive for oil producers to be vigilant. However, as a long-term regulatory mechanism, lawsuits had serious limits. Contamination had to occur before legal action could be initiated. With brine injection systems, long periods of time could pass before anybody discovered a problem. In Kansas, for example, most aquifers flowed toward distant discharge points at about ten feet per day. Two years might pass before the first slug of contaminated water reached freshwater wells one mile away. Hence, a significant amount of contamination could occur before anybody initiated legal action. In such cases, determining responsibility for contamination could be difficult.[54]

The fact that not all salt contamination of aquifers came from oil production also complicated matters. In 1945, in a project that had been put on hold during World War II, a team of state and federal investigators studied salt contamination in an aquifer that supplied the city of Medicine Lodge, Kansas, with over 500,000 gallons of water per day. By sampling water wells and drilling twenty-five test holes, they discovered two sources of contamination: an oil field about three miles northeast of the city and water-saturated sandstone that lay at a higher elevation than parts of the aquifer. Under certain conditions, the sandstone formation slowly discharged saline water into the aquifer.[55]

New problems also emerged in Texas, a state that depended on aquifers for 70 percent of its fresh water. In 1952, one of the three commissioners heading the Texas Railroad Commission noted that "hardly a day passes without . . . a complaint from some party within the state complaining of surface or underground pollution which invariably can be traced to improper plugging." With inspectors having a difficult enough time keeping track of the seven to nine thousand oil wells being abandoned each year, that same commissioner also pointed to a problem with "slim holes," small-diameter holes drilled just to extract geological information or to conduct seismic operations. No laws required anybody to document, much less plug, those holes.[56]

Furthermore, some oil producers in Texas still ran their salt water into disposal pits, increasing the risk of aquifer contamination. In the late 1950s, a Texas court ordered the operator of one saltwater disposal pit—a pit like thousands of others—to pay $22,000 in damages for polluting a freshwater aquifer. Upon upholding the decision, the court ruled that the operator of the oil well clearly "knew about the great amount of salt going into the pit and that the salt would not evaporate and knew it was going somewhere."[57] Oil producers took notice. One petroleum engineer familiar with the problem estimated that the industry produced "enough salt water to provide one hundred cubic feet of salt for every person in the United States," making it a truly large-scale disposal problem.[58] However, nobody knew what amounts of salt water went where. To find out more about disposal problems, the Interstate Oil Compact Commission, the agency founded to coordinate production levels in the major oil-producing states, conducted a nationwide study.[59]

When the Interstate Oil Compact Commission first requested information, officials in some states had yet to discover that they had a problem. For example, in Kentucky, the average concentration of chlorides in the Green River—which starts in Tennessee, cuts across the middle of Kentucky, and flows into the Ohio River—had an average chloride concentration of 10 ppm. That concentration soon increased. In 1958, wildcatters discovered that anyone willing to lift large quantities of salt water could extract oil from a shallow dolomite formation near Greensburg, Kentucky. Initially, operators in the field ran their waste salt water directly into streams, but the Kentucky Water Pollution Control Commission soon outlawed that practice. Operators then started running their wastewater into sinkholes that discharged into surface streams at lower levels. By 1959, investigators reported that chloride concentrations in the Green River frequently exceeded 1,000 ppm. In addition, springs and wells in the area that had previously yielded fresh water suddenly starting producing water with a chloride content of 10,000 ppm or more.[60]

Not surprisingly, the most serious problems existed in areas where the discovery of oil attracted people to previously unpopulated areas. For example, Texas officials reported that "barren areas . . . are becoming populated; agriculture is increasing; and the search for oil has made cities in former wastelands."[61] About 95 percent of the well water consumed in that state came from one of seven major aquifers, and most oil fields coincided with one of those seven aquifers. This overlap of aquifers and oil fields meant that oil field brines had the potential for contaminating an important supply of fresh water. To track the disposal of oil field brines more carefully, the Texas legislature created a Water Pollution Control Board and required the state's 70,000 operators of oil and gas leases to obtain disposal permits for their waste brine. As a result of the permit system, officials discovered that although oil producers injected about 1.5 billion barrels of salt water into subsurface formations each year, either for disposal purposes or as part of secondary recovery operations, they still ran about one-half billion barrels of salt water into unlined pits. About one-quarter billion barrels of salt water were still being discharged into surface waters. Another fourteen million barrels, less than 1 percent of the total, were being sprayed on dirt roads.[62]

Texas then outlawed all brine pits in the forty-eight counties overlying the Ogallala, a large aquifer that extends from the Texas Panhandle to parts of Oklahoma, New Mexico, Kansas, Colorado, and Nebraska. Thousands of years ago, streams flowing from the Rockies created the Ogallala by depositing ten trillion gallons of mountain water in subsurface gravel beds up to 300 feet thick. Although geological forces eventually diverted the flow of water away from the high plains, the vast reserve of accumulated water remained in place, undisturbed for millennia.[63]

In most areas, the first disturbance to the Ogallala came from farmers tapping into what they thought was a "hidden Mississippi" flowing beneath their feet. Although farmers believed they had access to an endless supply of water, scientists with the U.S. Geological Survey pointed out that the Ogallala had little natural recharge. Only about an inch of water trickled down through the soil each year. In effect, then, these farmers were "mining" a vast, but finite, supply of ancient water.[64] Of course, until farmers found ways to lift water faster and in greater quantities than they could with windmill-powered pumps, little reason for concern existed.

In parts of Texas and Oklahoma, the first disturbance to the Ogallala occurred when a drill bit passed through the water-saturated gravel and moved on to oil deposits below. In those areas, some of the salt water lifted by oil producers, which they discharged into pits or allowed to run free, eventually reached the aquifer. The slow movement of water in this immense pool prevented the saline contamination from dispersing too quickly, allowing it to remain as a body, undetected for years. In part, the Texas prohibition of saltwater disposal pits in fields above the Ogallala—a prohibition eventually extended to oil fields above other major aquifers—

was triggered by changes in the salt concentration of groundwater being withdrawn in the vicinity of oil fields.[65]

The "no pit" order did not always have the intended effect. Many operators responded to the order simply by injecting their salt water down old wells, some of which had corroded casings and poor cement seals. In some wells, salt water injected under pressure flowed behind the casing and still found a path—this time under pressure—to the aquifer being protected. Poorly plugged abandoned wells in the area further aggravated the problem.[66]

Texas was not unique. By the mid-1960s, oil field operators throughout the United States were lifting over 23 million barrels of salt water each day, 71 percent of which they injected back into subsurface formations. Twelve percent still flowed into unlined disposal pits and another 5 percent into streams and rivers.[67] Trade journals carried articles about the problem, but no one provided any good solution as to how producers could run a cost-effective secondary recovery operation and still verify that they were not contaminating aquifers in the process.[68]

The Bureau of Mines, which played a key role in the rapid dissemination of waterflood technology through its station in Bartlesville, Oklahoma, had no response to these new concerns. Engineers at the Bartlesville station did work on technology to help producers locate unrecorded abandoned wells, but the culture of demonstration did not work well for that technology.[69] The bureau simply could not frame efforts to prevent underground leaks as a problem in which eliminating waste would be good business. Controlling underground leaks clearly represented a net cost to the companies that undertook those efforts. Unless a regulatory body—industrial or governmental—could establish and enforce rules that forced companies to be vigilant, oil producers that invested capital to locate and control sources of brine contamination would be at an economic disadvantage. In the absence of regulations, oil producers simply did not have consistent incentives to search for and repair leaky wells.

Increasingly, the role of the Bureau of Mines as a nonregulatory educator to industry was becoming obsolete. Most oil companies no longer needed any prompting to make their operations more efficient. In the years since the Bureau of Mines established its first petroleum research station, oil companies had hired thousands of engineers trained in science and technology. Many had graduated from programs specifically designed to meet the needs of the oil industry, and the culture of the petroleum industry had been transformed in the process.[70] In addition, oil companies now invested heavily in research and development. As a result, technical managers were less inclined to either share their knowledge with engineers from the bureau or to seek guidance from them. Furthermore, the American Petroleum Institute, which had been in its infancy in the early 1920s, had since organized an elaborate system of engineering technical committees. Technocrats working for a governmental agency such as the Bureau of Mines no longer could expect to have as much influence in the industry as they once did.

A subtle indication that the bureau's position in the industry had eroded occurred when the Interstate Oil Compact Commission—and not the Bureau of Mines—took the lead in surveying problems associated with the contamination of fresh water by oil field brines. Funding problems also raised a red flag. In the early 1950s, budgets were especially tight for the Bartlesville research station, forcing administrators to secure more funds from outside sources.[71] Perhaps another indication of the bureau's changing status in the petroleum industry can be found in the comments of H. C. Fowler, the head of the Bartlesville station, after he attended the 1956 midyear meeting of the Interstate Oil Compact Commission. Fowler noted that the organization's committees were "losing their intimate touch" and that sessions were becoming crowded with "political types."[72] Fowler, and the Bureau of Mines, no longer commanded the respect they once did.

In a search for funded projects, Bartlesville courted the Atomic Energy Commission (AEC) and offered to research problems associated with injecting nuclear waste underground. In justifying this effort, administrators at the station cited their responsibility for ensuring the safety and health of those in the mining industry and for determining how to use the nation's mineral resources most efficiently. The station's initial interest in this area developed after several engineers from Bartlesville gained experience using radioactive isotopes to trace the flow of water through subsurface formations. In that project, engineers mixed a small amount of iodine-131 with the injected water and measured the radioactivity of samples from the producing wells.[73] The main purpose of the project was to develop more efficient waterfloods.

The idea of seeking funds from the AEC to study the subsurface disposal of radioactive waste came in 1957 after engineers from Bartlesville attended a conference presentation on the subject. The bureau's engineers noted that leakage from unrecorded holes might be a problem and observed that "it was apparent . . . that people are viewing the disposal problem as a sort of bogey-man that will go away if nobody looks too closely at him."[74] The AEC, however, turned to API's technical committees rather than to the Bartlesville station for advice in the disposal of radioactive wastes by deep well injection.

The AEC did ask for Bartlesville's help in designing an 85-foot-diameter spherical hole in a salt dome 3,000 feet below the surface.[75] Such cavities were not a new technology. Creating a storage cavity involved drilling a hole in a salt deposit to the desired depth, pumping in fresh water, and allowing the fresh water to dissolve the salt. After pumping out the salt water, a cavity existed into which hydrocarbon liquids or gases could be stored. Petroleum and petrochemical companies had constructed numerous storage cavities in salt domes since World War II, creating about 35 million barrels of storage capacity in the process. Despite the routine nature of this project, the head administrator at Bartlesville looked forward to working with the AEC and noted that "our bread cast upon the AEC's water finally may be coming home to us."[76]

Indeed, the station eventually became involved with the AEC-sponsored Project Gasbuggy, a program in which scientists and engineers hoped to stimulate gas production in a depleted oil field by detonating an atomic explosion in the oil sands. In addition, fellow researchers at the Bureau of Mines petroleum research station in Laramie, Wyoming, assisted in Project Bronco, which involved the use of an atomic explosion to help extract shale oil. As it turned out, the intense heat of the explosions sealed the test formations and prevented any gas or oil from flowing, bringing that line of research to a dead end by the mid-1960s.[77] These projects may have been in line with the bureau's mission. After all, if successful, they would have resulted in oil being extracted more efficiently than before. However, they eroded the notion that efforts to increase the efficiency of an industrial operation also addressed pollution concerns: the success of Project Gasbuggy would not have simultaneously addressed any pollution concerns. If anything, it would have created new pollution concerns.

Successful secondary recovery operations also created another problem. Waterflood operations required lots of water. Operators generally had to pump more water into the ground than they took out and, therefore, had to obtain additional water from other sources. If companies could secure a source of inexpensive fresh water, whether from a surface body or an aquifer, companies typically drew from that source. Otherwise, they had to lift salt water from another formation and treat it to avoid plugging the receiving formation with pore-clogging reactions between incompatible brines.[78] Therefore, some oil producers became major users of fresh water. Farms, ranches, and towns also put heavy demands on the available water, resulting in withdrawals from aquifers at a much higher rate than their natural rate of recharge. Oil producers, though, injected their fresh water into oil-bearing formations, thereby removing this reusable resource from the hydrological cycle—a practice at odds with the desire to use resources as efficiently as possible.

In summary, the petroleum industry initially experienced tremendous success in addressing concerns associated with the disposal of oil fields brines. By putting a waste product to use in increasing the efficiency with which they extracted oil, producers in the 1930s not only solved a serious disposal problem but also reinforced the notion that efforts to increase the efficiency of industrial operations also addressed pollution concerns. However, by the mid-1950s, pollution concerns associated with the disposal of oil field brines remained, yet companies had few incentives to address those concerns. Furthermore, the Bureau of Mines, the efficiency-minded government agency most closely associated with the success of brine injection systems, had no strategy for addressing the concerns that remained. Unless all interested actors could reach consensus on specific environmental objectives and put a regulatory system in place to meet those objectives little could be done to address the remaining concerns.

EIGHT **Eliminating Corrosion and Monitoring Flows**

IN THE 1920S AND 1930S, engineers in all sectors of the petroleum industry fought a never-ending battle against corrosion. This fight against corrosion—and hence against leaks, structural failures, fires, and explosions—represented another area in which engineering efforts to reduce waste and increase efficiency overlapped with efforts to reduce pollution-causing discharges. Oil companies also sought better ways to monitor the flow of oil from one point to another. Improvements in monitoring not only helped companies keep track of their oil but also helped them to identify breaks and leaks more quickly. Hence, success in both of these areas reinforced the link between efforts to make industrial operations more efficient and efforts to control the release of pollution-causing discharges.

In the battle against corrosion, pipeline engineers occupied the front lines. Companies that operated pipelines invested significant amounts of capital in constructing those lines, and they desired to protect their investment from the ravages of corrosion. Pipeline engineers also took the lead in efforts to monitor the movement of oil more accurately as millions of gallons of oil flowed through the nation's pipeline system each day, with much of the oil changing ownership in the process. Advances in corrosion protection and monitoring, of course, quickly spread to other sectors of the industry; the technical committees of the American Petroleum Institute were effective disseminators of innovation.

Because efforts to increase the efficiency of pipeline operations resulted in less oil being discharged to the environment, engineers and technical managers generally focused on the former without worrying too much about the latter. However, the wisdom of that logic only went so far. For example, protecting some equipment, such as small underground tanks, did not seem worth the cost. That steel often went unprotected, setting the stage for serious problems in the future. In addition, the industry stopped short of developing its monitoring technology to the point where small leaks from transportation and storage equipment could be remotely detected. Although developing such technology makes good sense from the perspective of pollution control, doing so was not necessarily "efficient" from the perspective of a firm attempting to maximize profit.

The Fight Against Corrosion

Scientists had understood the general chemistry of corrosion since the nineteenth century, recognizing that metals, when in contact with moisture, release metal ions. They also knew that when these positively charged ions go into solution, they leave negative charges—electrons—behind on the surface of the metal. If these electrons stay in place, the corrosion process—that is, the continual loss of metal ions—comes to a halt because an electric polarity prevents any more metal ions from going into solution. However, when electrons are drawn away, metal ions continually flow into solution, eventually resulting in a significant loss of metal. Indeed, scientists in the nineteenth century recognized the similarity between corrosion and the movement of ions in a galvanic cell having an anode and cathode.[1]

However, until the second decade of the twentieth century, no one could successfully explain the tendency of corrosion to devastate one section of metal while leaving a neighboring section untouched, which one scientist referred to as its "local and apparently capricious character."[2] Many scientists tried to find an elegant explanation for this phenomenon, but in practice, no single mechanism existed. Numerous factors affected the rate at which a metal corroded, including the homogeneity of the metal, its microscopic structure, variations in stress, stray currents, the type of chemicals in the corrosion-causing moisture, and whether it came into electrical contact with a different kind of metal. A scratch in the surface of a pipe could even serve as the anode of a galvanic circuit, with adjacent sections of the same pipe serving as the cathode. In some cases, metal ions flowed continuously from anode to cathode, creating a corrosive pit in the pipe's surface. In other cases, the products of corrosion formed a barrier to further corrosion.

The task, then, was not to search for one mechanism that explained all corrosion but to study the combination of factors that allowed the corrosion process to continue indefinitely. In 1911, in important research along these lines, engineers and physicists with the U.S. National Bureau of Standards began studying the effect of stray currents generated by electric traction systems on buried water and gas pipes (fig. 8.1). The researchers recognized that corrosion occurred only where current flowed out of the pipe, that is, when positively charged metal ions flowed from the pipe and into the surrounding soil. Investigators determined that pipes could be protected from the effects of stray traction current by securely bonding all pipelines to an appropriate electrical return (fig. 8.2). The investigators also noted something else of value. Pipes could be protected from the corrosive effects of electrolysis by keeping the pipes at a slightly lower potential than the soil. They described this form of protection as "cathodic protection."[3]

Over the next ten years, a handful of researchers continued to gain practical insight into the process of corrosion, mostly in areas unrelated to the petroleum industry.[4] For example, a researcher with the U.S. Bureau of Mines, Frank N. Speller, noted that pipes in hot-water heating systems did not suffer serious corrosion if the same water was circulated repeatedly through the pipes. However, when new water entered the system, the amount of corrosion increased. Speller attributed the differences in corrosion rates to the presence of dissolved oxygen in the make-up water. For closed systems, he theorized that hydroxyl ions of water [$(OH)^-$] combined with any iron ions that entered the water to form ferrous hydrate

Figure 8.1. Corroded pipe represented a major problem for water and gas utilities and oil companies in the early part of the twentieth century. (Smithsonian 99-2239)

[$Fe(OH)_2$]. The electrons left behind on the surface of the pipe then flowed to a section of the pipe that acted as a cathode, where they combined with hydrogen ions of water (H^+) to form small amounts of hydrogen gas. According to Speller, this gas, if not removed from solution, prevented the generation of any more hydrogen gas and halted the corrosion process. When new water entered the system, Speller speculated that any dissolved oxygen present in the water combined with the hydrogen gas to form water and allowed the corrosion process to continue. In a closed system, Speller asserted, the process of corrosion ceased when all the dissolved oxygen had been consumed.[5]

Speller recognized that he oversimplified. Furthermore, he did not see the presence or absence of oxygen as the only or even the main factor in determining the rate of corrosion. Rather, Speller and other engineers studying corrosion sought to identify any factor that either aided or hindered the process. For example, they knew that when zinc-plated (galvanized) steel got nicked, the bare steel always acted as the cathode and zinc as the anode, creating a situation in which zinc flowed from an undamaged section of the surface to the scratch, covering the exposed surface and preventing further corrosion. They also knew that one could inhibit corrosion by adding certain chemicals, such as arsenic salts, which tended to wet a metal surface and acted as a barrier to corrosion.

Such knowledge did not translate directly into strategies for protecting pipe, tanks, valves, and vessels. Each sector of the petroleum industry faced different problems and constraints. In the production sector, steel tubing used in wells not only had to withstand the corrosive effects of brine and sulfur but also had to have a high tensile strength. Protecting against corrosion by making pipe thicker was not an option as that approach only added more weight, and a mile or more of pipe hanging from a rig was already heavy enough. Transportation pipelines, which required less tensile strength, still had to withstand pumping pressures of 700 pounds per square inch or more. Moreover, when constructing a pipeline tens or even hundreds of miles long, the cost of the pipe became an important factor in the line's profitability. At refineries, pipes and vessels not only had to withstand high pressures but high temperatures as well. Oil

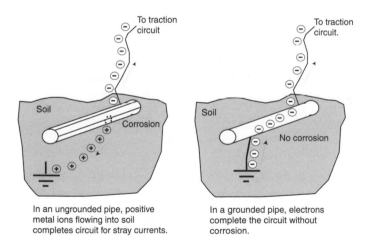

In an ungrounded pipe, positive metal ions flowing into soil completes circuit for stray currents.

In a grounded pipe, electrons complete the circuit without corrosion.

Figure 8.2. In 1911, researchers with the National Bureau of Standards emphasized the ability of stray currents to destroy ungrounded pipes. (Illustration by author)

tankers, attacked by salt water from the outside and by the corrosive impurities of crude oil from the inside, faced severe corrosion problems. Railcars that carried petroleum products also faced corrosion problems, sometimes losing their entire contents during transit through leaks.[6]

In 1926, Frank Speller published an in-depth and accessible treatise on the causes and prevention of corrosion. In this book, which helped establish his reputation as an authority on corrosion in the petroleum industry, Speller opened with a quotation from Commerce Secretary Herbert Hoover: "It is only through the elimination of waste and the increase in our national efficiency that we can hope to lower the cost of living, on the one hand, and raise our standards of living on the other. The elimination of waste is a total asset. It has no liabilities."[7] To Speller, who estimated that corrosion annually consumed about 2 percent of all steel in use, the battle against corrosion was a battle against waste.[8]

By the time Speller published his book on corrosion, steel companies and companies supplying chemicals and coatings marketed many products to engineers who hoped to gain the upper hand in their fight against corrosion. At one refinery, where engineers were actively looking for a metal to resist the corrosive effects of hydrogen sulfide, a salesman dropped off a pressure valve for them to test. They installed the valve with no plan other than to wait for it to deteriorate. Two years later, while technicians were upgrading all valves, one engineer recalled that the test valve was "just as shiny and pretty inside as you'd ever seen." The engineers immediately contacted the valve company to find out what kind of steel had been used. After being told the metal was Carpoloy, manufactured by the Carpenter Steel Company, the engineers began specifying "Carpoloy" for all critical components, refusing to consider anything else and giving Carpenter Steel several million dollars in business.[9]

As it turned out, Carpoloy was Carpenter Steel's brand name for a patented chromium-nickel steel alloy (stainless steel) developed by a German firm and licensed to U.S. companies. Carpoloy, of course, was not the answer to all corrosion problems. Although refineries could afford to use stainless steel for select components, no company could afford to use it for every application. Certainly, companies could not afford to use stainless steel for miles of pipe or for the acres of plate steel that went into building storage tanks and tankers. Second, numerous factors affected the perfor-mance of all materials. A steel that proved highly resistant to one chemical did not necessarily resist corrosion when in contact with another. Pres-sure, temperature, compression, and stress also affected a material's corro-sion resistance and performance. For example, high temperatures neutral-ized the effectiveness of galvanized steel by reversing the polarity of zinc relative to steel. Copper-steel alloys, which proved highly resistant to cor-rosive brines not containing sulfur, corroded rapidly in the presence of hydrogen sulfide.[10]

To help engineers sort out the conflicting claims of manufacturers and to exchange knowledge on the subject, in 1926 the API invited Speller to head a symposium on corrosion at its annual meeting, held that year in Tulsa, Oklahoma. The successful interchange of information at that sym-posium proved valuable to engineers, and undoubtedly reinforced the val-ue of holding technical sessions at API conferences.

The symposium focused first on the subject of pipeline coatings. One of the speakers, E. C. Kincaide of Gulf Oil, argued that investing 5 to 20 percent of a pipeline's initial cost in coating a line simply did not pay. With twenty years of experience in the industry, he did not see where coatings made any real difference. Instead, he believed that any significant reduction in corrosion depended on metallurgists developing inexpensive corrosion-resistant pipe. To justify his reasoning, Kincaide described two parallel lines 8 feet apart, one laid in 1907 and one in 1914. In his time with Gulf Oil, only 1 percent of the first line had to be replaced while 5 percent of the second line had been replaced, indicating to Kincaide that the dif-ference was in the steel. He concluded that the industry's understanding of "corrosion of underground pipelines is practically where it was when Columbus discovered America."[11]

Speller suggested that, by focusing only on better materials, Kincaide was limiting his options. He also doubted whether the older pipe could actually resist corrosion better than the newer pipe and wondered whether the old pipe was just a little thicker or whether the new pipe was buried a little deeper, just enough to bring it into contact with more water.[12] Speller also pointed to important improvements in the last few years. Joining pipe with welds instead of threads made a difference, because oil leaking through threads had a tendency to dissolve most coatings, exposing the metal around the joint.

Other speakers—including representatives from the Bureau of Mines, paint and chemical companies, and steel companies—agreed with Speller. They noted that although no general panacea had been discovered, engineers did know more about corrosion than their counterparts a generation ago. They also emphasized that the strategy used to fight corrosion depended on the application. Still, in general terms, only several options existed: use metal that was more corrosion resistant; apply protective coatings to the surface of the metal; use corrosion inhibitors in any fluid that came into contact with the metal; or, in the case of pipelines, modify soil conditions.

As it turned out, most engineers in the pipeline sector disagreed with Kincaide and saw coatings as the most appropriate technology. The chief engineer with the Prairie Pipeline Company noted that for trunk lines in harsh soil conditions and hard-to-reach areas, a coating of portland cement, which cost $3,000 per mile, might be more appropriate than a single ply of bituminous fabric, which cost $400 per mile. He did not recommend paint, which cost about $100 per mile, for use as anything but a primer. He also noted that red lead paint, which Kincaide used on his lines, had little adhesive strength and was not particularly suitable for underground use.[13]

Refinery engineers at the conference tended to focus on better steel as the most appropriate way to protect equipment subject to severe corrosion.[14] For one thing, the consequences of a failed component tended to be more serious at a refinery than in a pipeline. Not only could the failure of a critical component shut down the entire refinery, but a failure also could cause a disastrous fire or injure workers or both. For example, in the late 1920s, when engineers at one refinery upgraded the distillation process to operate at higher pressures, they neglected to replace old cast-iron valves in the receiving house, where condensing vapor flowed through glass-covered look boxes so that operators could inspect the condensate. Two failures of cast-iron valves soon occurred. The first incident suffocated six men. In the second, the vapor ignited, killing seven men.[15]

Engineers working in the production sector faced a different set of circumstances than those faced by either pipeline or refinery engineers. A producing well—with concentric rings of steel tubes and various types of screens, tools, and pumping mechanisms in contact with crude oil containing salt water and other corrosive impurities—posed an eclectic mix of corrosion problems. Hence, preventing corrosion in this sector required an eclectic response. Although engineers could specify the use of special alloys for valves and some downhole tools, they could not do so for the miles of casing, tubing, and sucker rods required for most wells. Coatings, which crude would quickly attack, were not an option either. Instead, operators routinely pulled and inspected production tubing and rods and, if necessary, replaced corroded segments.[16]

Although oil well operators expected a certain amount of corrosion, maintaining a corrosion-plagued well could become expensive. Therefore, they did seek to diagnose and solve the most serious problems. For example, one operator found that the strainers he placed at the bottom of a hole to prevent loose sand from entering the well often failed within a few months. Only after looking into the matter did he discover that different parts of the screen were made of different metals, an invitation to corrosion.[17] Other operators discovered that their use of an air-lift mechanism—an alternative to mechanical pumping—accelerated the action of corrosion too much to be economically justified.[18]

Chemicals that inhibited corrosion also had a place in the production engineer's arsenal. Some operators protected their drill pipe from rapid corrosion by adding alkalis to their drilling mud.[19] Production engineers with Cities Service experimented with corrosion inhibitors in the early 1920s, but the use of these chemicals did not become common until after companies started acidizing wells after 1929.[20] When acidizing a well, crews injected a solution of hot hydrochloric acid into the well at high pressure and held that pressure for several hours. Ideally, the acid dissolved the limestone around the well and widened the fractures through which oil flowed. Unfortunately, the acid also attacked any steel with which it came into contact, requiring the companies that promoted the process—Pure Oil and Dow Chemical—to develop more effective inhibitors.[21] Subsequent efforts to fight internal corrosion in pipelines and to protect brine injection systems from corrosion stimulated additional interest in inhibitors.

The cost of corrosion justified the attention petroleum companies paid to the problem. By the late 1920s, the industry had about 45,000 miles of pipeline in service and corrosion of those lines cost the industry about $100 million each year. That figure did not include costs associated with the loss of product, shutdown time, or damage to property and groundwater. Neither did it include any costs associated with corrosion in oil wells, storage tanks, tankers, and refineries.[22] In 1928, the API initiated a research project aimed at cataloguing existing corrosion problems and comparing the strategies of different companies.[23] Several years later, that organization also sponsored half of a $132,000 effort to test pipeline coatings, with the remaining funds coming from manufacturers of coatings.[24]

Government agencies also conducted research in corrosion prevention. The Bureau of Mines, through its Petroleum Research Station, focused on the corrosion of storage tanks and oil field equipment.[25] Though not directly interested in oil pipelines, researchers with the National Bureau of Standards, in an ongoing project costing about $26,000 per year, studied the corrosion of underground pipelines by burying and periodically examining 17,000 segments of pipe in various types of soil.[26]

The only significant method of corrosion prevention not explicitly discussed at the 1926 API meeting was cathodic protection. Although this form of protection had been discovered over a decade earlier, nobody applied cathodic protection on a large scale until 1929. Then, engineers seeking to protect new gas mains in New Orleans successfully employed such a system.[27] Engineers there used a low voltage rectifier to keep the pipeline at a slightly lower voltage than the surrounding soil. In 1956, when engineers reviewed the history of that pipeline system, then consisting of about 475 miles of pipe in southern Louisiana, they discovered that no leaks caused by corrosion had occurred in its twenty-six years of service.[28]

Companies also used cathodic protection to protect large storage tanks. In such cases, engineers could use sacrificial anodes made of magnesium in place of rectifiers. When the structure being protected was bonded to the block of magnesium, any local galvanic circuits that formed drew all positively charged ions from the block of magnesium and consumed the sacrificial anode instead of the metal structure. The results could be dramatic. At one tank farm operated by Humble Oil, maintenance crews repaired about seventy-five leaks annually in the years before 1940, including leaks in pipelines running between tanks. After the company implemented a program of cathodic protection, the number of leaks dropped significantly, reaching zero by 1950.[29] By that time, the newest storage tanks had sacrificial anodes integrated into their design.[30]

Engineers, on the lookout for better products and techniques, discovered that fighting corrosion was a full-time job. Their interest in the matter led to the establishment of corrosion committees within all technical divisions of the API. That interest also led, in the mid-1930s, to the formation of groups such as Mid-Continent Cathodic Protection Association. In 1943, engineers in the petroleum industry were also active in founding the National Association of Corrosion Engineers. Within two years, that organization had 750 members, a third of which were employed by pipeline companies. Two years later, the National Association of Corrosion Engineers started publishing its own journal, *Corrosion*.[31]

Advances in Pipeline Systems

The growing confidence that pipeline companies had in their ability to construct robust pipelines is reflected in decisions to transport products such as gasoline by long-distance pipelines. The first major pipeline built to transport gasoline long distances, completed in 1930, ran about 700 miles from a Sun Oil refinery on the Delaware River to a distribution terminal near Pittsburgh, Pennsylvania.[32] To protect the line, the company encased the most vulnerable portions of the pipeline in concrete and asphalt. Cathodic protection, which emerged in the 1930s as a successful technology, only increased the integrity of such pipelines.

Transporting gasoline long distances through a pipeline would not have been practical ten years earlier. First, too much of the valuable product would have leaked through the threaded joints then in use, which had since been replaced by welded joints. Second, a large amount of gasoline would have escaped as vapors. In the early 1920s, all oil entering and leaving a station still flowed in and out of storage tanks, which resulted in large vapor losses. By the 1930s, improvements in pumping technology allowed oil to flow through an entire line without being diverted to temporary storage tanks, reducing vapor losses and making product lines more practical.

Three years after the product pipeline went into service, engineers with Sun computed that 99.66 percent of the gasoline put into the line reached its destination. Few of the losses were due to leaks or evaporation. The pipeline had yet to suffer significant deterioration and the oil did not come into contact with air while in the pipeline. An unknown fraction of the losses were due to thieves who made an illegal tap into the line.[33]

As pipeline engineers made significant advances in pipeline technology, the industry still had to maintain the 45,000 miles of old pipeline that had been laid with a minimum of protection. Maintenance departments found themselves routinely reconditioning stretches of old pipeline and, by the late 1930s, could do so proficiently. For example, when a twelve-inch pipeline owned jointly by the Texas Company and Cities Service developed corrosive leaks at points where it ran along and crossed the Neches and Angelina Rivers in Texas, engineers with Sun Oil took notice. Sun operated a parallel line and quickly developed a plan to recondition their line before the dry season ended. In August 1938, the company put out bids for over ten miles of line to be "pulled out of the ground, placed on skids, thoroughly cleaned with a W-K-M portable cleaning machine, carefully inspected for pits, patch welded where pitting is excessive, coated with hot Barett enamel, and wrapped with J & M 15# asbestos felt." Furthermore, they expected the subcontractor performing the job to finish before November and do so without ever taking the pipeline out of operation.[34]

Leaks or breaks in pipe on the bottom of a river posed the biggest challenge for repair crews. However, by the 1930s, companies had developed a fairly routine procedure—as routine as such efforts can get—for making such repairs. For example, in 1932, when a small leak occurred in a line under the Mississippi River, crews with Standolind Pipe Line Company first displaced the oil in the line with pressurized air. The air kept water from entering the pipe and generated a stream of bubbles that allowed divers from a repair barge to locate the leak. Crews on the barge then lowered a suction line, which the divers used to remove silt covering the leaking segment of pipe. Finally, the divers placed a stuffing box around that segment of pipe and used a hose from the surface to fill the stuffing box with cement.[35]

Good incentives to prevent leaks existed, of course. Not only did companies want to prevent the loss of petroleum and, in a growing number of cases, petroleum products such as gasoline, they also wanted to prevent damage claims associated with leaks and spills. According to one pipeline engineer speaking at an API session on pipelines held in 1953, the expense of damage claims had become a serious concern.[36] Some leaks, however, were small and difficult to detect, and companies relied on line walkers making routine visual inspections. A few companies started using airplanes in the 1930s, but many people doubted whether aircraft could replace a person on foot.[37]

Pipeline owners were especially concerned that leaks in lines transporting gasoline, a thinner and more volatile fluid than crude, would be too difficult to see. In one case, pipeline engineers tested the skills of their observation pilot by dumping a drum of gasoline along the pipeline right-of-way. The pilot later reported seeing a fluid on the right-of-way but noted that "all indications are that somebody emptied out a drum."[38] In the years after World War II, pipeline companies integrated the use of airplanes into their operations on a more systematic basis. By that time, field crews also had access to mobile radiotelephones, which allowed maintenance crews to receive instructions and reach serious breaks more quickly than crews a generation earlier.

Pipeline companies also continued to reduce their vapor losses from storage tanks. By the 1950s, the largest tanks had become highly engineered structures designed to minimize corrosion damage and vapor losses, with many aspects of their design covered by API standards. Indeed, about a third of all storage tanks owned by pipeline companies were equipped with floating roofs that moved up and down as the level of the petroleum changed, eliminating much of the vapor losses associated with filling empty tanks.[39] In addition, by using internal floats capable of measuring the level of petroleum automatically, engineers also eliminated the need to access tanks for gauging, which further reduced opportunities for vapor to escape.[40]

Pipeline companies also took steps to reduce losses from their fixed-roof tanks. Some companies vented the vapor space of their fixed-roof tanks to a centralized vapor space, which allowed their tanks to breathe without the loss of hydrocarbons. Other companies reduced vapor losses by floating microballoons—tiny air-filled plastic balls—on the surface of any petroleum exposed to a vapor space. Although pipeline companies operating in the 1950s still experienced some vapor losses (as anybody who lived near a tank farm could testify) those losses were much lower than the 3 to 6 percent of throughput that companies had experienced in the early 1920s.[41] One study in the 1950s showed that losses from a 55,000-barrel tank with a fixed roof ranged from .4 percent of throughput for winter gasoline (a highly volatile gasoline for use in colder temperatures) to less than .02 percent of throughput for fuel oil.[42]

Engineers employed by pipeline companies also reduced the amount of material that settled to the bottom of large storage tanks. They did so by installing mechanisms to prevent settling. For example, in 1954, engineers with Interstate Oil Pipe Line Company designed a system in which a shaft protruding through the wall of the tank drove propellers inside the tank. Workers equipped with portable motors capable of driving these shafts kept tanks free of bottom sediment by occasionally stirring up the contents.[43] Other companies reduced the accumulation of sediment and water by using suction lines with inlets near the lowest point of the tank floor.[44] By preventing the settling of sediment, pipeline companies shifted the problem of removing sediment to the refineries they served.

In part, pipeline companies began shifting the sediment problem to refineries because they increasingly played the role of common carriers with less direct ties to refineries. A change in the size of pipelines encouraged this restructuring of the industry. The first step toward this change came in the late 1920s, when steel mills developed the technology to manufacture thin-walled, twenty-four-inch-diameter pipe capable of withstanding high pumping pressures. However, no single company commanded the traffic necessary to justify using that pipe for transporting anything but natural gas. After all, a single twenty-four-inch line could transport about 250,000 barrels of crude each day, or as much as nine eight-inch lines. For a decade, most new pipelines for moving crude remained under twelve inches in diameter.[45]

The motivation to use the larger pipe for transporting petroleum came in the first six months of 1942, when German submarines sank fifty-five oil tankers carrying crude to the Atlantic coast. Shipments to eastern re-

Figure 8.3. When constructing the "Big Inch" pipeline during World War II, contractors coated pipelines with a primer using specialized machinery before wrapping with asphalt paper, as seen on this segment of the line being laid north of Little Rock, Arkansas. (LC-US-W3-9323-D)

fineries dropped by four-fifths. To get petroleum from the mid-continent to the eastern seaboard, the U.S. government subsidized the construction of two pipelines: first, a twenty-four inch line known as "Big Inch," and, later, a twenty-inch line called "Little Inch" (fig. 8.3). Although the federal government eventually sold Big Inch and Little Inch to a natural gas transmission company, companies moving crude recognized the value of using large-diameter lines. Innovators began consolidating traffic into large-diameter pipelines jointly owned by several companies, resulting in a looser corporate connection between pipeline operators and the refineries they served.[46]

Pipeline companies also made significant advances in monitoring the flow of oil from point to point. Their main incentive for doing so came from an interest in automating control. In the 1920s and 1930s, tracking the flow of petroleum through pipelines typically required an army of clerks, all communicating over telephone and telegraph lines, to make sure that the right batch of fluid arrived at the right place at the right time.[47] As engineers took hesitant steps toward automation, they also placed more attention on losses from the system and, hence, on efforts to reduce those losses.

Although the most dramatic advances in automation would not come until the 1950s, pipeline companies made some progress toward their goal in earlier decades. For example, in 1930, engineers with the Empire Pipe Line Company designed the first pumping station operated by remote control. An operator at a neighboring station started and stopped the remote pumps by sending signals over a telephone line.[48] Stations of any complexity still could not be controlled remotely, but partial automation certainly was possible. Relay circuits and new instruments such as automatic tank gauges, positive displacement meters that measured the flow of oil, and galvanic chart recorders allowed operators to control an entire pumping station from a central control panel. Engineers also used basic control technology to shut down pumps automatically when changes in pressure indicated that either a blockage or a break had occurred.[49]

In the mid-1930s, improvements in the design of positive displacement meters, along with new techniques for superimposing carrier currents on telephone wires, made "telemetering" feasible. Companies, though, first had to create a standard code of practice for using the new meters in commercial transactions. Engineers started work on such a code in 1938 but did not complete the final ASME-API Petroleum Meter Code until after World War II. Telemetering improved the ability of operators to detect discrepancies between bulk receipts and deliveries and, therefore, potential leaks.[50] Although operators could neither pinpoint the location of leaks nor necessarily detect them immediately, consistent discrepancies raised a red flag.

Efforts to automate pipeline operations accelerated in the 1950s.[51] The development of electronic regulators, magnetic amplifiers, sophisticated electromechanical devices, and microwave communication made the operation of remote pumping stations more practical.[52] Advances in technology also allowed pipeline companies to automate the transfer of oil from leases. The main impetus for automating this task came from the increasing number of offshore leases served by pipeline companies. Traditionally, pipeline companies gauged transfers of petroleum by measuring the fluid level in storage tanks before and after the transfer. Representatives from both the pipeline company and the producer had to be present for the transfer, an event that was both time-consuming and subject to delays due to missed appointments and changes in schedules.[53] The logistics of transferring oil from offshore leases complicated matters, providing greater incentives to automate the task.

Automating the transfer of oil from leases involved more than developing reliable meters. It also required equipment that could automatically measure temperature, specific gravity, and sediment and water content. By the 1950s, this technology was available and pipeline companies began installing "automatic custody transfer equipment."[54] In addition to being convenient, this equipment also reduced vapor losses. Tests showed that by getting the oil into the pipeline more quickly and by eliminating the need for gauging, less gasoline had a chance to evaporate, resulting in gains of "API gravity" between .5 and 1.5 degrees, which translated into one or two cents per barrel per degree of API gravity.[55] Improvements in efficiency, engineers could argue, continued to result in less material being lost to the atmosphere.

Diminishing Incentives

In the period between 1925 and 1955, pipeline companies made significant advances in reducing the leaks and spills that occurred while moving petroleum and petroleum products through pipelines. Simply protecting pipelines from corrosion reduced the frequency of leaks. The routine use of "pigs" (mechanical devices that traveled inside a pipeline) also helped to keep the inside of pipes maintained (fig. 8.4). The shift to large-diameter pipelines also made a difference, allowing the industry to move much larger quantities of petroleum without dramatically increasing the length of line to be inspected and maintained. Finally, better methods of monitoring, communication, and transportation allowed crews to find and repair most leaks sooner.

Figure 8.4. Pipeline companies maintained the inside of their lines by running "pigs," such as the device shown here, through them. (Library of Congress LC-US-W3-9569-D)

As engineers improved the efficiency of pipeline systems, they had little reason to ask specifically about the level of environmental quality they should respect. They generally assumed that efficiency-related advances would continue to address pollution concerns. Damage and nuisance suits from leaks and spills indirectly specified measures of environmental quality that had to be respected, but damage claims had limits as incentives to continued improvement. Indeed, by the 1950s, the economic incentives to further reduce the risk of leaks and spills was diminishing rapidly. Yet problems remained.

Small pipeline leaks, for example, still occurred, and any that went undetected could slowly contaminate groundwater. Furthermore, when line walkers discovered a leak, companies had no incentive to determine the extent of contamination. If nobody complained, why go looking for problems? And if someone complained, paying damage awards cost less than developing the science and technology necessary to clean up an aquifer.

Large spills also presented problems. When breaks in a pipeline occurred, such as when erosion shifted a line or when heavy construction equipment accidentally tore through a pipe, operators could shut down a line fairly quickly. In some cases, the pressure drop was enough to trigger action. However, the spilled oil still had to be cleaned up. The size of the spill, of course, depended on the time taken to close the valve feeding the broken line, the distance of the closest valve to the break, and the size of the pipe. If there were any delay in shutting down a line and the break occurred a good distance from the nearest valve, spills from a large-diameter line could be significant.

When such breaks occurred, crews first recovered any oil they could by pumping it into tank trucks. Oil that reached streams and rivers posed the largest problem. In fast-moving rivers, crews typically let the current carry away the slick. In slower-moving streams or in areas where obstacles trapped pools of oil, they removed the oil in the most convenient fashion—by skimming it from the surface, by burning it off, or by spraying it with carbon-coated sand. The carbon-coated sand formed a heavier-than-water mixture with the oil and sank to the bottom.[56]

Companies hoping to minimize potential damage suits had good reason to spray Carbosand on any oil that reached a stream. The Carbosand helped to sink the oil before it could do further damage. However, in 1950, the authors of a report published by the Ohio River Valley Water Sanitation Commission explained that "oil deposited with Carbosand will remain toxic to aquatic life, and its use, therefore, should recognize this limitation."[57] Companies, though, had little incentive to find a better strategy. Neither did they have an economic incentive to install more remotely controlled valves simply for the purpose of reducing the size of spills. After all, no one knew where a spill would occur, which meant that large numbers of remotely controlled valves would have to be installed to be effective.

Other segments of the industry also had little incentive to detect and prevent leaks. For example, incentives to protect small underground tanks from corrosion were also minimal, that being less of a pipeline system concern than an industry-wide concern. Tens of thousands of such tanks were already buried at gasoline stations throughout the United States, and few people expressed serious concern over the possibility of their leaking. In 1950, the API's Division of Marketing did release a bulletin describing

"Recommended Practices for Bulk Liquid Loss Control in Service Stations," but the focus was on keeping accurate statistics on their inventory, not pollution control.[58]

The bulletin included a few pages on losses from small leaks, which the authors emphasized could add up. According to the bulletin, a leak of two drops per second represented a loss of 1,360 gallons per year. Drops breaking into a steady stream would result in a loss of 8,760 gallons per year if allowed to continue. The bulletin also provided station owners with a way to determine if their tanks had a leak. However, it provided little advice as to what one should do if a leak was found. In practice, repairing a tank buried under cement or asphalt could be more expensive or, at the very least, more of a hassle than allowing small leaks to continue. Furthermore, even when station owners tested a tank and found no leak, it did not mean that no problem existed. All tanks not adequately protected against corrosion would eventually corrode.

NINE

Creating a Pollution Control Manual

IN THE PERIOD FROM 1925 to 1955, engineers in the refining sector were forced to address pollution concerns more directly than their counterparts in the production and pipeline sectors. Indeed, in the 1930s, a technical committee sponsored by the API produced a manual for the disposal of refinery wastes, which became the standard against which individual refineries compared their own disposal efforts. The technical committee that created and maintained this manual also provided a forum for refinery engineers to discuss waste disposal practices. For over two decades, the directors of the API could point to this committee and its manual as evidence that the industry was serious about addressing pollution concerns.

The industry had good reason to demonstrate its interest in controlling pollution. In 1926, the Army Corps of Engineers, in a report required by the Oil Pollution Act of 1924, concluded that the prohibition of oily discharges from ships should be extended to industrial facilities such as refineries.[1] When a bill to amend the Oil Pollution Act was introduced in House of Representatives, the directors of the API acted quickly to convince legislators that the industry could regulate itself.[2] They did so by initiating a new study of pollution concerns and by establishing a technical committee to recommend remedial actions.[3]

The API's pollution control committee produced its manual on the disposal of refinery wastes after another bill to regulate oily discharges from industrial facilities was introduced in 1930.[4] At the hearing for that bill, the head of the API committee—Robert Hand, a veteran at testifying before Congress—argued the same point he emphasized at all previous hearings: discharging "crystal clear water or water absolutely free from oil, emulsion, sludge, or particles of carbon" was a technical impossibility.[5] Therefore, if a law that prohibited the discharge of oil passed, refineries would always be in violation. He also noted that domestic sewage contained "crank-case drainings" and "gear lubricants," which would put the operators of municipal sewers in violation as well. Therefore, rather than pass legislation that would interfere with the "operation of sewers," Hand suggested that the industry create a code of good practice that would "accomplish more than any law."[6]

Soon after the hearings ended, the API Committee on the Disposal of Refinery Wastes released the first volume of a multivolume manual on the disposal of refinery wastes, *Waste Water Containing Oil.* The committee issued two more volumes, *Waste Gases and Vapors* and *Chemical Wastes,* in 1931 and 1935. It also wrote a tentative draft of a fourth volume, *Solid Wastes,* which the API distributed in 1932 but did not release as an official part of the manual until 1963.[7] And, once again, the effort to regulate the discharge of oily effluent from land-based facilities failed.

For two decades, the API's claim of self-regulation through industry standards was more than mere window dressing: refineries did reduce their discharges of pollution-causing wastes in this period, as measured per barrel of crude processed. To some extent, the increasing scale of refinery operations, which new technology encouraged, dictated some changes in disposal practices. A refinery processing 260,000 barrels of crude each day simply could not be as lax in its waste disposal practices as a refinery processing 26,000 barrels.

The API's ability to advance the state of the art in refinery waste disposal practices eventually stalled. Many of the reductions in pollution-causing discharges that refineries secured in the 1930s and 1940s overlapped with efforts to increase the efficiency of refineries. By the 1950s, pollution concerns remained, yet refiners could no longer count on addressing those concerns through process improvements or simple changes. To further reduce pollution-causing discharges into the environment required significant capital investment that did nothing to increase the efficiency of production. Only the threat of new legislation stirred the industry into action.

The API Pollution Control Manual

The authors of the API manual emphasized three strategies for reducing the amount of oil refineries released: more efficient oil separators, more sophisticated sewer designs, and the recovery of spent chemicals. The three strategies were related. Many refineries had only one sewer, and wastes from all processes, along with storm water, entered that sewer. Spent chemicals and sludges that reached the sewer often caused oil to emulsify—that is, to mix so intimately with water that gravity-based separators were ineffective. Left untreated, emulsified oil simply flowed through oil-water separators. Furthermore, at refineries with a single sewer, storm water often overwhelmed what treatment processes existed.

To gain control over the different waste streams, the API manual recommended that refiners maintain separate sewers for storm water, reduce the amount of spent chemicals and sludges entering the sewer, and treat wastewater from some processes even before it reached the main sewer. The manual specifically noted that acid sludge and waste caustic were emulsifying agents that could be significantly reduced by recycling the spent acid and caustic. It also described a gravity-based oil-water separator—essentially a series of concrete settling pools equipped with skimming devices—capable of meeting the demands of a refinery with few emulsions in its waste stream.

Given the importance of keeping spent chemicals and sludges out of the main waste stream, engineers on the API committee for the disposal of refinery wastes emphasized the recovery of spent material as the obvious first step a refinery should take. Most refiners, however, did not need a manual to bring this to their attention. Although they did not always pay much attention to the details of what entered and exited their sewers, they were very aware of any material that had to carted away or replaced. Indeed, the operators of many refineries had already taken significant steps in this direction, though not necessarily to improve the quality of their effluent.

For example, before 1921, all refineries disposed of their fuller's earth—a claylike material used to filter color-imparting impurities from refined products—after it became clogged with tarry material. Some companies had tried burning off the impurities, but fuller's earth regenerated in this way performed poorly. Hence, tons of clay, along with any absorbed hydrocarbons, ended up being dumped in convenient areas around refineries. To eliminate this waste of material, the head of Humble Oil's refinery in Baytown, Texas, initiated a project to regenerate fuller's earth without destroying its filtering capacity. Researchers with Humble soon discovered that the furnaces used to burn off the oily impurities were too hot. The heat fused the microscopic surfaces of the clay together, reducing the surface area available for filtering. The refinery then experimented with a furnace typically used for roasting ores that allowed operators to control temperatures more precisely.[8]

After Humble Oil achieved success in regenerating fuller's earth with ore-roasting ovens, the practice quickly spread. The manufacturer of the oven facilitated dissemination of this innovation by marketing its equipment to other oil companies for the same task.[9] By 1926, most refineries of appreciable size made use of such ovens to regenerate their spent fuller's earth. Operators first "washed" the material with naphtha and then "steamed" it to remove as many impurities as possible. Conveyors then moved the fuller's earth to the ore-roasting oven where the temperature was raised enough to vaporize the remaining hydrocarbons without fusing the clay.

This particular effort to recycle material did nothing to improve the quality of a refinery's effluent. Although the process reduced the amount of oil-saturated clay being dumped by refineries, it merely shifted some of that waste into another form. "Roasting" the spent clay, for example, increased the amount of unburned hydrocarbons and sulfur compounds being released to the atmosphere. In addition, the cleaning process added more contaminated wastewater to the sewer.

Refiners also sought better ways to recycle or dispose of the acid sludge left over after they washed refined products with sulfuric acid. When the Corps of Engineers performed its 1926 survey of pollution-causing discharges, the largest refineries were generating about 150 tons of acid sludge each day. Investigators with the corps indicated that these plants recovered most of the acid in that sludge, losing only 2 to 4 percent of their acid each day.[10] Although these investigators probably underestimated the losses, any recovery effort represented an improvement over previous practices.

Refineries that recycled their acid sludge did so in two steps. First, they separated out the liquid acid, leaving a thick, oily sludge that some refiners burned as fuel. Second, they concentrated the recovered acid for reuse. Again, manufacturers of equipment helped disseminate the process. For example, the De Laval Separator Company, which manufactured centrifugal oil-water separators, began marketing their separators as a way to separate the acid from the oily sludge. Other companies such as the National Lead Company and the Chemical Construction Corporation marketed equipment for concentrating the recovered acid.[11]

Many smaller refineries, though, continued to dispose of their acid sludge without recovery. Some simply diluted their sludge with water, neutralized it with spent caustic, and discharged the mixture to their sewer system. The authors of a 1933 textbook on refinery processes recognized that this practice still occurred, but considered it appropriate only in "isolated places, as large quantities of acid cannot be discharged into rivers or seas without affecting animal life." The authors of this textbook also frowned on small refineries that dumped their acid sludge into pits, with the acid seeping into the ground. The authors noted that this method resulted in "a gradual contamination of the ground" and could "injure the vegetation of surrounding territory and bring consequent liability or damage suits."[12] An alternative was to burn the acid sludge for fuel, but the corrosive action of the acid on the furnace—not to mention the ill will caused by releasing tons of sulfur dioxide into the air—generally exceeded the value of the acid as fuel. Still, many refineries, including large ones, continued to burn some portion of their acid sludge well into the 1950s.[13]

The API manual on the disposal of refinery wastes also discussed measures of pollution other than oil concentration, including the pH of the waste water, its oxygen-depleting capacity, the presence of taste-imparting and odor-imparting impurities, the color and turbidity of the final discharge, and the concentration of reactive chemicals.[14] In suggesting that refineries use these parameters as indicators of effluent quality, the API committee on the disposal of refinery wastes took a step toward a more sophisticated discussion of effluent standards. However, the committee shied away from recommending explicit standards. In practice, the quality of effluent deemed acceptable depended on the location of the nearest intakes for a public water supply. If foul odors and strange tastes in that water triggered concern, tighter standards were needed.

Hence, when the city of Chester, Pennsylvania, complained to the Pennsylvania State Water Board about "objectionable tastes in its water supply," the state engineer contacted the API and suggested that a member of the API committee on refinery wastes accompany him on inspection trips to three local refineries. The head of that committee, D. V. Stroop, then contacted the refineries. However, the managers of those facilities felt no need to show that their discharges met specific standards. Rather they simply demonstrated that wastes from their plants could not "possibly get back to the City of Chester water intakes," which were upstream from their refineries.[15]

As with wastewater standards, the authors of the API manual did not expect all refineries to meet the same emissions standards. Instead, the authors emphasized the importance of taking local conditions into consideration and suggested that engineers first determine what could be emitted without violating local ordinances or causing a nuisance. Then, technicians at the refinery could be trained to take samples and analyze whether those samples met the refinery's own disposal standards. As an aid, the API manual included appendices describing how to design instruments for making the necessary measurements.

The manual suggested that all refineries install gas collection systems and that they think of them as sewer systems for vapors and gases. For this reason, the authors frowned upon the disposal of acid sludges in lagoons and pits. When dumped, sludges tended to release vapors in an uncontrolled fashion, including a sulfuric acid mist full of malodorous compounds that was corrosive to nearby steel structures. For refineries too small to justify an acid recovery plant, the authors of the manual recommended burning the acid sludge in a special type of furnace that resisted corrosion.[16] The manual included plans for such a furnace. However, because burning sulfur compounds produced sulfur dioxide (SO_2), which the authors noted was "an irritant gas that in relatively high concentrations is harmful to plant and animal life" and was "open to legislative control," the authors of the disposal manual suggested that refineries in urban areas had to be more careful about the quantity of sulfur-containing material burned than did other refineries.[17]

The authors of the manual assumed that all refineries already captured and flared hydrocarbon gases released during distillation or created during cracking. Therefore, they focused on the control of poisonous, irritant, and malodorous gases. Hydrogen sulfide (H_2S), which was released when sulfur compounds broke down during cracking and other processes, posed the most problems. Also included in the manual were warnings about vapors containing aromatics such as xylene, toluene, and benzene. A chart specified the concentrations, in parts per million, at which these and other chemicals were acutely toxic.[18] At one refinery, some workers did not need numbers representing toxic thresholds to avoid the vapors of aromatics; according to one employee, the men "somehow got the opinion" that if they "got [aromatics] on them or breathed the vapors, it would affect their manhood."[19] What he meant by "affecting their manhood" is not clear, but the concern encouraged these particular workers to avoid these toxic vapors.

As for ways to capture particulates from stacks, the manual suggested using cyclone collectors, which used centrifugal force to separate particles from a gaseous stream, or electrostatic precipitators, which diverted particles by placing an electric charge on them. Captured particulates were classified as solid wastes, along with waste tars, coke, tank bottom deposits, ashes, and spent clays that could not be reclaimed. The manual recommended that operators mix any combustible waste material with fuel and burn the mixture in a furnace, even if the waste material did not have much heat value. Ashes, captured particulates, and any asphalt impractical to burn could be dumped or used as fill.[20]

What effect did the manual have on the industry? Few refiners moved too quickly in adopting the practices recommended by the manual be-

Figure 9.1. This view of an independent refinery in Oklahoma City from 1939 shows that older and smaller plants did not always live up to the standards of the API manual. (Library of Congress LC-US-F34-34061-D)

cause many of the practices—such as building a separate sewer for storm water, designing a coordinated system of gas collection, and monitoring the entire disposal system—involved significant expense without promise of monetary return. Hence, ten years after its release, many facilities still fell short of the goals set by the API manual (fig. 9.1 and fig. 9.2). For example, the manual suggested that a well-operated refinery should not release water containing more than .003 percent oil (30 ppm). That concentration of oil, the manual noted, did not even leave a noticeable film on pleasure craft.[21] However, in the early 1940s, E. F. Eldridge, a leading authority on the treatment of industrial wastes, estimated that the effluent of most refineries, when operating as designed, had an oil concentration of about 100 ppm.[22]

In addition, many refineries met those concentration levels primarily by dilution. Given that refineries of the time diluted their effluent with copious amounts of cooling water, they could discharge relatively large amounts of oil without causing a visible problem. For example, a refinery that discharged 20 million gallons of water daily could release about 600 gallons of oil every day without ever exceeding the concentration suggested by the API manual. Furthermore, because some of the oil in a refinery's effluent formed emulsions with the water being discharged, the oil was less visible in the receiving body than it would have been if all the oil immediately rose to the surface, allowing refineries to discharge even greater concentrations of oil without calling too much attention to themselves.

In general, then, the API manual held up the disposal methods practiced by the newest refineries as the state of the art, though no refinery necessarily incorporated all features. Furthermore, despite significant weaknesses in the manual created and distributed by the API Committee

Figure 9.2. As this view of a refining facility from the early 1940s shows, some companies still allowed drips and leaks from storage tanks to seep in the ground or be carried away by storm water. (Library of Congress LC-US-W3-10086-D)

on the Disposal of Refinery Wastes—including poor organization, poor writing, poor graphics, an abundance of hedging, and a lack of details in important areas—it did establish a standard of disposal practice where none had existed before.[23] In doing so, it provided refinery engineers with a rough guide for judging the pollution control performance of their own refineries.

In addition, significant changes lay on the horizon. A major expansion of the nation's refining capacity that occurred in the 1940s and early 1950s allowed refineries to integrate better sewers and pollution control equipment into their plants at a pace faster than would have otherwise occurred. In the effort, refineries moved closer to the practices recommended by API manual.

Technological Advances Affecting Disposal Practices

The pressing demand for petroleum products during World War II motivated much of the expansion that occurred in the 1940s. Indeed, during the war, the federal government financed the construction and expansion of numerous facilities, including those producing aviation gasoline, toluene, and special gases required in the manufacture of butyl rubber. In the years after the war, the availability of new process technology and steady increases in the demand for gasoline provided addition incentives to expansion.

New refineries accounted for some of the increased capacity. One of the largest wartime projects involved the construction of a facility in Lake Charles, Louisiana, which was completed in 1944 and fueled the advance of the Allies' Western Front with 130,000 barrels of petroleum each day. The facility also included 10 million barrels of steel storage and 2 million barrels of underground storage in artificially created salt caverns.[24] However, most of the increased refining capacity came from the expansion of existing facilities (table 9.1). For example, after Humble Oil expanded its refinery in Baytown, Texas, that plant could process twice as much petroleum as the Lake Charles facility (fig 9.3). By 1950, the total capacity of refineries in the United States exceeded 6 million barrels per day, up from about 2.3 million in 1927. In that same period, the number of refineries dropped from 508 to 355.[25]

Figure 9.3. This 1944 view of Humble Oil's Baytown refinery shows new equipment installed as part of the wartime expansion. (Library of Congress LC-S89-17827)

This steady expansion of refining capacity kept an army of engineers and scientists at work developing and redesigning processes and equipment. In designing and installing this wave of new equipment, engineers increased both the efficiency of refineries and the sophistication of their controls (fig 9.4). The expansion also facilitated the dissemination of innovations.[26] For example, in the late 1920s, a few refiners started using more efficient and specialized processes to extract impurities from products such as lubricating oils. These processes typically involved circulating a solvent, such as benzene or phenol, through a closed loop of extraction and separation. This closed loop resulted in relatively few losses of the working solvent. As companies modified and expanded their facilities, they made better use of these closed-loop processes of solvent extraction.[27]

Table 9.1. Major Refineries in the United States, 1927 and 1950

Location	1950 Owner	1927 Capacity bbls/day	(rank)	1950 Capacity bbls/day	(rank)
Baytown, Texas	Humble	50,000	(10)	260,000	(1)
Baton Rouge, Louisiana	Standard Oil (La.)	60,000	(7)	245,000	(2)
Port Arthur, Texas	Gulf	45,000	(13)	230,000	(3)
Port Arthur, Texas	Texas Company	102,000	(1)	190,000	(4)
Whiting, Indiana	Standard (Indiana)	52,000	(8)	175,000	(5)
Bayonne, New Jersey	Standard Oil (N. J.)	88,000	(4)	173,000	(6)
Linden, New Jersey	Standard Oil (N. J.)	77,000	(5)		*
Beaumont, Texas	Magnolia	45,000	(12)	150,000	(7)
Marcus Hook, Pennsylvania	Sun Oil	15,000	(39)	140,000	(8)
Richmond, California	Standard Oil (Ca.)	100,000	(3)	138,000	(9)
Lake Charles, Louisiana	Cities Service	—		130,000	(10)
El Segundo, California	Standard Oil (Ca.)	100,000	(2)	117,000	(11)
Philadelphia, Pennsylvania	Atlantic Refining	50,000	(9)	117,000	(12)
Wood River, Illinois	Roxana/Shell	30,000	(21)	115,000	(13)
Texas City, Texas	Pan-American	—		114,000	(14)
Houston, Texas	Shell Oil	—		110,000	(15)
Philadelphia, Pennsylvania	Gulf	—		107,000	(16)
Torrance, California	General Petroleum Co.	—		100,000	(17)
Wilmington, California	Richfield	35,000	(16)	91,000	(18)
Martinez, California	Tidewater	60,000	(6)	85,000	(19)
Houston, Texas	Sinclair	16,000	(36)	85,000	(20)
Tulsa, Oklahoma	Mid-Continent Petroleum	40,000	(14)	45,000	(36)

Sources: "Refinery Locations and Crude-Oil Capacities," supplement to *Oil and Gas Journal* 48 (Sept. 21, 1950); Standard Oil (N. J.), Marine Dept., "Oil Refineries and Principle Oil Fields of the United States," January 1, 1927.

*The 1950 capacity of the Standard Oil plant in Linden, New Jersey, is included in the 1950 capacity for the company's plant in Bayonne, New Jersey.

One technology that took the industry by storm was the application of catalysts to cracking, which rendered thermal cracking obsolete. Designing successful catalytic crackers depended not only on new scientific knowledge but also on technological advances in the design of pressure vessels, heat exchangers, turbo compressors, motorized valves, control instrumentation, and other types of basic process equipment. For example, when using catalysts to facilitate cracking, a layer of carbon tended to form on the porous surface of the catalysts, rendering the catalyst ineffective. To design a practical catalytic cracking plant, designers had to figure out a way to remove the carbon, thereby "regenerating" the catalyst, without disrupting the flow of product through the plant.[28] In the first commercially successful catalytic cracking process, installed in 1937 at Sun Oil's plant in Marcus Hook, Pennsylvania, inventor Jules Houdry solved the problem by placing two units in parallel. In that way, the refinery could crack hydrocarbons in one unit while regenerating the catalyst in the other. Switching the flow of petroleum from one unit to the other and making use of the heat produced by the regeneration cycle required a use of heat exchangers, compressors, process control mechanisms, and automatic valves as sophisticated as any application up to that time.[29]

In an effort to get around the licensing requirements of the patented Houdry process, a consortium of companies led by Jersey Standard and

Figure 9.4. By the late 1940s, advances in control technology allowed many refinery operations to be controlled from a central panel, such as in this refinery in Oklahoma City. (Library of Congress LC-US-W3-10761-D).

the M. W. Kellogg Company then put approximately one thousand scientists and engineers to work developing a more efficient catalytic cracking process.[30] This team soon developed a process in which powdered catalyst flowed along with the petroleum, as if part of the fluid. As the mixture of powdered catalyst and hydrocarbon vapors left the reaction chamber, cyclones separated the mixture into two streams. The stream of hydrocarbons continued on while the stream of catalyst flowed back through a regeneration chamber. Upon exiting that chamber, the "fresh" catalyst mixed in with a new stream of incoming oil. While a single Houdry unit could crack no more than 19,000 barrels per day, a single fluidized unit could crack over 100,000 barrels per day. The first fluidized catalytic cracking unit, installed in Standard's Baton Rouge refinery, went on stream in 1942.[31]

This successful use of catalysts encouraged researchers and engineers throughout the industry to develop other catalytic processes capable of manipulating hydrocarbons. For example, alkylation plants constructed during World War II produced high-octane aviation fuel by combining smaller molecules—light gases—into larger molecules. With alkylation, light gases, including some created during catalytic cracking, could now be transformed into a valuable product. Another innovation, catalytic reforming, allowed refiners to bend straight molecules into cyclic molecules. Initially used to produce toluene for explosives, this process could also convert low-octane gasoline molecules into higher-octane molecules. Soon after, refiners developed various types of hydrotreating, which allowed refiners to stabilize unsaturated hydrocarbons that sometimes gummed up fuel systems. Hydrotreating could also be used to remove sulfur compounds by converting those compounds into hydrogen sulfide, a gas that could be extracted from the product stream using a solvent. All of these new processes allowed refiners to convert more of their crude oil into refined products and to do so with fewer chemicals and a smaller percentage of waste.

A much smaller number of engineers worked more directly with pollution control, and the general expansion of the nation's refining capacity in the 1940s allowed them to experiment with new ways to break emulsions, recover spent material, and otherwise treat wastes. In general, these engineers freely shared their knowledge with engineers from other companies during sessions sponsored by the American Petroleum Institute's Division of Refining. By the 1950s, these sessions covered topics that were far more sophisticated than those discussed in the 1930s. For example, at the API meeting in 1954, one could attend sessions such as "New Developments in the Disposal of Refinery Waste Gases," "The Chemical Flocculation of a Refinery Waste," and "Planning and Executing of a Refinery Waste-Stream Survey."[32] W. B. Hart, in charge of the "gas, acid, and drainage" department at the Atlantic Refining Company refinery in Philadelphia, credited those meetings—and the API committee on the disposal of refinery wastes—with changing industry practices.

Safety considerations also encouraged some new practices. For example, the new alkylation processes used hydrogen fluoride as a catalyst. Any of this material that leaked out and combined with moisture in the air formed a steamlike mixture capable of causing severe chemical burns. Therefore, designers of these plants typically placed all equipment over limestone neutralization pits with drain lines that led to a secondary system of neutralization before allowing any fluid to enter the main sewer system. Process designers also included systems for capturing any vapors that formed.[33] Such designs also encouraged engineers to think about their use of sewers and release of vapors in new ways.

Engineers in charge of waste disposal also began using spent solutions associated with one process to help break emulsions or treat wastes elsewhere. For example, Roy F. Weston, one of Hart's colleagues at Atlantic Refining, used aluminum chloride as a coagulant in a chemical flocculation treatment plant because he could obtain that material from a spent catalyst sludge generated at the refinery. For a weighing agent, he used ground-up spent clay.[34]

Some advances in refinery technology created new problems. For example, in releasing emissions into the air, the regeneration chambers of cracking units lost a small portion of their catalyst in the form of dust particles. Given that one "cat cracker" could contain over 2 million pounds of catalyst, even a small fraction translated into a significant loss. Furthermore, in the initial wave of fluidized cracking units, the problem increased over time. Engineers discovered that the fast-moving catalyst chewed up cyclones, forcing engineers to reduce the speed at which the gas flowed. The lower flow rates also reduced the ability of these devices to separate the stream of catalysts from other streams, increasing the amount of catalysts escaping. In this case, the expense of the catalyst encouraged refineries to redesign their cyclones and to add electric precipitators capable of capturing more of the catalyst dust that escaped.[35]

Keeping releases of sulfur under control proved to be one of the more challenging tasks for refineries. Refiners had developed numerous methods for removing sulfur from distillates or, at the very least, for "sweetening" those distillates by rendering sulfur compounds less obnoxious. However, the most common method of removing impurities involved washing the distillate with sulfuric acid, a process that created substantial quantities of acid sludge. Even at refineries that recycled their sludge, a lot of this sulfur escaped, in one form or another. New methods of removing sulfur compounds from petroleum were being developed, but even then the sulfur had to go somewhere. For example, by the 1930s, engineers recognized that they could use bauxite to facilitate the decomposition of sulfur compounds into hydrogen sulfide.[36] The development of catalytic hydrotreating in the 1940s accomplished the same task more efficiently. But refineries still had to dispose of the highly toxic hydrogen sulfide gas by flaring it, thereby releasing huge quantities of sulfur dioxide. Although these technologies simplified the job of removing impurities and reduced the need for acid washes, they left a major pollution concern unaddressed.

Changes in scale also mattered. For example, Sun Oil succeeded in cutting per-barrel losses at its refineries in half, from 2 percent of throughput in 1932 to 1 percent in 1945.[37] However, in the same period, the capacity of the company's main refinery increased almost tenfold, meaning that the absolute amount of oil being lost from all sources increased fivefold. Hence, despite significant success in reducing the quantity of pollution-causing wastes released per barrel of petroleum, waste disposal engineers continued to face challenges in preventing wastes from causing a nuisance. Indeed, in the early 1940s, the Public Health Service proposed funding the Bureau of Mines to study the character of refinery wastes so as to determine "possible changes in plant operation to so alter the character of the resulting wastes." The proposed budget for the study ($12,250), though, was far too low to represent a serious effort.[38]

Therefore, by the 1950s, refiners had reduced the amount of material they discharged as emissions, effluent, and solid wastes. In addition to using more efficient process technologies, refineries had achieved these reductions by installing segregated sewer systems, recycling their spent chemicals, making better use of closed-loop extraction processes, and operating better oil-water separators. However, pollution concerns still remained. Yet, further reducing the amount of oil and chemicals being discharged would require significant effort. For example, engineers at Humble Oil's Baytown refinery estimated that drips from the thousands of valves at the refinery added up to about 17,000 barrels of oil per year.[39] Small spills originating from all sources—including spills at times of unit shutdown and equipment repair, accidental spills and overflows, and tank bottom drawoffs—could amount to as much as 2 percent of a refinery's annual charge of crude.[40] Not all of those spills reached a refinery's main sewer, and putting the infrastructure in place to capture these small distributed losses was difficult to justify.

Some refinery engineers predicted that the pressure on large refineries to control their pollution-causing releases would only grow stronger. For example, in the late 1940s, Hart, an active member of the API's committee on the disposal of refinery wastes, predicted that as urban areas constructed more sewage treatment facilities, oil pollution would stand out like a sore thumb. Calling the petroleum industry one of the "four horsemen of pollution," Hart doubted that the public, after spending lots of money for sewage treatment, would allow the petroleum industry to continue discharging wastes at its present level.[41]

Addressing New Pollution Concerns

As it turned out, Hart was right. In the late 1940s, new concerns with industrial discharges into surface bodies of water emerged as urban areas put more effort into the treatment of domestic sewage. In some areas, local officials put pressure on refinery managers to go beyond API-recommended practices.

Local officials who desired to reduce the quantity of municipal and industrial waste entering major rivers and bays faced a daunting task. Many wastes came from facilities over which local officials had no control. For example, industrial waste dumped into the Ohio River near Pittsburgh caused as many problems for cities and towns downstream, such as Wheeling, West Virginia, and Cincinnati, Ohio, as it did for residents of Pittsburgh. In such cases, downstream states could not accomplish much by unilaterally enacting and enforcing local- or state-level water pollution laws. At the same, industrial leaders and officials in heavily industrialized states argued that federal legislation would be too broad, unable to take into account the specifics of the local situation.

The campaign to clean up the nation's interstate rivers gained momentum with the passage of the federal Water Pollution Control Act of 1948. This act, by approving the use of interstate compacts to coordinate pollution control efforts in watersheds, sidestepped the question of whether federal or state agencies should take the lead in coordinating pollution control efforts. In the years following that legislation, states in various watersheds formed organizations such as the Ohio River Valley Water Sanitation Commission and the Delaware River Valley Compact Commission. Although these interstate organizations identified industrial wastes as a problem and encouraged companies to adopt cleaner practices, many of their early efforts went toward reducing the amount of raw sewage being discharged by towns and cities.[42]

As some refinery engineers recognized, decreases in discharges of raw sewage placed more attention on industrial wastes. To make the job of the waste disposal engineer easier, W. B. Hart of the Atlantic Refining Company hoped to find a simple indicator of water quality, such as a type of bacteria that could survive in clean water but not in polluted water. Then, refinery technicians could check a sample of their refinery effluent by seeding it with the test bacteria. In an API-sponsored effort to find a type of bacteria that would serve as the test organism, Hart led a team that studied water samples from 171 streams. However, in 1950, Hart reported that the "bacteria forms found in these samples were just as pronounced in the polluted water as in the clean water."[43]

Hart's discovery came as no surprise to microbiologists. Municipal sewer treatment plants in large cities had been using activated sludge—sludge containing bacteria to break down organic wastes—for several decades.[44] Refineries, though, did not make any systematic use of biological action until the early 1950s, when engineers with the Dow Chemical

Company demonstrated the feasibility of removing phenols through biological oxidation. Extremely small quantities (parts per billion) of phenols could impart a foul taste to chlorinated drinking water and removing these organics from wastewater was difficult. Hence, refinery engineers at facilities under pressure to control their phenol discharges began experimenting with the biological oxidation of phenols. Some refineries secured their initial source of bacteria from Dow facilities in Midland, Michigan, but engineers eventually discovered that activated sludge from municipal plants served just as well.[45]

Not everybody was comfortable with this line of experimentation. Some people expressed concern "that a phenol-tolerant strain of pathogenic, or disease-causing bacteria might develop in the unit" and, since phenol was widely used as a bactericidal agent in medicine, "a phenol-resistant strain could become a serious health menace." Investigations showed, however, that the bacteria that flourished in the process were nonpathogenic.[46]

By 1958, several refineries, including two new refineries that Shell Oil and Mobil Oil constructed on Puget Sound in Washington, established biological treatment as a practical form of wastewater treatment. Through the use of activated sludge systems, trickling filters, and oxidation ponds, other refineries also began exposing their effluent to mixtures of oxygen and bacteria.[47] Meanwhile, federal legislators reaffirmed their interest in cleaning up the nation's rivers by heavily subsidizing the construction of municipal sewage treatment plants through the Federal Water Pollution Control Act of 1956.[48]

In some areas, local officials put more pressure on refineries to reduce their emissions into the air. By far the most pressure came from officials concerned about smog in Los Angeles, California. At first, the effort to address the smog problem looked as if it would be yet another example in which technological advances associated with using resources more efficiently would address the concerns at hand. However, it soon became apparent that this approach would not be sufficient. In the end, the effort to address the smog problem placed the focus on measuring and monitoring pollution-causing releases and their effects on the shared environment.

Earlier in the century, health officials in many urban areas expressed concern with smoke pollution but addressed the problem in piecemeal fashion. However, in the 1930s and 1940s, some cities began to address the concern more systematically. Municipal leaders in St. Louis, for example, significantly reduced that city's smoke problem by the start of World War II. Inspired by the success of St. Louis, civic leaders in Pittsburgh, then known as the "smoky city," initiated a clean air campaign that achieved significant results by the late 1940s and early 1950s.[49] In 1948, news of a "death smog" in Donora, Pennsylvania, reinforced the importance of such campaigns. In Donora, temperature inversion trapped fumes from a zinc smelter, making one-third of the population ill and killing over twenty people in a twenty-four-hour period.[50] Officials in Los Angeles, concerned with the smog that settled over its city, also initiated a campaign to identify and eliminate the source of its smog.

The problem in Los Angeles first became noticeable in the early 1940s. At that time, decreases in visibility occurred that could not be explained by any climactic change. During the war years, residents noticed that this hazy mist appeared more regularly and more intensely than before. Complaints that the smoggy air damaged crops and irritated eyes increased as well. Immediately after the war, the County Board of Supervisors hired a director of air pollution control. The California state legislature reinforced that action with the creation of pollution control districts.[51]

The first director of the Los Angeles Air Pollution Control District, Louis C. McCabe, previously had worked as an air pollution control expert with the Bureau of Mines. McCabe knew that refineries released large amounts of sulfur dioxide and estimated that, in 1948, refineries in the Los Angeles basin were emitting about 380 tons of this gas each day.[52] McCabe also knew that refineries generated much of this sulfur dioxide when they flared gases containing hydrogen sulfide, and that technology existed to convert hydrogen sulfide into either sulfuric acid or elemental sulfur. Indeed, as early as 1935, the Atlantic Refining Company's refinery in Philadelphia had succeeded in converting hydrogen sulfide into a usable acid.[53] Therefore, when McCabe issued his first pollution control regulation in 1947, he instituted limits on the amount of sulfur dioxide that refineries could release into the air. McCabe's focus on refineries, along with photographs of sprawling, steam-spewing refineries in newspapers, quickly made these facilities popular scapegoats for the entire smog problem.

Although the operators of refineries complained that the action was hasty, arbitrary, and inequitable, they took corrective action. Within several years, the total sulfur dioxide emissions from all refineries dropped to 80 tons per day, about one-fifth the level being released three years earlier. Initially, Richfield Oil, the Texas Company, Standard Oil of California, Shell Oil, and General Petroleum—all of which operated refineries in the area—accomplished this reduction by installing equipment capable of absorbing hydrogen sulfide from gas streams. They sent this hydrogen sulfide gas to nearby chemical plants for conversion into elemental sulfur. Union Oil went a step further. In addition to installing absorption units, the Union refinery installed its own "Claus" recovery plant, a move that other refineries eventually followed.[54]

These Claus sulfur recovery plants transformed hydrogen sulfide into elemental sulfur using a version of the process developed by C. F. Claus in the late nineteenth century. In that process, about a third of the hydrogen sulfide was first burned to produce sulfur dioxide. The sulfur dioxide and the unburned portion of hydrogen sulfide then reacted in the presence of a catalyst to form elemental sulfur and water vapor.[55] At the time, a high demand for sulfur as a raw material allowed refineries to recover much of their expense for installing absorption units and Claus recovery plants.[56]

Some refineries also installed hydro-desulfurization plants to convert sulfur-containing impurities directly into hydrogen sulfide, which complemented their use of Claus plants. Hydro-desulfurization, a form of catalytic hydrotreating, took place in an atmosphere of hydrogen and facilitated processes in which sulfur-containing hydrocarbons released their sulfur molecules to bond with hydrogen. Hydro-desulfurization not only transformed sulfur impurities into a form that could be sent to Claus plants, but also reduced the need for large quantities of sulfuric acid and caustic solutions during refining.[57]

Here then seemed to be another case in which a solution to a pollution problem could be framed in terms of eliminating waste and increasing efficiency. New technology had allowed refiners to convert sulfur-based impurities found in crude oil into hydrogen sulfide, which could then be turned into elemental sulfur and sold. This technology supplemented older methods of removing sulfur, reducing the amount of sulfuric acid sludge and spent caustic that would otherwise be generated. Furthermore, by stripping hydrogen sulfide from gas streams, the remaining gas could be used as fuel rather than flared directly to the atmosphere. If this solution proved successful, any need to explicitly discuss standards of environmental quality and to take action to maintain those standards would have seemed less urgent.

As it turned out, reducing sulfur emissions did not eliminate the smog problem in Los Angeles. This did not come as a surprise to A. J. Haagen-Smit, a biochemist whose research indicated that unburned hydrocarbons from automobiles and refineries contributed more to the creation of smog than sulfur dioxide. He came to this conclusion after the Los Angeles Chamber of Commerce asked him to investigate the source of damage to local crops. In that investigation, completed in the late 1940s, Haagen-Smit identified photochemical processes, which involved hydrocarbons reacting with other chemicals in the presence of sunlight, as the source of the smog problem in Los Angeles.[58]

Investigators with the Stanford Research Institute, a nonprofit laboratory not connected with Stanford University, disagreed with Haagen-Smit's conclusion. Their work, funded by the Western Oil and Gas Association, suggested that chemicals from thousands of sources were involved and that no single class of chemical could be held responsible. However, Haagen-Smit's hypothesis was one that could be tested. He was talking about chemical reactions that, under specific conditions, happened in a matter of hours. For example, in 1949, a traffic jam in and around a football stadium north of Los Angeles produced the same smoglike conditions that Los Angeles typically experienced. Such events helped to strengthen the link between smog and automobiles.[59] Haagen-Smit also pointed to other sources of hydrocarbon vapors as problems, including refineries and service stations.

Researchers at Stanford Research Institute soon moved closer to Haagen-Smit's position. They concluded that the Los Angeles basin had the topography and climate of a giant reaction chamber. Mountains that surrounded Los Angeles on three sides and warm air masses which settled over the basin prevented polluted air from escaping. After studying the problem further, they agreed that nitrogen oxides and unburned hydrocarbons reacted in the stagnant atmosphere, usually with the help of sunlight, to create other compounds that settled over the city in a chemical mist. According to their estimates, automobiles released approximately 1,000 tons of unburned hydrocarbons into the air each day. Hydrocarbon vapors from refineries, tank farms, distribution terminals, and filling stations accounted for another 440 tons. A third major source of pollutants came from the burning of paper, grass clippings, and wood by individual residents of the county.[60]

In light of this consensus, the Los Angeles Air Pollution Control District sought to have refineries in the area reduce their vapor losses. However, engineers in the petroleum industry believed that they already had reduced vapor losses as much as they could. Indeed, the effort to eliminate vapor losses had been one of the great successes in the conservationist fight to reduce waste and increase efficiency. At some refineries, storage tank losses had dropped to .025 percent of throughput each year.[61] In addition, over the last several years, in voluntary cooperation with the Los Angeles Air Pollution Control District, refineries already had cut their hydrocarbon emissions in half, mainly by replacing old storage tanks with new ones having floating roofs and vapor recovery systems. Incentives to reduce hydrocarbon emissions any further would involve more costly action, such as putting roofs on oil-water separators and reducing vapor leaks from thousands of valves (fig. 9.5).[62] Small amounts of vapor came from many different sources, and the value of those vapors did not justify their recovery. Refineries were not likely to reduce their vapor emissions much further without further action from county officials.

The API also turned its attention to the problem. In 1952, the directors of the API created a "smoke and fumes" committee and charged it with "erecting centers of information throughout the country." By 1955, this committee was funding about a dozen researchers for a total expenditure of about $250,000 annually. These researchers—at institutions such as Stanford Research Institute (Pasadena), the Armour Research Foundation (Chicago), the Industrial Hygiene Foundation (Pittsburgh), the Franklin Institute (Philadelphia), and Kettering Labs (Cincinnati)—studied the chemical reactions involved in smog formation and developed specialized

Figure 9.5. Refineries house large numbers of valves from which small amounts of hydrocarbons could escape. (Library of Congress LC-US-W3-10336-D)

equipment for taking samples and making measurements.[63] However, if the members of the API's Smoke and Fumes Committee desired to control the research agenda associated with smog-related concerns, they failed. As evidence mounted identifying hydrocarbons as a smog-causing pollutant, the research funds available from other sources also increased. By the mid-1950s, the API-funded researchers represented only a fraction of the scientists studying the formation of smog and developing instruments to measure and monitor chemicals in the atmosphere.[64]

First, civic leaders in Los Angeles organized an Air Pollution Foundation to study the smog problem in more detail. In a joint project with the Los Angeles Air Pollution Control District, the foundation established ten sampling stations throughout the basin. These stations routinely measured the concentration of various pollutants in the air, including hydrocarbons, sulfur dioxide, and nitrogen oxides. The Air Pollution Foundation, with an annual budget approaching a million dollars, received and welcomed assistance from a wide variety of sources. The U.S. Navy and the Goodyear Rubber Company even helped researchers obtain air samples at high altitudes, with Goodyear putting a blimp at the foundation's service. The Automobile Manufacturers Association also presented the Air Pollution Foundation with $100,000.[65]

Second, the federal government, through the Air Pollution Research and Technical Assistance Act of 1955, authorized the Public Health Service to spend $5 million per year to help state and local agencies investigate the problem of air pollution. Most of these funds, whether directly or indirectly, went toward advancing the agenda set by researchers in Los Angeles.[66] One line of research had to do with sampling automobile exhaust, which investigators saw as the most troublesome source of pollutants. Some federal funds also went toward studying emissions from refineries. For example, when engineers with the Petroleum Research Station in Bartlesville, Oklahoma, secured funding to study the content of automobile exhaust, they were also asked to assist officials in Los Angeles with sampling emissions from thirty-five refinery emission points—ranging from valves and flanges to waste gas flares—and analyzing the content of those samples.[67]

Not surprisingly, papers on instrumentation and procedures attracted significant interest at a conference on air pollution sponsored by the Air Pollution Foundation in 1955. Two industrial scientists drew the most attention presenting a technique for using gas chromatography to identify organic compounds in a sample of air.[68] As attendees to the conference recognized, this technology made it easier to identify the hydrocarbon molecules present in automobile exhaust. In the technique presented, researchers first passed their sample through a column packed with an adsorbent. Differently sized molecules then moved through the column at different rates, and as each group of molecules exited the column, they triggered a signal that could be recorded on chart paper. Mass spectrometers, first used by physicists in the 1930s to measure the atomic weights of important isotopes, provided industrial scientists with another way to analyze the hydrocarbon content of a sample. In using mass spectrometers, researchers first ionized a portion of the sample and then passed those ions through an electromagnetic field. The electric field deflected the molecules by an angle proportional to their weight. The angle of deflection associated with different ions provided an indication of a sample's constituents.[69]

These two technologies for identifying the molecules present in a sample were not altogether new. Researchers hoping to identify the constituents of crude petroleum had reduced to practice the basic technique of gas chromatography in the 1940s. However, gas chromatography required considerable application-specific expertise and skill, especially in preparing samples, choosing adsorbents, and interpreting results.[70] Mass spectrometers, first used by physicists in the 1930s to measure the atomic weights of important isotopes, had been around somewhat longer but were also in an early stage of commercialization and standardization.

Therefore, soon after the 1955 conference, scientists working for the Los Angeles Air Pollution Control District decided to purchase their own mass spectrometry and gas chromatography equipment and to develop their own testing procedures. Over the next several years, the Air Pollution Control District—in cooperation with Western Oil and Gas Association, the California Bureau of Air Sanitation, and the Public Health Service—completed its study of refinery emissions and summarized the results in a report titled "Atmospheric Emissions from Petroleum Refineries: A Guide for Measurement and Control."[71] This guide represented a significant change from the type of monitoring and measuring technology described in the API manual.

Indeed, by the mid-1950s, the status of the API manual had changed. Managers of refineries had discovered that, even after installing equipment recommended by the API's manual on the disposal of refinery wastes, emissions and effluents coming from their plants were still identified as a concern. People knowledgeable about the petroleum industry could say, without fear of contradiction, that developments in waste disposal had "run significantly ahead of the [API] Manual," in the words of A. Berk, the head of the U.S. Bureau of Mines Industry Water Lab.[72]

In many ways, the group responsible for the API manual was simply unprepared to address the new wave of concerns. In 1952, that committee released a new edition of the volume on waste gases and vapors, in which they predicted that "an enlightened, discerning, and sympathetic public will understand that atmospheric pollution is a joint industry-community problem, and will not regard the industry as a menace to the public health."[73] However, the changes to this volume fell short of a serious response. The main change from the previous edition involved the addition of two supplements: one on smokeless flares and another on air pollution surveys. In the supplement on smokeless flares, the authors emphasized the need to mix air with the escaping gas to ensure complete combustion. They also advised engineers to place flares, if possible, on the ground, noting that elevated flares attracted attention and generated complaints. The section on air pollution surveys provided some general advice on why and how to perform a pollution survey. Among other things, the authors suggested taking air samples at various points around the refinery to determine if air pollution equipment was working properly and for use as evidence in lawsuits. The API Committee on Disposal of Refinery Wastes also released two new volumes: one on the sampling and analysis of wastewater and one on the sampling and analysis of waste gases and particulate matter. Both volumes were little more than a collection of separate procedures assembled in one binder.[74]

Though periodically updated, the API manual on the disposal of refinery wastes fell further behind advances in technology and, by the late 1950s, no longer described the state of the art in pollution control. A new perspective on pollution control had begun to emerge. For example, in a paper presented at the API's 1960 meeting, John Frame, an engineer with Cities Service's Research and Development Division, raised the question, "Can refinery pollution control have a payout?" He concluded that "in the usual sense of the word, we believe the answer would have to be no." The payout, he noted, came in the form of better public relations.[75]

Hence, while holding on to his faith in self-regulation based on the bottom line, Frame recognized that sustaining a certain standard of environmental quality was neither free nor the inevitable result of making operations more efficient. But what standards of environmental quality should refineries be expected to meet? And who should define those standards? Nuisance law certainly suggested that some minimum standards applied, but the law did not always define those standards. The API Committee on the Disposal of Refinery Wastes, which had been setting standards for two decades, could no longer expect to establish these standards either.

The lack of standards also represented an obstacle to local officials hoping to place controls on industrial pollution. The experience of Health Department officials in Houston highlighted the problem. If a "water pollution" equivalent of Los Angeles smog had existed, the Houston Ship Channel probably would have been involved. Few places in the nation had as great a concentration of industrial plants disposing such a wide variety of wastes into a single channel. When W. A. Quebedeaux assumed his position in 1954 as the first director of the Harris County Health Department's air and water pollution control section, over one hundred industrial plants, including three of the largest refineries in the world, lined the ship channel along its twenty-five-mile route from Houston to Galveston Bay.[76] The three main refineries—operated by Humble, Shell, and Sinclair—processed 520,000 barrels of crude each day. Several other average-sized refineries contributed another 70,000 barrels.[77]

Previous to his appointment, Quebedeaux had been the air and stream pollution director for the Champion Paper Company. Before that, he operated his own chemical analysis laboratory in St. Louis, Missouri, where he provided what he described as "consultant and engineering services especially connected with litigation."[78] Upon assuming his position, Quebedeaux immediately embarked on a campaign to reduce the amount of wastes being discharged into the channel, but the effort never had the same public support as the Los Angeles campaign to eliminate smog, perhaps because nobody had to breathe the water.

But water quality problems clearly existed. A water quality survey performed by Humble Oil in 1955 indicated that the channel from the "San Jacinto monument" to the "turning basin," a distance of about sixteen miles, was devoid of dissolved oxygen because sewage and industrial wastes consumed all the oxygen available.[79] Few people, however, expressed any real concern with the channel's poor water quality. For example, in 1961, when one member of the Texas Railroad Commission retired after two decades of service, his associates honored him with a party held on a boat that sailed down the ship channel.[80] The industrial plants lining the shore, not an afternoon of fishing and swimming, served as the main attraction.

Without public support or sufficient resources, Quebedeaux could do little. Described by a local newspaper as a "prophetic voice crying in the wilderness," Quebedeaux routinely collected and analyzed air and water samples but met with numerous frustrations when he tried to get companies to change their practices.[81] This list of frustrations included companies that waited for lawsuits before responding to him and laws that prevented local officials from inspecting a plant at will. Also near the top of that list stood cases lost in court by a prosecuting attorney "out of law school probably less than one year" going against an industry-hired "defense lawyer with twenty or thirty years' experience."[82] Little changed during his tenure with the Harris County Health Department. Quebedeaux's small staff was simply unable to take consistent and systematic action against companies not willing to cooperate.[83]

Still, if nothing else, the limited pressure Quebedeaux's staff placed on industrial plants provided engineers who wished to install better waste treatment equipment with a better case for doing so. And some changes did occur. For example, in the 1950s, engineers with Humble Oil installed coalescing equipment to remove and recover oil from wastewater streams at several process units, a new master oil-water separator, filters to remove emulsions, equipment to strip hydrogen sulfide from various waste streams, cooling towers to increase the reuse of cooling water, and a 20,000-gallon-per-minute pump at the point of discharge to increase the amount of dissolved oxygen in the water.

The pump that the Humble refinery installed also had the effect of diluting the refinery's effluent with another 28 million gallons of water each day. However, in this case, dilution may not have been that beneficial. Engineers at the refinery claimed that, between 1949 and 1960, the quality of their effluent "showed more than a 90 percent improvement," making it "generally of better quality than that of its receiving body, the Houston Ship Channel." In the early 1960s, the company also installed a three-pond lagooning system capable of holding water for forty-five days, with the extended retention time permitting enough bacteriological action to reduce pollutants in the effluent by an additional 70 percent.[84]

Discharging effluent "generally of better quality" than water in the Houston Ship Channel may not have been a particularly lofty goal, but determining how clean industrial effluent should be was no simple matter. Struggling pollution control officials such as Quebedeaux recognized that the wide variety of wastes being discharged into waterways, especially wastes from chemical and petrochemical plants, prevented any easy analysis of the problem. Indeed, traditional measures of water quality—such as those associated with dissolved oxygen, pH, mineralization, odor-causing impurities, and oil content—were not always sufficient.

The need to establish quality standards eventually led to the federal Water Quality Act of 1965. The legislation gave states until June 30, 1967, to formulate water quality standards or to accept those furnished by the Department of Health, Education, and Welfare.[85] Nobody knew the extent to which those standards would affect refineries, but many in the petroleum industry believed that oil companies had already made substantial progress toward eliminating refineries as a source of water pollution. For example, the editors of the *Oil and Gas Journal* felt confident that "odor" was "the primary remaining water-pollution offense of refiners." This problem, they maintained, was "subtle and difficult to solve" but not "the kind of flagrant pollution that poses health hazards."[86]

An Attempt to Reestablish Leadership

Unlike the editors of the *Oil and Gas Journal*, the directors of the API did not assume that the industry would be unaffected by the Water Quality Act of 1965. Setting standards of environmental quality was clearly a political issue. Previously, they had proceeded on the assumption that voluntary action based on solid scientific knowledge, with industry leading the way and determining the pace of change, would satisfy legislators. However, those efforts had failed.

For over a decade, W. L. Stewart, Jr., the first chairman of the Smoke and Fumes Committee, had argued that the smog problem was a scientific issue, not a legal issue. Laws, he emphasized, would not cure pollution. Industry, he noted, had to learn more about the problem and voluntarily take steps to solve that problem.[87] Even after Stewart's committee failed to define the research agenda, he continued to treat the problem as a technical one and integrated the API's research program with the work of other scientists. This strategy, though, failed to deflect regulatory initiatives. If anything, the new knowledge created by ongoing air pollution research only raised more questions about hydrocarbon vapors, automobile emissions, and the effects of sulfur dioxide.

The API's lobbying arm had not been dormant in the 1950s: it had been busy fighting never-ending charges that the industry fixed prices.[88] The roots of the most recent wave of charges reached back to 1938, when the Roosevelt administration created the Temporary National Economic Committee (TNEC) to investigate the concentration of economic power in the United States. After collecting thousands of pages of testimony related to the petroleum industry, the committee concluded that price stabilization measures associated with production controls and the vertical integration of major oil companies undermined competitive forces.[89] The Department of Justice eventually charged twenty-two oil companies and the API with fixing prices, a suit that came to be known as the "Mother Hubbard suit" because it contained so many charges. It was laid aside during the war years.[90]

Although the government eventually dismissed the Mother Hubbard suit, concerns associated with the concentration of power in the petroleum industry remained. In the late 1940s, a shortage of home heating oil and higher prices in the northeastern markets further fueled that concern. The API then launched a publicity campaign to allay public distrust of the oil industry, in which "suspicion, misunderstanding, hostility, investigation, and regulation" were positioned as an ugly shadow darkening the potential for "enduring progress."[91] Soon after, though, Eugene Rostow, a prominent economist, put the API on the defensive again by publishing a widely read book, *A National Policy for the Oil Industry,* in which he argued that the federal government should disperse the economic power of the petroleum industry.[92]

The lobbying arm of the API also devoted significant effort toward shaping legislation associated with taxes, tariffs, and import quotas. On these issues, though, different oil firms had different strategies and interests, making it impossible for one organization to speak for the entire petroleum industry. Although these differences between firms undermined the API's ability to represent the interests of all constituents, they also served as evidence against charges of monopolistic control.[93]

While focused on pricing regulations and charges of monopolistic control, the API made little effort to influence political decisions associated with pollution control. Even as state after state passed some form of air pollution statute, the lobbying arm of the API generally remained uninvolved. The directors of the API continued to treat the issue as a technical one.[94] The Clean Air Act of 1963, though, caught their attention. That act promised further legislation aimed at regulating emissions from motor vehicles and power plants, and these regulations had the potential to significantly affect petroleum products.[95] The Water Quality Act of 1965, which required states to formulate water quality standards, finally moved the API to take a more active role in the political process.

Despite the shift in strategy, the API did not let up on its efforts to learn more about air and water pollution. Indeed, after Congress passed the Clean Air Act of 1963, the API's directors set up a new division to study the relationship between its products and air pollution, budgeting $1.8 million for this effort in its first year.[96] The advice that W. L. Stewart originally gave in arguing for research funds was still valid. More than a decade earlier, Stewart had argued that the industry should come to "know more about our [air pollution] problem . . . than the policeman—because he is coming, believe me!"[97]

T E N **The Ocean Ignored As
Tankers Grow**

THE WAY IN WHICH OIL TANKERS disposed of their oily wastes
changed little in the three decades following passage of the Oil Pollution
Act of 1924. That law prohibited ships from discharging any oily wastes
within three miles of shore, and captains typically complied by flushing
their tanks further out to sea. Doing so, however, only delayed the day of
reckoning.

From the start, various people complained that oil discharged anywhere in the sea would still cause problems. At the preliminary international conference on oil pollution convened by the United States in 1926,
several speakers called for a complete prohibition on the discharge of oily
wastes into the sea, regardless of how far from shore a ship emptied its
tanks.[1] In the end, shipowners undermined the need for a formal treaty by
informally committing to keep routine discharges of oily water as far away
from the shore as possible, usually fifty miles but sometimes more. For
most representatives at the conference, that distance was enough.[2]

For three decades, despite occasional efforts by Great Britain to revive
interest in a complete prohibition, few changes occurred in how tankers

discharged their oily wastes. Great Britain finally succeeded in convening another conference in the mid-1950s. By that time, several new factors—such as increases in the size of oil tankers, increases in the amount of oil being transported by sea, a more complex pattern of tanker ownership, and a move toward offshore production—had further complicated matters.

Out of Sight, Out of Mind

A typical oil tanker in the years before World War II contained between twenty and thirty cargo compartments, each capable of holding about 3,000 barrels of oil. To maintain stability in an empty ship, crews kept about a third of those compartments filled with seawater. By moving ballast water from one compartment to another, all the while keeping the ship balanced, the officer in charge could schedule each cargo tank to be cleaned by the time a ship arrived at its destination. The extent to which each tank had to be cleaned depended on a variety of factors, including the type of petroleum or petroleum product to be loaded next and whether the tank was due for maintenance, but flushing tanks with seawater sufficed for most routine trips.[3]

Shipowners did make a few technological changes associated with removing oily residue from cargo tanks, but those changes only allowed crews to clean tanks more efficiently. All oily residue still ended up in the sea. For example, to prepare a compartment for repairs, crews first had to remove all loose scale from the walls. Otherwise, hydrocarbons absorbed by the scale would serve as a continual source of dangerous vapors that could suffocate maintenance workers or explode. To clean the compartment, crew members wearing breathing apparatus and protective gear entered the tank and washed down its walls with salt water from high-pressure hoses. In the years after World War II, most tanker owners simplified this task by installing rotating nozzles that sprayed high-pressure jets of water around the inside of the tank, eliminating the need for anybody to enter a tank until most of the oil had been removed. Use of the high-pressure jets also allowed routine cleanings to be more thorough than before.[4]

Occasionally, such as when unexpected repairs or schedule changes upset a tanker's normal routine, crews found themselves having to clean a ship's tanks within fifty miles of shore. In such cases, they were supposed to pump any emulsified slop they produced into an empty compartment. In theory, that compartment would serve as a huge oil-water separator, with the oily emulsions gradually breaking down and the oil rising to the top. Crews could then discharge whatever oil-free water they found at the bottom of the tank and store the remaining slop until reaching a marine terminal with an oil-water separator. In practice, though, a thick layer of emulsions remained between the water and the oil, especially if the mixture had only a short time to settle, and crews often pumped much of that emulsified oil into the sea close to shore.[5]

Captains hesitated to store their oil-water emulsions for unloading at a land-based facility because that transaction complicated matters. Not only did unloading oily water and sludge take time, but if the owner of the refinery and the owner of the tanker were not the same—a situation that became more common over time—some negotiation was necessary.[6] To avoid unloading any oily slop while docked, many captains discreetly discharged their oily wastes en route, usually outside the three-mile limit but well within the industry standard of fifty miles. Ships coming in for unexpected maintenance also faced the question of what to do with their oily wastes while close to shore.[7]

Accidental releases of oil due to collisions or groundings also occurred, but the Oil Pollution Act of 1924 did not apply to accidents. Most people regarded accidental discharges of oil from ships as something beyond the control of shipowners. Hence, little discussion as to how to prevent such accidents took place.[8] For example, in 1930, just before hearings on a bill to amend the Oil Pollution Act, the steamship *Edward Luckenbach* ran aground near Nantucket, Massachusetts, and released enough oil to spoil beaches and kill a large number of birds. Even with the accident fresh in mind, the chairman of the API pollution committee could say, without fear of reproach, "Take the *Luckenbach* steamship, you could have 100 laws and it would not stop such a case as that. It was a wreck."[9] Not long after, another tanker collided with a cargo ship in Massachusetts Bay. Oil from the tanker spread, enveloped both vessels, and ignited.[10]

In practice, a gray area separated willful discharges from accidents. As authors of an API report on oil pollution noted, if a hose breaks while loading or discharging petroleum at 10,000 barrels per hour, "it can be readily appreciated that a considerable quantity of oil will enter the water before the pumps can be stopped and the valves closed."[11] But to what extent should companies or individuals be held liable for such accidents? Should an oil company be fined for using poorly maintained hoses or for continuing to load at full capacity when a tank was nearly full?

In the mid-1930s, officials at oil ports in Texas and Louisiana began pressuring oil companies to take responsibility for such spills. In one case, a crew member loading a cargo tank continued to operate the pump at full capacity even after the compartment being loaded was full, causing crude to overflow onto the deck. The first mate then pulled the plugs out of several scuppers, which allowed the oil to drain off the deck and into the harbor. Port officials witnessed the scene from an adjoining ship and made a report to the district attorney, who charged the shipowner with violating the Oil Pollution Act of 1924. The district attorney argued that pulling out the scupper plugs represented a willful discharge of oil.[12]

In another case, residents around Lake Charles, Louisiana, complained to officials that an oil tanker headed toward the Gulf leaked crude. The tanker, the *Bidwell,* had been involved in a major explosion several years earlier, and in the words of one inspector, "did not enjoy a good reputation as being a tight boat." Sometime later, when loading oil in Lockport, Louisiana, officials arrested the captain of the ship and charged him with violating the Oil Pollution Act of 1924. The shipowner attempted to fight the violation on grounds that loose rivets did not constitute a willful discharge of oil, but the Circuit Court of Appeals in New Orleans held that the tanker was "a leaking tanker of long standing" and that operating the ship represented a willful discharge of oil.[13]

Around the same time that officials in several U.S. ports began cracking down on ships that violated the Oil Pollution Act, officials in Britain, under the auspices of the League of Nations, began organizing another international treaty on oil pollution. Significant amounts of oil, presumably carried by ocean currents, were still washing onto English beaches. Hostilities in Europe, though, put that effort on hold. The problem, of course, only grew worse during and after World War II due to increased oil shipments to Western Europe. Hence, in the early 1950s, the British government commissioned a major study of the problem. After performing laboratory analysis of samples from beaches, oil-stained boats, harbors, and piers, the investigating committee concluded that "crude oil residue from oil tankers" represented the main source of contamination.[14]

The report published by the United Kingdom Ministry of Transport led to the long-delayed international conference on oil pollution. Held in London, the 1954 International Convention for the Prevention of Pollution of the Sea by Oil drew representatives from twenty-two nations. British attendees initially proposed that tankers be prohibited from discharging oily slop or ballast containing over 100 ppm of oil anywhere in the ocean at any time, but they found little support for their proposal. In particular, United States representatives believed that whatever problems existed in the past had been solved. Other attendees, including those from Germany and the Netherlands, expressed just as little concern and argued that oil discharged in the open ocean evaporated or biodegraded before reaching shore.[15]

In many ways, the protocol did little more than reinforce the informal agreement reached at the "preliminary conference" held twenty-eight years earlier. Participants even raised the old call for the development of an efficient shipboard oil-water separator. One new provision did represent a step toward establishing tanker discharges as a concern to be monitored and regulated: tanker captains had to maintain an official record of all ballasting, deballasting, and tank-cleaning operations. Despite the weak language and expectations of the protocol, four years would pass before ten governments—the number needed before any signatory nation was expected to enforce its provisions—signed the protocol. Even then, the United States was not among the signers.[16]

Although the 1954 conference did not produce an effective international agreement, it did focus some attention on the issue of tanker discharges. Following the conference, the United Nations collected, analyzed, and disseminated information about the availability of port facilities for the reception of oily residues in forty-two countries. From the United States, researchers received reports from fifty-five shipyards, 104 oil terminals, and sixteen companies in the business of cleaning tanks. The authors of the report also noted that the Oil Pollution Panel of the U.S. Coast Guard indicated that conditions along the U.S. coast had been satisfactory since 1935.[17]

The UN report showed that although some changes had taken place over the last three decades, no systematic method for receiving and treating oily wastes from ships had emerged. In the United States, most loading terminals and shipyards identified some method for handling oily wastes from tankers, but many of those facilities were minimal, especially those in remote areas. For example, the operators of some loading terminals simply filtered any oily ballast water they received through bales of hay. Others ran the water into settling pits, from which they then bled the water and burned whatever remained. In still other cases, barges picked up oily sludge and scale from tankers but then deposited that material in dumps along the shore or in deep water. In general, then, some facilities did exist to handle oily water and wastes that tankers carried to shore, but these facilities varied widely in quality. Furthermore, tankers were still flushing large quantities of oily water from their tanks in open water, far from shore.[18]

In 1954, the API did fund an effort to improve ballasting and cleaning operations, but the goal was to reduce the rate at which cargo tanks corroded, not to reduce oily discharges. Engineers had discovered a direct correlation between cleaning with high-pressure nozzles and a tank's corrosion rate. Therefore, the API's board of directors approved funding for a ten-year tanker corrosion study. In the course of this project, researchers did examine a technology that would later (in the 1970s and 1980s) play an indirect but key role in addressing the problem of oily ballast: flue gas inerting systems. These systems, installed on tankers operated by Sun Oil since the late 1930s, reduced the chance of explosion by filling empty compartments with oxygen-poor exhaust gas from the ship's flues. Researchers examining corrosion expressed interest in the ability of these systems to control oxidation.[19] Unpredictably, these systems would also eventually cut the quantity of oily wastes discharged by tankers by allowing owners to install residue-removing crude oil washing systems.

New Concerns Surface

A complication not present at the 1926 international conference on oil pollution emerged at the 1954 conference. Increasingly, oil tankers, the terminals at which they loaded and unloaded, and the oil they carried were owned by different companies. In addition, the nations under which ships were registered, the nation in which oil companies were incorporated, and the nations in which terminals were located could differ as well. Any agreement had to be acceptable to all parties involved.

In the earlier period, oil companies owned about 85 percent of the tankers that moved petroleum and petroleum products from one port to another in the United States. Independent shipping companies owned the remaining 15 percent of U.S.-registered oil tankers, chartering those tankers to oil companies for a specific time period or number of voyages. That independent companies even owned 15 percent of the U.S. tanker fleet was due to a decision by the federal government to sell government-financed ships, including tankers, constructed for World War I. To provide an incentive for U.S. companies to purchase these ships and to keep them available for national defense in the future, Congress passed the Merchant Marine Act of 1920. Also known as the Jones Act, this law required any ship carrying cargo, including petroleum, between two ports in the United States to be U.S. built, registered, and staffed. Independent companies ended up purchasing many of the tankers available.[20]

In the years after World War II, a similar opportunity to purchase government-owned tankers again presented itself. During the war, the federal government contracted with U.S. shipyards to construct almost five hundred tankers, most of them based on a standard design. These sturdy "T-2 tankers," which could carry about 16,500 tons each, were not much larger than the tankers already in service but could travel at higher speeds and had a well-designed system of pumps, pipes, valves, and controls. This time, international companies secured many of the ships.

As with the tankers built during World War I, Congress considered these ships vital to U.S. national interests and, after the war, passed legislation requiring that they be sold to U.S. companies. Although enough firms expressed interest in purchasing the T-2 tankers, many planned to use them for moving crude from the Middle East to Europe, which had become the most heavily traveled tanker route in the world. These plans to use the T-2 tankers throughout the world quickly embroiled the purchase program in international politics. After some debate, 133 of the ships ended up being sold to foreign interests.[21] In addition, two entrepreneurial Greek shippers, Aristotle Onassis and Stavros Niarchos, each obtained twenty more of the T-2s by forming companies incorporated in the state of Delaware.[22]

World War II also encouraged "flags of convenience." Before the United States officially entered the war, the U.S. Maritime Commission allowed firms to transfer U.S.-flagged vessels to Panamanian registry, which provided a convenient way around the official U.S. policy of neutrality.[23] After the war, some companies continued using flags of convenience to avoid certain taxes and rules imposed by the United States and other maritime nations. For example, after acquiring T-2 tankers, Onassis and Niarchos promptly chartered the ships to corporations in Panama, which then rechartered those ships to oil companies. Both individuals also pioneered convenient use of the Liberian flag.[24]

The increasing prominence of independent tanker owners made reaching consensus within the industry difficult. Although vertically integrated oil companies did business with independent shipping firms, the two groups were not always on the best of terms. Indeed, around the time of the 1954 conference, Onassis, in a secretive deal with the Saudi government, attempted to monopolize all oil traffic leaving Saudi ports. In response, the major oil companies tried to drive Onassis out of the tanker business by not chartering his ships. By 1956, half of the tankers in Onassis's fleet were idle.

Ironically, the availability of Onassis's fleet worked to his advantage. When hostilities—and a line of scuttled ships—closed the Suez Canal in 1956, all tankers traveling between Europe and the Middle East had to take a long detour around Africa. In the hectic spot market that followed, Onassis managed to charter all of his tankers at premium prices, making between $60 and $70 million. This windfall financed his next round of tankers.[25]

By 1959, vertically integrated oil companies owned less than half of the 493 tankers that sailed under the U.S. flag and less than 35 percent of the 2,614 tankers in the world fleet. Therefore, almost 1,700 tankers were available to be chartered for a single trip or for a specific period of time. And neither the API nor any other organization could speak for the eclectic mix of shipowners that operated the world's tanker fleet.[26] In addition, the

United States had since become a net importer of petroleum. Although most tankers carrying petroleum to the eastern seaboard still came from ports in Texas and Louisiana, an increasing number came from ports in foreign countries, mainly Venezuela. Those tankers did not have to meet the standards of U.S. registry, which allowed more independents to secure a piece of this long-distance international business. Hence, by the time new concerns emerged in the late 1950s, the tanker business had become significantly more complex than it had been in the 1920s.[27]

Furthermore, more information about tanker pollution would soon become available. By the late 1950s, many scientists studying the matter had reached a consensus: oil persisted in the ocean too long for the 1954 strategy of disposal to be a long-term solution. Indeed, nations bordering the Mediterranean Sea discovered this firsthand. The Mediterranean, wide enough for ships to legally discharge oily water in its central shipping channels, saw larger amounts of oil wash ashore each year.[28] When nations bordering the Mediterranean expressed a new interest in oil pollution, the International Maritime Organization (IMO), the UN agency established to manage maritime treaties, scheduled a conference to revisit the 1954 agreement.[29] In 1961, the United States—still not one of the signatory nations—became one so as to become a full participant in the upcoming discussions.[30]

New studies had influenced the official position of the United States. In 1957, in response to reports of oil pollution along the coast of southeastern Florida, the API Committee on Tank Vessels and the API Oil Pollution Abatement Group initiated a year-long project to investigate the problem. To head the research team, the API retained John Dennis, a former employee of both the National Park Service and the U.S. Fish and Wildlife Service. Dennis had been trained as both an ornithologist and geologist.[31]

One portion of the API research team monitored three beach sites, each one hundred feet long, on the Atlantic coast of Florida. These researchers knew that exposure to sun, sand, shells, and waves could change the physical characteristics of oil, resulting in oil washing ashore in the form of a soft, sticky tar or, with longer periods of exposure, a hard solid. Therefore, they recorded all forms of oily material they observed. In the end, they noted that some incoming oil, though not always enough to cause a nuisance, appeared on 341 of the 355 days monitored. Other members of the team surveyed beaches along the entire Atlantic Coast and found some oil at most sites they visited. In Provincetown, Massachusetts, they reported seeing "an obvious discharge from a ship close to shore" that resulted in about 33 pounds of oil being recovered from every hundred feet of beach. However, at most beaches, they found less than 6 ounces of oil per hundred feet, with beaches in New Jersey being the exception. There, the average beach contained 1.3 pounds of oil per hundred feet.[32]

Investigators could not determine the origin of most oily material they found, mainly because the oil had weathered enough to lose its original character. As possible sources, Dennis mentioned all types of oily discharges, including natural oil seeps on the ocean floor, the "careless disposal of used crankcase oil," and "oil washed from streets into sewers." He also listed tankers that had been sunk by German submarines during World War II as a possible source. After all, German submarines had sunk sixty-one fully loaded tankers in U.S. coastal waters, and gradually, as compartments corroded, those tankers released their cargo. According to Dennis, the worst such releases occurred in February 1952, when the sunken tankers *Fort Mercer* and *Pendleton* broke apart off Cape Cod, releasing over 150,000 barrels of crude in areas heavily populated by wintering seabirds. The spill resulted in a huge bird kill. Dennis emphasized the vulnerability of birds to oil, saying that "a small patch of oil about an inch in diameter" was sufficient to "break down the insulation of plumage and thereby expose the bird to cold and pneumonia."[33]

Ultimately, researchers on the API-sponsored project concluded that some oil—in whatever form—floated off shore. Whenever the weather conditions were right, that oil washed onto beaches. Furthermore, they concluded that most of this oil came from distant sources, not from discharges or spills close to shore. As evidence supporting this conclusion, the report noted that the Coast Guard recorded only two major offshore oil slicks in the twelve months during which the API team made their observations. Therefore, the final report suggested that oil being discharged in the open ocean by tankers might be reaching the nation's beaches.[34]

When the next international conference on oil pollution convened in

1962, the British received more support in their call for stricter international standards. Specifically, participants backed the British proposal that new tankers over 20,000 tons be prohibited from discharging water containing over 100 ppm of oil, regardless of the tanker's location. Theoretically, then, the next generation of tanker would discharge much less oil into the sea than existing tankers.

From the perspective of oil companies, the best strategy for complying with the 1962 amendments came from companies that practiced what came to be known as "load-on-top" procedures. That strategy required no new equipment or facilities. Instead, as empty tankers headed back to oil ports, crews pumped any oily slop they collected while cleaning tanks into an empty compartment, allowing gravity to separate as much oil and water as possible. Then, while still en route, crews pumped any clean water that settled underneath the oily slop into the ocean. Ideally, the tanker arrived at port with nothing but dewatered slop in the bottom of a single compartment. At the loading terminal, crews simply pumped the new crude right on top of the oily slop, eliminating the need to discharge any oil in the open water or to unload any wastes at a remote oil-loading port.[35]

Load-on-top procedures made the most sense for vertically integrated companies that operated tankers on regular routes. When the cargo owner differed from the tanker owner, problems arose. Many independent tankers received payment only for the amount of oil loaded onto a tanker. Any oily slop already on board before loading—that is, the slop left over from the previous voyage—represented dead space. Therefore, owners of independent tankers still had an incentive to discharge all slop before arriving in port. To many of these independent tanker owners, treaties prohibiting them from discharging oily wastes were not likely to matter. Onassis, for example, already had established a poor track record in respecting international treaties. In the late 1930s, he instructed his whaling ships to harpoon whales protected by international treaty even while other fleets looked on and registered complaints.[36] Hence, expecting such shipowners to respect a treaty prohibiting the discharge of oily water anywhere in the ocean was optimistic, at best.

Furthermore, cargo owners chartering an independent tanker did not want their crude to be contaminated with oily slop from previous trips. Hence, many cargo owners required that the tankers they chartered be cleaned before taking on crude. That requirement occasionally encouraged independent tankers to discharge oily slop close to shore. As a result, some oil-producing nations, none signatories to the international protocol, required tankers to arrive at port with tanks that already had been flushed clean, defeating the purpose of the protocol.[37]

Finally, the load-on-top strategy did not always work. Tankers switching from crude to a refined product could not follow load-on-top procedures because the slop would contaminate the refined product. In addition, on many voyages, rough weather prevented much of the oil and water from separating, especially when the specific gravity of the oil was close to that of water. In such cases, tankers had to discharge water containing a significant amount of oily emulsions. Tankers on short runs faced a similar problem.[38]

Enforcement proved difficult as well. Captains still could legally discharge water containing less than 100 ppm of oil. Given the lack of devices to monitor discharges, crews relied on visual inspection of the water to verify compliance, something that was difficult—or, depending on one's point of view, easy—to do at night. Furthermore, many governments in the best position to verify compliance, those of oil-producing nations, did not support the treaty. In addition, companies operating in those oil-producing nations typically did not add facilities to receive oily slop from tankers.[39]

In practice, then, extending the prohibition of oily discharges to the entire ocean did not have much effect on tanker practices. At best, the main strategy for satisfying that prohibition, the load-on-top strategy, represented a partial solution to a complex problem.[40] Meanwhile, two new concerns appeared on the horizon. First, shipowners began to order larger tankers. Hence, the potential damage from spills due to collisions and grounding, while not yet a major public concern, grew as well. Second, the industry made its first serious efforts to develop offshore fields. With offshore production came a potential new source of oil pollution: petroleum spewing from out-of-control offshore wells.

While the dangers of offshore drilling prompted some debate, the growth of tankers initially generated little concern. The size of an average oil tanker changed little between 1920 and 1950. Large tankers built during World War II were about the same size as large tankers built in the early 1920s. In 1921, eighty-one ships were added to the U.S. tanker fleet, with the average ship capable of carrying about 12,000 tons of petroleum (about 84,000 barrels of crude). The largest vessel, constructed for Standard Oil of New Jersey, came in at 22,000 deadweight tons (dwt), meaning that it could carry 22,000 tons of oil (154,000 barrels). Over the next sixteen years, only forty-four more tankers were added to the U.S. fleet, none capable of carrying more crude than the largest ships already in the Jersey Standard fleet.[41]

By the late 1940s, shipbuilders had the knowledge and technology to build larger ships, but shipowners faced an upper limit on the size of tankers they could operate. World-class tankers had to fit through the Suez Canal to avoid the long detour around Africa. Indeed, in 1955, even the world's largest tanker, the *Spyros Niarchos* (47,750 dwt), could still fit through the canal, though without much room to spare.[42] This limit on the size of the largest tankers soon disappeared.

When hostilities closed the Suez Canal in 1956, the need to carry oil around Africa suddenly increased the demand for tankers. Independent shipowners with tankers available to be chartered offered their services at a premium. Some of these independents, such as Onassis, invested their windfall by placing orders for tankers not limited by the size of the canal. The largest tanker ordered by Onassis was capable of carrying 100,000 tons of oil, or more than six T-2 tankers could carry. Other tanker owners soon placed orders for even larger ships.[43]

Good reasons to build larger tankers certainly existed: the cost of transporting each barrel of oil decreased with tanker size. First, one could employ roughly the same size crew regardless of tanker size. Second, the cost of designing a tanker and its auxiliary systems did not increase in proportion to size. The amount of steel required to build the hull and the cost of fuel required to operate the tanker did not increase in proportion to size either. Finally, Japanese shipyards, reestablishing themselves in the years after World War II, proved willing and capable of constructing these tankers at a reasonable cost.

However, as the size of tankers increased, their strength decreased. In addition to decreases in robustness due to increases in scale, a more theoretical approach allowed naval architects to reduce the margin of safety previously designed into the ship. One measure of robustness—the ratio of a tanker's weight when empty to the amount it could carry—dropped from .36 for a T-2 tanker to .13 for the largest tankers eventually constructed.[44] The size of cargo tanks also increased, which increased the potential for larger spills due to a single puncture.

At the same time, the view that tanker accidents and collisions could not be reduced by government-enforced regulations still held sway. People reasoned that shipowners and insurance companies already had enough incentives to prevent accidents if they could. What more could laws accomplish? By the early 1960s, though, that position was difficult to defend. Tankers capable of carrying 130,000 tons (910,000 barrels) had been launched, and even larger ships were under construction. One large tanker accident could now do significant damage to a fragile ecosystem, and nobody would be liable for those costs. Furthermore, insurance spread the costs of any damage over a large pool of shippers, encouraging the industry to reduce its risks only so far.[45]

Drilling wells in oil fields that lay offshore also complicated the issue of oil pollution in coastal waters. By the mid-1950s, oil companies had developed the technology necessary both to locate offshore fields and to drill from rigs surrounded by water. In addition, many legal issues surrounding the leasing of submerged land had been settled, allowing development to proceed in a reasonably systematic fashion.

Companies had been drilling wells in submerged land for at least a half century before the systematic development of offshore fields began in the 1950s. By 1906, for example, operators in Santa Barbara County, California, had drilled over two hundred wells from piers extending out over the water (fig. 10.1).[46] By that same year, some offshore drilling was attempted off the coast of Peru but was not widely publicized due to the Peruvian government being opposed to the effort.[47] Oil companies also gained some "offshore" experience by drilling in shallow lakes and marshes. In 1924, Standard Oil of New Jersey drilled the first of many wells in Lake Maracaibo, a body of water in Venezuela having an average depth of sev-

Figure 10.1. In Santa Barbara, producers had been drilling wells in shallow water since the turn of the century. (Library of Congress LC-US-Z62-16232)

enty-five feet.[48] Companies drilling in the lakes, bayous, and bays of Louisiana also faced many of the same difficult logistical challenges faced by crews drilling offshore.[49] Companies that gained such experience gradually ventured further out into the Gulf of Mexico. First, in 1934, the Texas Company constructed a drilling barge for use in shallow waters just off the coast of Louisiana. Then, four years later, Pure Oil and Superior Oil successfully completed several producing wells from a pier extending about a mile offshore, encouraging other operators to venture farther from the coast. However, the next wave of efforts resulted in dry holes, temporarily dampening the enthusiasm that had begun to emerge.[50]

Advances in directional drilling also improved access to underwater deposits. For example, in the mid-1930s, operators in Huntington Beach, California, started using directional drilling to reach offshore oil deposits there. They typically drilled four or five closely spaced holes from a single lease, each hole curving out to drain different portions of a reservoir three-quarters of a mile offshore.[51] This technology would eventually prove useful when drilling multiple holes from a single offshore platform.

One barrier to development revolved around the question of who owned the continental shelf. In general, as one ventures away from the shore, the floor of the ocean slopes gently down until reaching a depth of about 600 feet (100 fathoms). After that point, the floor of the ocean drops off rapidly, falling a mile or more in a relatively short distance. The width of the shelf varies widely, ranging from only a few miles wide to over one hundred miles wide along portions of the Texas and Louisiana coast. In total, the area of the continental shelf off the entire United States amounts to approximately one-tenth of the country's area, making it a valuable asset to whoever could claim ownership.

The question of ownership first emerged in the 1920s when California started leasing the submerged land off its coast to oil companies that wished to drill from piers. Texas and Louisiana began leasing some of their offshore lands in the 1930s. However, in 1937, the federal government laid claim to that land, triggering a legal dispute with coastal states that would last almost thirty years. Initially, this fight over the question of ownership did not have much of an effect on the strategy being pursued by oil companies. The pressing demands of World War II, which included shortages of material and submarine attacks on ships and tankers, temporarily discouraged oil companies from exploring most offshore areas.[52]

In the meantime, other people raised questions about the wisdom of drilling oil wells in offshore areas. One of the more significant debates occurred in connection with the development of oil fields underlying Galveston Bay. By 1936, geologists had located six "oil domes" beneath the bay. The Texas State Land Commission, with definite authority over the bay because it lay within state boundaries, then began leasing rights to drill. Some of those leases, however, lay in the path of the thirty-two-mile-long channel that connected the Houston Ship Channel to the Gulf of Mexico. All ships headed to Houston from the Gulf of Mexico had to follow this channel across the shallow bay. Once across the bay, those ships still had to travel another twenty-five miles up Buffalo Bayou, which had been dredged to make Houston accessible to oceangoing ships.

Keeping the entire ship passage open, including the portion in the bay, was serious business to the hundred or so chemical and manufacturing companies that operated plants along the ship channel.[53] Therefore, upon hearing of the plans to drill in Galveston Bay, officials with the Port of Houston expressed concern that waste oil, wild wells, and construction debris would disrupt traffic through the channel. To press their case, they contacted the State Land Commissioner, the U.S. Army Corps of Engineers, and the State Game, Fish, and Oyster Commission and argued that any well drilled in the bay threatened navigation, fishing, and recreational interests. In particular, port officials expressed concern about what would happen "should an oil well adjacent to the ship channel blow out and cover the Bay with oil and then become ignited."[54]

The Army Corps of Engineers had the power to prohibit drilling if engineers in the corps believed that it threatened navigation. However, the corps' official reply came from the division engineer stationed in San Francisco. He noted that in California, where similar protests had been raised against wells "drilled in the Pacific Ocean adjacent to the shore," no real problems had been experienced.[55] For that matter, he could have pointed to wells that earlier producers had drilled in Galveston Bay, along the shore bordering the Goose Creek field.[56] In the end, the corps decided not to prohibit drilling.

Port officials then narrowed their focus to the strip of bay surrounding the 400-foot-wide channel. They asked the U.S. Engineers Office in Galveston to prohibit any company from drilling "closer than 2,500 feet from the nearest edge of the ship channel." Port officials suggested that, where necessary, directional drilling be used to reach oil underneath the prohibited zone.[57] Eventually, they secured this restriction on drilling. In part, port officials succeeded in securing even this restriction because they could point to offshore wells that recently had blown wild. Photographs of one incident, published in *Look* magazine, showed a well in Lake Maracaibo, Venezuela, blowing out of control as its derrick collapsed.[58] In early 1941, two offshore gas wells blew out of control in the Gulf, one in Louisiana and one near Corpus Christi. Then, a few months later, drillers in Galveston Bay also struck a pocket of natural gas under high pressure, causing gas, sand, and salt water to shoot freely from the well.[59]

Figure 10.2. The first generation of offshore drilling platforms, such as this 1950 structure located off the shore of Louisiana, were fixed structures constructed in relatively shallow water. (Library of Congress LC-US-Z62-124119)

After the war, companies cautiously resumed their exploration, gradually moving further into the Gulf of Mexico. In 1947, in shallow water about twelve miles offshore, the Kerr-McGee Oil Company struck oil from a fixed platform supported by pilings driven over one hundred feet into the sea floor. Neither the depth of the water, which was less than twenty feet, nor the type of platform represented an advance in offshore drilling. However, the distance from shore—and the distance from the nearest port, which was over fifty miles away—marked a significant step toward the sea. Drilling, completing, and operating the well presented Kerr-McGee with a logistical nightmare. Engineers had to address numerous issues nobody needed to address on shore. What was the best way to get crews to and from the platform? What was the best way to get cement trucks out to the well? Where did one store the oil that the well produced and what was the best way to get that oil to shore? The company successfully addressed these logistical challenges and began lifting oil from that well on a production basis.[60] In the five-year period following the completion of the Kerr-McGee well, oil companies drilled 233 wells in shallow waters off the shore of Louisiana, resulting in the discovery of twenty-seven new fields (fig. 10.2).[61]

However, offshore development in that state soon slowed. In 1947, the Supreme Court affirmed the federal government's right to all offshore land and the federal government asserted those rights, triggering new court battles that temporarily brought development to a standstill in 1952.[62] In 1953, after numerous hearings and debates, Congress passed the Submerged Lands Act, which gave states jurisdiction over submerged lands three miles from shore (except in the Gulf of Mexico, where some boundaries extended ten miles from shore). Numerous issues, though, remained unsettled. For example, California owned a series of islands thirty to seventy miles off its coast. Who owned the portion of the continental shelf between the shore and those islands? Did the submerged land owned by California extend three miles past the islands or did the state simply own a three-mile circle of land around each island?

As such questions were being addressed, companies resumed their exploration and development in the Gulf of Mexico.[63] By 1958, companies had drilled 1,610 wells from about 200 platforms in 125 offshore fields, some in water over one hundred feet deep.[64] By the early 1960s, oil companies had developed the technology to drill in slightly deeper waters off the shore of California, bringing California's claim to that land to the fore once again. In 1965, after the U.S. Supreme Court ruled that California controlled a three-mile band around each island and no more, offshore development in that area began on a larger scale.[65]

Differences in the Perception of Risk

In general, people concerned about oil pollution caused by offshore wells approached the matter differently than people concerned with tanker discharges. With tankers, the focus was on intentional discharges.

With offshore wells, people focused on accidental releases. Indeed, when Congress passed the Outer Continental Shelf Lands Act of 1953, which gave the Department of the Interior the authority to regulate offshore production, legislators made little mention of routine discharges. Like the officials who were concerned about keeping the Houston Ship Channel clear, they focused instead on accidental releases of huge amounts of oil.

In Texas, the rules and regulations governing drilling and producing operations in coastal waters also reflected this emphasis on accidents. In 1965, those regulations specified (1) the required casing program for submerged wells, including the quality of the casing and the amount of cement used; (2) the need for "two dual-control hydraulic ram-type blowout preventers in combination with a master valve"; and (3) the need to remove all oil that leaks from a well or underwater pipeline before any repairs were begun." However, routine discharges, such as oil discharged with the waste brine, were ignored. Although the regulations specified that all operations be performed in a manner to "prevent such pollution of the waters as will destroy fish, oysters, and other marine life," the rules did not indicate what quantity of oil represented a threat to marine life.[66]

By the mid-1960s, problems associated with discharges of oily water into the sea had grown worse, not better, even when measured in per-barrel terms. As tanker traffic increased, the lack of any serious effort to address oily discharges, which previously had been disguised by releasing oily wastes further from shore, could no longer be ignored. In a tank maintained to prevent adhesion, approximately .4 percent of a tank's content remained behind after unloading. This percentage translated into about 2,800 barrels of crude in a 100,000-ton tanker, the largest at the time.[67] The potential effects of accidents involving these large tankers also presented new concerns, as did the dramatic increase in the number of offshore wells. These concerns would soon become major issues.

PART III

Regulating Industrial Activity to Maintain Environmental Quality

ELEVEN   # Crude Awakening

IN THE FIRST HALF OF THE TWENTIETH CENTURY, efforts to optimize the efficiency of industrial operations served as a general guiding ethic for many engineers and technical managers, even in their efforts to address pollution concerns. From their perspective, pollution-causing wastes from industrial facilities would cease to be a problem after they optimized the efficiency with which firms transformed resources into valuable products. Setting, or even discussing, long-term environmental objectives did not seem necessary. As for short-term concerns, damage and nuisance suits defined the limits of what was and was not acceptable.

In the petroleum industry, many efforts to increase the efficiency with which companies extracted, transported, and refined petroleum did overlap with efforts to address pollution concerns. The success of these efforts reinforced the faith that many technical experts placed in the ethic of utilitarian conservation and undermined the arguments of anybody who supported strong pollution control regulations. After all, if economic incentives associated with eliminating waste were sufficient, why add another layer of complication?

But efficiency-minded incentives turned out not to be sufficient. By the mid-1950s, despite significant reductions in the quantity of pollution-causing discharges released per barrel of oil, pollution concerns remained. Increases in the scale of operations overwhelmed whatever per-barrel reductions had been achieved. Furthermore, the remaining pollution problems were typically difficult to solve and could not be framed in terms of using resources more efficiently. In the absence of regulations, the actions necessary to reduce pollution-causing discharges represented costs that contributed nothing to the bottom line.

For example, to a refiner, preventing the small quantities of hydrocarbons escaping from thousands of valves was not worth the cost. Yet those vapors contributed to the formation of low-level ozone or smog. Similarly, oil producers using brine injection wells could not economically justify searching for and repairing poorly plugged abandoned wells that might provide a path for the injected brine to move between geological strata. Yet, if not detected, such paths could result in the long-term contamination of an aquifer. In the transportation sector, companies operating tankers were not likely to invest in strategies to reduce the risk of accidents beyond that demanded by insurance associations. Yet, a spill from one of the many large tankers being constructed in the 1960s could wreak havoc on fragile ecosystems.

Meanwhile, public expectations concerning use of the shared environment were changing.[1] By the early 1960s, most people knowledgeable about the sources of industrial pollution recognized that new regulatory mechanisms were necessary to meet rising public expectations associated with air and water quality. But what kind of regulations were needed? Who should enforce them? What should be measured and monitored? Legislation such as the Water Quality Act of 1965 and Air Quality Act of 1967 specifically attempted to answer such questions. However, neither piece of legislation successfully established either the standards or the regulatory mechanisms necessary to change industrial practices.

In the years after 1967, events unfolded in a way that few people—especially decision makers in the petroleum industry—could have predicted. A tanker spill off the coast of England, an incident involving the escape of oil from an offshore well in California's Santa Barbara Channel, and the decision to build a pipeline across the frozen tundra of Alaska soon entangled the petroleum industry in a national debate over pollution, environmental quality, and the use of resources. These incidents raised questions that the industry was not prepared to address, and led to legislation that dramatically changed the world of pollution control, not only for the petroleum industry, but for all industries.

Toward Strong Federal Pollution Control

The American Petroleum Institute (API) took seriously both the Water Quality of Act of 1965 and the debates leading up to the Air Quality Act of 1967. In response to the water quality legislation, the trade organization established a new committee, the Coordinating Committee on Air and Water Conservation, to "review major legislative issues" and "to analyze and interpret the great and growing mass of data on air and water conservation so as to enable the industry to make the best possible factual case whenever and wherever necessary."[2] The API also jointly sponsored a similar committee, the Coordinating Research Council, with the automotive industry.[3]

The Air and Water Conservation Committee served as the petroleum industry's voice in the effort to develop air and water quality standards. For example, the Water Quality Act required that each state submit water quality standards to the Federal Water Pollution Control Administration—a newly created agency in the Department of the Interior—along with a plan for achieving those standards. To help the various states prepare these standards, the API offered its technical assistance and generally found state agencies to be cooperative. In the end, the directors of the API saw the proposals set forth by the various states as setting realistic goals to be achieved in a reasonable time frame.[4]

The chairman of the Air and Water Conservation Committee expressed far more concern with ongoing congressional debates over air quality legislation, which he feared would result in a "virtual takeover by the federal government of all power to regulate air pollution."[5] The legislative roots of the proposed Air Quality Act stretched back to the early 1950s when officials in California first took steps to address the smog problem in Los Angeles County.[6] California's actions influenced a series of federal air pollution laws, starting with legislation in 1955 that awarded grants and technical assistance to states. A decade later the Clean Air Act of 1963 set in motion efforts to define air quality standards, with amendments in 1965 directing the Department of Health, Education, and Welfare (HEW) to establish national standards for automobile emissions. Encouraged by the plan to set national standards for automobile emissions, advocates of strong action then lobbied to establish national emissions standards for stationary sources such as power plants and refineries.[7]

To have technical information and alternatives ready when making their case, the directors of the API committed over $10 million to studies associated with atmospheric chemistry, automobile emissions, the health effect of various pollutants, and instrumentation for measuring emissions. They saw these studies as providing "a factual basis for debating the effects of these components with public agencies," which they ended up doing frequently. In the first nine months of 1967, representatives from the API's Air and Water Conservation Committee also came before congressional committees six times.[8]

To the directors of the API, one of the more worrisome issues concerned a recommendation by HEW that oil companies keep the sulfur content of their fuel oil below .3 percent.[9] Industry spokespeople challenged this provisions on two fronts. First, they questioned the ambient air quality standard—.1 part per million averaged over a 24-hour period— that the Public Health Service advocated as being necessary to protect the public health. Second, they argued that even if this standard could be justified, one did not need to reduce the sulfur content of fuel that low to achieve that standard. Even in New York City, which was the city for which HEW first made its calculations, they asserted that wind currents and stack heights kept ambient levels at an acceptable level.[10]

At first glance, the controversy over reducing the sulfur content of fuel oil seems surprising. After all, oil companies had a long history of manufacturing products to meet various specifications, dating back to early state-level requirements that kerosene meet specific sulfur and flash point standards. Expecting refiners to produce low-sulfur industrial fuel did not represent a dramatically new type of demand. The attention that the federal government gave the issue also seems, at first glance, out of proportion. In the mid-1960s, when Congress first charged HEW with determin-

Table 11.1. Sulfur Dioxide Emissions, 1966

SO_2 Emission Sources	SO_2 Emissions (tons)
Combustion Sources	
Utility coal	11,925,000
Non-utility coal	4,700,000
Non-utility fuel oil	4,386,000
Utility fuel oil	1,218,000
Natural gas	3,500
Total	22,232,500
Industrial Processes	
Ore smelting	3,500,000
Petroleum	1,583,000
Sulfuric acid manufacturing	550,000
Coke processing	500,000
Refuse burning	200,000
Miscellaneous	75,000
Total	6,408,000

Source: Oil and Gas Journal 68 (May 11, 1970): 64.

ing the concentration of sulfur dioxide in ambient air that represented a health hazard, most sulfur dioxide came from the burning of coal. Industrial fuel oil accounted for less than a fifth of all sulfur dioxide being released into the air (table 11.1).

The issue received as much attention as it did because many electric power plants located in the congested urban strip between Philadelphia and Boston burned residual oils, the heavy fraction left over after refiners removed the more desirable hydrocarbons. Eighty-five percent of this fuel came from the residual of Venezuela crude, which contained an average sulfur content of 2.5 percent.[11] Consolidated Edison, the utility servicing New York City, burned about 70,000 barrels of this fuel oil daily, which meant that its power plants released about 250 tons of sulfur dioxide into the air each day.[12] After health officials in New York City identified high concentrations of sulfur dioxide as a major cause of the city's air pollution problem, they sought to reduce the quantity of sulfur dioxide entering the air from plants burning industrial fuel oil.

Therefore, in January 1967, officials in New York City prohibited industrial plants from burning fuel oil having more than 2.2 percent sulfur by weight. They also announced that they planned to reduce that amount incrementally to 1 percent by May 1971. Representatives of oil companies regarded this goal "as not too unreasonable from the time viewpoint."[13] Refineries supplying power plants with fuel initially lowered the sulfur content of this fuel by blending it with low-sulfur residuals from North Africa.[14]

However, blending residuals to produce a low-sulfur product depended on steady supplies of low-sulfur crude, which could not be relied upon in the long term. Low-sulfur crude, typically defined as oil having less than .5 percent sulfur, was becoming less available and more expensive. In the long term, then, companies supplying industrial fuel to refineries on the East Coast recognized they would have to remove some of the sulfur from their heaviest cuts before selling it as industrial fuel. However, even this was not a major concern because technology for reducing the sulfur content of residuals existed.

To remove sulfur from cuts of petroleum, refiners could pass those cuts through a catalytic hydro-desulfurization process. In this process, which took place in an atmosphere of hydrogen, sulfur compounds reacted with the hydrogen to form hydrogen sulfide, a gas that could be delivered to a sulfur recovery plant.[15] Although the process could reduce the sulfur content of its feedstock to .1 percent, much less than any proposed standard, refiners could not pass their entire residual through the process. The heaviest impurities—metals—tended to remain in the residual, and these metals could deactivate the catalysts. To remove some sulfur from their residuals while preventing metals from deactivating any catalysts, several refineries first used a vacuum process to vaporize about a third of the residual. They then desulfurized that portion before mixing it back with the unprocessed residual, reducing the total sulfur content of the final blend to about 1 percent.

These refineries saw their plans to reduce the concentration of sulfur in fuel oil as an opportunity to secure business. For example, soon after officials in New York City enacted its requirements for low-sulfur fuel oil, Shell Oil announced that it was adding a hydro-desulfurization plant to its new 335,000-barrel refinery in the Caribbean and that it planned to sell one million barrels of desulfurized fuel oil to Consolidated Edison each year.[16] Cities Service announced that its refinery in Lake Charles, Louisiana, already had such a process up and running.[17]

However, plans to lower the allowable sulfur limit below 1 percent worried even these refiners. According to the designer of the process used by the Lake Charles refinery, getting the sulfur content of the entire batch of residual lower than 1 percent "would be prohibitively expensive."[18] U.S. refiners who were not major suppliers of the residual fuel also had a reason to be concerned. After all, they often reduced their residuals to coke in a process that produced additional lighter oils. They sometimes sold that coke, which could have a high sulfur content, as fuel.[19] Hence, in 1967, when the interstate agency responsible for air quality in the region encompassing New York City and parts of New Jersey adopted a recommendation by HEW to reduce the sulfur content of fuels to .3 percent, many in the petroleum industry condemned that requirement as "needlessly severe."[20] They also lobbied against suggestions by HEW that the standard be extended to other geographic areas.[21]

Debates over federal standards for vehicle emissions also concerned the directors of the API. Those debates brought tetraethyl lead, the additive refineries used to boost the octane number of gasoline, under attack on two fronts. Not only was lead a health concern in its own right, it also interfered with the catalytic pollution control devices that automobile manufacturers were experimenting with.[22] These devices reduced automobile emissions of carbon monoxide and smog-causing unburned hydrocarbons by facilitating oxidation reactions that transformed these compounds into carbon dioxide and water vapor. If forced to abandon its use of tetraethyl lead, refiners would have to install expensive equipment—four billion dollars' worth according to API estimates—to raise the octane of its gasoline in other ways.[23]

Automobile manufacturers had been familiar with the technology underlying these devices for some time. Indeed, in 1955, at the Third National Air Pollution Symposium held in Pasadena, California, a representative of the Automobile Manufacturers Association noted that "catalytic mufflers" were one of several methods being considered for reducing the quantity of unburned hydrocarbons released.[24] Research activity followed but soon tapered off.[25] In 1967, the managing editor of a leading trade journal serving the petroleum industry saw a trend "toward modifying engines rather than installing some system of air injection or catalytic muffler to control [hydrocarbon] emissions."[26]

The Coordinating Research Council, the committee jointly sponsored by the API and the Society of Automotive Engineers downplayed the health effects of lead and encouraged more research. In a press release, that committee pointed to a report showing "that human and atmospher-

ic lead levels in Los Angeles, Cincinnati, and Philadelphia had not increased in 10 years—despite the increase in automobiles in that time." In
the way of research, the API initiated a joint program with the Public
Health Service to study special groups whose work put them in close daily
contact with automobile traffic.[27]

In the end, the Air Quality Act of 1967 established a rather complicated
procedure by which the federal government set up "air quality criteria"
that states would use in establishing control and abatement programs in
interstate air quality control regions.[28] In addition to being complicated,
the law failed to place any limitations on the time it took for states to meet
the criteria set by HEW. Hence, the responsibility for enforcement fell on
the states, and few states, perhaps fearing an exodus of industry, took the
lead in enacting tough air pollution laws. Setting air quality standards
proved equally difficult. Industrial groups successfully slowed down the
process by taking advantage of a cumbersome procedure to solicit information from industry, mainly by always insisting that industrial researchers be allowed to study a problem thoroughly before responding to
requests for information.[29]

Three years later, Congress would find itself considering changes to this
piece of failed legislation. In the meantime, however, public interest increased dramatically in having the federal government control pollution,
especially oil-related pollution. Environmental groups also gained a
greater say in decision-making processes in that same period, further undermining the ability of industry-sponsored committees to dominate decision-making processes.

A New Perception of Accidents

Before 1967, lawmakers paid little attention to accidental oil spills. Indeed, in 1966, lawmakers passed an amendment to the Oil Pollution Act of
1924 that allowed prosecutors to enforce the law only when a tanker discharge was "grossly negligent." This amendment weakened the original act
and reinforced the notion that no one should be held responsible for accidents. The representative from Texas who inserted the amendment,
James C. Wright, Jr., explained that his purpose was to protect from heavy
fines "the poor little devil who might have done it [discharged oil from a

ship] accidentally."[30] This perception of accidents changed dramatically over the next three years.

On the morning of March 18, 1967, the *Torrey Canyon*—a tanker carrying Kuwaiti crude, flying the Liberian flag, owned by the Barracuda Tanker Corporation of Bermuda, chartered to the U.S.-based Union Oil Company, rechartered to British Petroleum Company, and operated by an Italian crew—approached the southern tip of England carrying 840,000 barrels of crude. No foul weather or unusual traffic threatened the tanker. The trip was routine, and the tanker was traveling at full speed, approximately fifteen knots. The captain, headed for Milford Haven on the west coast of Wales, planned to pass between the southern tip of England and a group of islands, the Scilly Isles, that lay about thirty miles offshore.[31]

As the *Torrey Canyon* approached the Scilly Isles, the tanker's heading took the ship perilously close to a rocky shoal known as the Seven Stones. The captain ordered the helmsman to alter course by a few degrees. Minutes later, with the tanker still moving at full speed, another fix on the ship's position showed that the dangerous rocks still lay directly ahead. The captain ordered a sharper change in direction. Again, the ship did not respond. Upon investigation, the captain discovered that a switch determining whether the ship was under manual control or under the control of a gyroscope had not been correctly set. The captain adjusted the controls but the change in direction came too late. As the huge tanker responded to the request, a submerged rock jutting up from the shoals caught the hull of the tanker.[32]

The *Torrey Canyon*—almost one-fifth of a mile long, carrying 120,000 tons of cargo, and moving at full speed—was a massive but fragile structure. By the time the tanker came to a stop, rocks had ripped open a line of compartments on the starboard side of the ship. Crude oil flowed into the sea. Over the next few days, as a wide ribbon of oil moved eastward toward the English Channel, a portion of that oil caught the southern tip of England and flowed northward along the coast of Cornwall. The remainder of the slick continued eastward, eventually splitting into two, with one ribbon flowing south toward the Bay of Biscay and the other north into the English Channel. Both branches of the slick then converged on the coast of Brittany, France.

British authorities assumed that the British Navy could handle such a spill. Seven years earlier, a small spill of about 250 tons—one-fiftieth of the *Torrey Canyon* spill—had set in motion a series of studies aimed at determining how best to contain oil spills. The researchers concluded that detergents, which had been used to disperse the 250-ton spill, represented the best alternative. The detergents broke the oil into small globules, which then emulsified and sank. Subsequent studies conducted at Swansea University determined that the combination of detergent and oil had killed about 30 percent of the marine life in the vicinity of the spill, but the recommendation to use detergents in the case of an emergency held.[33]

Therefore, when the navy commander responsible for containing the spill arrived on deck of the *Torrey Canyon,* barely two hours after the ship came to a halt, he already had given orders for tugs loaded with detergents to move toward the spill. Marine biologists argued that the detergent was being sprayed on a scale never envisioned by its designers, and they feared that the chemical would kill oyster and mussel beds in the area without stopping the bulk of the slick from reaching shore. Indeed, ships spraying detergent—eventually over 10,000 tons—confronted the slick, but the sheer quantity of oil overwhelmed them. The stench of crude oil soon wafted over the resort towns of southern England, with the oil itself soon following.[34]

Meanwhile, when tugs attempted to refloat the damaged tanker, the hull pivoted, broke in two, and released more oil. To reduce the amount of oil reaching shore, naval authorities then ordered eight British jets to drop forty-two high-explosive bombs on the ship, which set the remaining oil on fire. In total, the *Torrey Canyon* released 840,000 barrels of crude, with 140,000 barrels flowing toward the Cornish coast, 210,000 barrels entering the English channel, 350,000 barrels flowing toward the Bay of Biscay, and 140,000 barrels being consumed by fire. In addition, biologists later showed that the chemical-based carrier for the dispersant contributed significantly to the toxicity of the spill.[35]

Most people evaluated the immediate response to the spill as a complete failure, demonstrating the inability of national governments to protect their coastlines from the damaging effects of large oil spills. The question of liability also emerged as a central issue. At the time, the law of the sea limited the liability of shipowners to the value of their ship and its cargo. In the case of the *Torrey Canyon*, that liability translated into the value of a few lifeboats. Business owners in resort towns, people in the fishing industry, and governments that financed the clean-up bore the economic brunt of the spill.

The *Torrey Canyon* spill also made news in the United States. Among other things, the event raised the question of what would happen if such a spill occurred off the shore of North America. An incident in which oil from an unknown source fouled thirty miles of shore on Cape Cod in Massachusetts further increased interest in the issue. President Lyndon Johnson then ordered the secretary of the interior and the secretary of transportation to "determine how best to mobilize the resources of the Federal Government" to prevent similar catastrophes from occurring in the United States.[36]

At a press briefing announcing the initiative, a reporter asked whether anybody had determined the source of "this thing on Cape Cod." Stewart L. Udall, the secretary of the interior, indicated that some people believed the spill to have been caused by "old World War tankers" breaking up. Udall, though, did not see the source of the oil as relevant. The plan, he emphasized, was to recognize that accidents such as the *Torrey Canyon* disaster were going to happen and to put materials and resources in place "to do something about it."[37]

Officials with the Department of Transportation and the Department of the Interior were charged with putting the necessary contingency plan in place. They soon discovered that tankers twice the size of the *Torrey Canyon* were already on order. Given the potential damage a spill from one of those tankers could cause, they emphasized the need for preventative action as well as contingency plans. The cost of preventative measures, they noted, "must be weighed against the costs, both tangible and intangible, that arise from disastrous spills." But exactly how to measure those intangible costs and benefits was not clear.[38]

The report that the secretary of the interior and the secretary of transportation eventually submitted to the president contained much of the same information about oily discharges to the sea as previous reports prepared by the U.S. Bureau of Mines, the U.S. Army Corps of Engineers, and the API. New information and new concerns included the results of an investigation in which the navy inspected four sunken World War II tankers (three were empty, one had some oil left); concern about the transportation of synthetic organic chemicals produced by twelve thousand chemical companies, much of which had unknown effects on aquatic and human life; the dramatic increase in tanker size and traffic in the last decade; and the drilling of six thousand wells in the Gulf of Mexico since 1960.[39]

The report also contained a list of thirty-four action items, most not particularly controversial. Among other things, the report recommended performing "a feasibility study of shore guidance systems;" developing "more effective on-board systems for oil-water separation;" and designing "equipment for improved shore protection." One of few controversial recommendations suggested that the Oil Pollution Act of 1924 be amended to include accidental discharges of oil. The investigators identified the need for such a change because they did not believe that companies were using existing technology as effectively as possible. Authors of the report also raised, but did not answer, questions having to do with who should be held responsible for accidents, how long that party would be given to remove the oil, and to what extent that party could be held liable for damages caused by such accidents.

More accidents followed. In the year after the *Torrey Canyon* spill, seven more tankers lost about 380,000 barrels of oil in groundings and collisions, including an incident in the waters off Puerto Rico.[40] Although the total amount of those seven spills represented half that released by the *Torrey Canyon,* that amount kept attention on the contingency plan then under development. In September 1968, Johnson announced its completion: a "National Multi-Agency Oil and Hazardous Material Pollution Contingency Plan" specified how the federal government would mobilize needed equipment and manpower and have them on the scene of an emergency in the least possible time. It was also to serve as a model for the development of state, municipal, and industrial plans.[41]

The API responded to the call for contingency plans by encouraging oil companies to organize local oil spill cooperative programs.[42] For example, the Massachusetts Petroleum Council, consisting of representatives from fourteen companies that operated oil terminals in the Boston area, jointly purchased several motorboats, 12,500 feet of boom, and dispersants and stored these supplies at several sites in the area. Oil terminal operators in Portland, Oregon, incorporated a "third-party service organization" to manage their response to a major spill. In Pensacola, Florida, the operators of oil terminals arranged to have the local fire department serve in that same capacity. Within two years, oil companies had organized twenty-three such local cooperatives in port cities throughout the United States; another twenty-seven were in the planning stages.[43] Until they were put to the test, though, nobody could say for sure whether these efforts would be effective in controlling large spills.

As it turned out, officials did not have to wait long for a test. A drilling incident off the shore of Santa Barbara soon sent oil flowing toward local beaches. Over the previous decade, offshore drilling had evolved into a major component of the petroleum industry; companies had drilled approximately 8,000 holes into the U.S. continental shelf, some up to 20,000 feet deep. The tentative structures and secondhand barges from which crews drilled offshore wells had evolved into large, highly engineered platforms. Indeed, each platform served as the base for numerous wells and represented a major capital investment.[44] Oil production in the Santa Barbara area was nothing new either. Sailors in the nineteenth century noticed that oil occasionally seeped up from the ocean floor in the area, an observation that encouraged early oil producers to drill wells along the shore and from piers extending into the Santa Barbara Channel.[45] By the mid-1960s, however, oil companies were ready and able to drill in deeper waters from platforms.

Therefore, in 1968, when the federal government began leasing blocks of submerged land in the Santa Barbara Channel, they received bids totaling $603 million for 363,000 acres, including a $64.1-million bid from Union Oil, Gulf Oil, Texaco, and Mobil for a jointly leased block of 5,200 acres. That September, in 180 feet of water, Union Oil crews tipped into place a drilling platform owned jointly by those four oil companies.[46] Residents of Santa Barbara had succeeded in getting state and federal officials to prohibit drilling any closer than five miles from shore, so the consortium placed the platform, capable of supporting approximately sixty wells, just over five miles from shore.[47]

Union Oil already had considerable experience with offshore wells. The company also had some experience with wells blowing out of control, with the most memorable incident being the "Wild Tiger of the Gulf," a well six miles off the coast of Louisiana that blew in 1957. In that case, unexpectedly high pressure encountered in a deep formation caused a flange in the blowout preventer to leak. A mixture of gas and sand soon chewed away the metal, and the escaping gas eventually cratered the area around the well and toppled the platform. To prevent the potentially explosive gas from hindering containment efforts, navy aviators flew over, released a tank of gasoline, and ignited the mixture with tracer bullets. Over the next five months, with the gas safely consumed in an island of flame, crews drilled another well one-third of a mile away. They made a hydraulic connection with the first hole two miles below the surface and ended the blowout by flooding the hole with water, mud, and cement.[48]

Twelve years after the Wild Tiger, Union Oil crews began drilling from the Santa Barbara platform. Within three months, Union Oil crews had drilled four wells, each slanting away to a different portion of the submerged lease. The borehole associated with each well first passed through a shallow, low-pressure reservoir holding a "brownish" oil. The holes then went down another 3,000 feet, eventually reaching a high-pressure reservoir that held a "green" oil. In these four wells, a string of casing isolated the two reservoirs from one another.[49]

The drilling crew at Santa Barbara encountered its first problem on January 28, 1969, just after drilling a fifth hole to its target depth of 3,500 feet. In that hole, still uncased below 240 feet, crews relied on tons of drilling mud to prevent the flow of fluid between the two reservoirs. As crew members began removing the drill stem, gas started blowing out of the top of the stem. The crew then dropped the pipestring back down the hole and triggered the blind rams of the blowout preventer, which sealed off the well and eliminated the threat of a blowout. However, soon after they closed the blowout preventer, a brownish oil started appearing several hundred feet from the platform.[50]

Apparently, high pressure from the "green" reservoir, hydraulically communicated to the shallow "brown" reservoir, forced "brown" oil to seep out of an old fault. An estimated 500 to 5,000 barrels of oil leaked through the fissure each day.[51] Looking back at the events that were about to unfold, one investigator concluded that, even "without the excuse of hindsight," Union Oil should have cased off the hole immediately after passing through the first reservoir so as to prevent the pressure of the lower reservoir from "feeding a known oil seep" in the higher reservoir.[52] Over the next ten days, crews injected mud and cement into the uncased hole so as to sever the connection between the two oil reservoirs. In the process, they filled the hole with cement from top to bottom, leaving 2,500 feet of drill pipe and collars frozen in place. A smaller amount of crude, between 20 and 200 barrels per day, continued to seep from the ocean floor for some time afterward, presumably until the pressure of the shallow "brown" reservoir dropped to its previous level.

In the well-publicized cleanup effort that followed, it became clear that neither officials nor oil companies were prepared for a spill even of this limited size. The total amount of oil released into the Santa Barbara channel represented about 5 percent of the amount spilled by the *Torrey Canyon*. That quantity of oil made it a serious spill to be sure, but nowhere near the worst-case disaster that local and national contingency teams supposedly were prepared to address with a few oil booms and some chemicals. Furthermore, the weather cooperated with the cleanup effort. Although debris from a recent storm became oil-fouled and added to the visual impact of the spill, the sand on the beach, being water-wet, repelled more of the oil than it would have otherwise (fig. 11.1). Hence, the task of absorbing and removing that oil was less complicated than it could have been.[53]

The cleanup effort also served as fertile ground for testing ways to remove oil from beaches. The Coast Guard, which took the lead in coordinating the cleanup, limited the use of dispersants so as to reduce the toxic effects of those chemicals on aquatic life. However, booms, the next line of defense in most contingency plans, proved to be of limited value in the choppy water. Much of the oil reached shore. Crews then attempted to remove the oil using various absorbent materials, including flour and ground styrofoam. In the end, the Coast Guard found hay to be the most effective absorbent. Volunteers eventually spread and collected over 170,000 bales of hay.[54]

Figure 11.1. The spill resulting from the blowout at Santa Barbara demonstrated the ineffectiveness of existing contingency plans. (Smithsonian 99-2235)

Efforts to clean dying birds also became a focus of media attention. However, of the sixteen hundred birds going through the labor-intensive cleaning process, approximately thirteen hundred eventually succumbed to the oil. Also capturing significant attention were spokespeople for a group called GOO. Standing for "Get Out Oil," GOO collected 60,000 signatures in a petition to ban further oil development in the area. Later in the year, the group also led efforts to block a new drilling platform from being set into place, which they did by placing boats where the platform was to be anchored. Although the group failed to stop drilling operations, its efforts attracted national attention and increased public awareness of oil spills.[55]

The Santa Barbara incident also reinforced the need for more research on the legal and technological means for preventing and containing major oil spills. One response came later in the year when the API and the Federal Water Pollution Control Administration jointly sponsored a conference to assess and discuss the technology available for preventing and controlling oil spills. In his opening remarks, Kerryn King, the chairman of the API's Committee on Air and Water Conservation, asserted that the "cost of environmental conservation will be high" and that the industry "will be required to allocate an ever-increasing share of our economic resources to it." Such remarks demonstrated a recognition that maintaining a specific level of environmental quality had a certain cost associated with it, suggesting that participants no longer assumed that economic incentives to increase efficiency and reduce waste were sufficient incentives for industry in addressing pollution concerns. However, the notion of "conserving the environment" implied that they still framed the entire issue as an optimization problem that technical experts could solve on their own terms.[56]

L. P. Haxby, the chairman of the newly formed API Subcommittee on Oil Spills Cleanup, also spoke, announcing that the API had appropriated $1.2 million toward the "development of new methods of cleanup." He noted that "in an age where we can go to the moon, we should be able to do better" than spread "straw as an absorbent and pick it up with rakes." As a first step toward this goal, the API contracted with the Battelle Institute to evaluate all available "sorbent materials, sinking agents, combustion aids, dispersants, and biological agents." In another API-funded project, researchers with the consulting firm Arthur D. Little studied factors that might affect the decision to use dispersants, such as the location of recreation and fishing areas, water conditions, local flora and fauna, and legal restraints.[57]

The next speaker at the conference, a representative from the Federal Water Pollution Control Administration, Alleb Cywin, announced a $10-million package of government-financed research efforts. These efforts included seventeen separate research projects on preventing releases of oil, eleven associated with detecting spills, fourteen on the control of spills, eight associated with studying the effects of a spill, and three on the restoration of beaches. Each project was being performed by a different firm or university. Over half of the federal funds went to four projects: $1.7 million to Amoco for studying the chemical and biological treatment of refinery effluents; $1.5 million to the National Oil Recovery Corporation for developing technology to remove oil-water emulsions from industrial wastes; $1.5 million to the Armco Steel Corporation for developing technology associated with re-refining used crankcase oil; and $737,000 to the city of Buffalo, New York, for developing technology to "detect, sample, contain, and collect oil" that entered the nation's rivers.[58]

Officials at the Federal Water Pollution Control Administration expressed hope that such projects would help develop a pool of experts trained in pollution control. In an earlier report, they noted that research grants, in addition to supporting specific projects, had the "long-term effect of increasing the supply of highly trained scientists and engineers" interested in a particular line of research.[59] Indeed, the $11.2-million commitment of private and public funds for water pollution research, and other funding efforts like it, represented a new level of funding for research projects associated with studying water pollution and the effects of oil pollution.

The Santa Barbara spill also placed the issue of liability back in the spotlight. One month after the spill, the House Committee on Merchant Marine and Fisheries met to discuss amending the Oil Pollution Act of 1924 to require those who caused oil spills to clean up the mess. The proposed amendments also gave the federal government authority to take over a cleanup effort if the responsible party did not take sufficient action. The responsible party would still be liable for the cost of the cleanup efforts.[60] In the bill that eventually became law as the Water Quality Improvement Act of 1970, legislators based the liability provisions on an insurance plan that tanker owners voluntarily put into place immediately after the *Torrey Canyon* disaster. Known as TOVALOP for "Tanker Owners Voluntary Agreement Concerning Liability for Oil Pollution," the plan's participants agreed to fund any cleanup effort associated with a tanker spill for an amount up to $10 million per vessel or $100 per ton of vessel capacity, whatever was less. The new legislation included provisions that made the owners of vessels strictly liable for cleanup costs up to $14 million or $100 per gross ton of the vessel, whichever was less. The liability of onshore facilities was limited to $8 million.[61]

In the debates preceding passage of the bill, much of the discussion revolved around the liability of onshore facilities. How high should it be? If too low, the threat of liability might not be sufficient to encourage any real change in industrial practices. If too high, insurance might not be affordable. In either case, what costs should responsible parties be liable for? The API's vice president for environmental affairs, Peter N. Gammelgard, argued that the liability cap for waterfront facilities (initially $10 million) was "unrealistically high" and would prevent smaller firms from obtaining the liability insurance necessary for them to stay in business. Gammelgard suggested that Congress did not wish "to penalize anyone simply because his activities may entail a risk of spilling oil" and argued that companies be liable for cleanup only if their actions were "willful or negligent."[62]

However, many in Congress did wish to penalize firms that operated poorly designed facilities and, therefore, made the owners of all facilities strictly liable for any oil spills that occurred. By "strictly" liable, Congress meant that companies were still liable even if the spill was accidental and involved no negligence or willful intent. Still, ambiguities and loopholes remained. For example, the final legislation did not hold firms responsible for spills that occurred due to an act of God, an act of war, or the negligence of a third party. Hence, if a 300-ton vessel rammed into a 300,000-ton tanker, causing both vessels to spill cargo, the larger tanker would not be held liable for any cleanup costs while the smaller vessel's liability would be capped at $30,000.[63]

Despite such loopholes, the Water Quality Improvement Act of 1970—with its condition of strict liability, a $35-million revolving fund for cleanup efforts, and its application to onshore facilities—represented a stricter piece of legislation than the Oil Pollution Act of 1924. Indeed, it served as the model for later legislation passed to clean up chemical spills and abandoned waste sites, the legislation that eventually came to be known as Superfund.[64] The Water Quality Improvement Act also gave the Coast Guard a larger role in preventing and containing oil spills, including a charge to undertake research toward those ends. The act also required anybody responsible for a spill to notify the Coast Guard, or an agency that would do so, or face a $10,000 fine. This last provision resulted in the Coast Guard accumulating data associated with oil spills that otherwise would have been lost.

The Trans-Alaska Pipeline and Environmental Assessments

Matters would soon become even more complicated for petroleum companies. In January 1970, President Richard Nixon signed the National Environmental Policy Act (NEPA); its stated purpose was to encourage "productive and enjoyable harmony between man and his environment; to promote efforts that will prevent or eliminate damage to the environment or biosphere and stimulate the health and welfare of man; [and] to enrich the understanding of the ecological system and the natural resources important to the Nation. . . ."[65]

Not expected to be much more than a symbolic piece of legislation, NEPA had consequences that, according to one scholar, "outran the wildest imagination of its sponsors and most enthusiastic supporters."[66] The unexpected consequences sprang from a provision that required federal agencies to assess the impacts of federally funded projects that could affect the environment in a significant way. For example, the framers of NEPA expected the U.S. Army Corps of Engineers to assess the environmental impacts of any new dam or large construction project they planned, demonstrating that they were aware of how a particular project might affect the environment and that they had considered alternative approaches.

Few people expected environmental assessments to do more than raise questions that might not otherwise have been asked. As it turned out, the provision allowed groups normally outside the decision-making process to enter that process by raising questions that the authors of an environmental impact statement had failed to consider. Indeed, soon after passage of NEPA, groups outside the normal circle of industrial committees and government agencies began using its assessment provision as a wedge into the decision-making processes of numerous projects. One of these was the planning of the Trans-Alaska pipeline.

The territory of Alaska became a state in 1959, one year after oil companies developed Alaska's first commercial oil field. However, that first field, located on the Kenai Peninsula southwest of Anchorage, was relatively small.[67] Petroleum geologists felt confident that much larger oil fields could be found on the state's remote north coast. Nineteenth-century explorers to northern Alaska had noticed oil floating in the Arctic Ocean, which indicated that an oil deposit could be found in the area. Subsequent geological expeditions reinforced that hope. Indeed, in 1923, the federal government set aside twenty million acres on the North Slope as a naval petroleum reserve.[68] In the 1930s, several companies drilled exploratory wells in various parts of the territory, but World War II placed most of this exploration on hold. Companies resumed their search in the 1950s, leading to the discovery of oil in the Kenai Peninsula. The wild scramble for leases that followed caused groups promoting the preservation of the wilderness to oppose further oil development. However, the search for new fields in Alaska continued, reaching the Arctic by the mid-1960s.[69]

Any oil discovered north of the Brooks Range, an east-west string of mountains about 150 miles above the Arctic Circle, would create a major transportation problem. All roads, including the Alaska Railroad, ended in Fairbanks, which lay about 250 miles south of the Arctic Circle. Transporting crude from any field discovered on the North Slope would require a major investment. Still, a huge discovery might make that investment worthwhile, and two companies, Sinclair and British Petroleum, took the lead by drilling six exploratory wells near Point Barrow, the northernmost point in Alaska. None of the six holes they drilled tapped into an oil formation.

Another team consisting of the Atlantic Richfield Company and Humble Oil (a subsidiary of Exxon) then drilled an exploratory hole in the same area, but that effort also failed to locate any oil. In 1967, these crews moved two hundred miles to the east and drilled another hole near Prudhoe Bay. This time, the hole showed signs of oil. Another hole, drilled a year later and seven miles away, verified the existence of a reservoir containing at least ten billion barrels of oil, the elephant all had hoped to find.[70]

The question of how to transport Prudhoe Bay crude, which could come out of the ground at temperatures over 150 degrees Fahrenheit, now had to be answered. Various alternatives were discussed, such as an all-tanker route through the Arctic Ocean and an all-pipeline route across Canada, but companies owning the main block of leases in Prudhoe Bay decided on a pipeline to the southern coast of Alaska.[71] Therefore, on February 10, 1969, Atlantic Richfield, Humble, and British Petroleum publicly announced plans to build a forty-eight-inch-diameter line running eight hundred miles from Prudhoe Bay to Valdez. With 500,000 tons of the pipe already on order, the entity responsible for the project—the Trans-Alaska Pipeline System, known as TAPS—estimated construction costs to be $900 million and planned to start pumping crude in 1972.[72]

The announcement to build the pipeline came at a bad time: the Santa Barbara oil spill was still making national headlines. Hence, tough questions by groups concerned about how the pipeline might affect Alaska reached a sympathetic national audience. What unique features would be required in a pipeline moving hot crude across frozen land? How would oil companies clean up any oil spill that occurred? How would the project affect the fishing port of Valdez?[73]

Anybody familiar with the region knew that constructing a pipeline from Prudhoe Bay to Valdez would not be easy. The proposed route crossed long stretches of permafrost—particles of soil and rock suspended in ice. In some areas, this permafrost had no real strength when melted and was known to be over 1,000 feet deep. In many places, the top few feet thawed each year, but a fragile vegetative crust held the surface together. However, if anything destroyed that vegetation, the spring thaw turned the top few feet into a gooey mess that left long-lasting scars on the land.[74] The potential for a pipeline carrying hot oil to cause havoc with such land was obvious. Potential damage to the permafrost was not merely a theoretical concern. In 1944, drivers in the first convoy of trucks to use a 400-mile highway from Anchorage to Fairbanks learned about permafrost firsthand. One hundred miles out of Fairbanks, the entire convoy came to a halt when the summer sun turned a section of the road into muck.[75] Given that each mile of a forty-eight-inch pipeline could hold 50,000 barrels of crude, a pipeline break in this region could be a major ecological disaster.

As it turned out, engineers with TAPS were given longer to think about the design of the pipeline than they expected. The federal government controlled 97 percent of the state's 365 million acres, and under the conditions of statehood, Alaska had the right to select over 100 million acres of federal land not already claimed by the Native Alaskans.[76] That selection process, which had not seemed too urgent in the years before the Prudhoe

Bay strike, suddenly became a highly charged affair. Hence, the outgoing secretary of the interior, Stewart Udall, placed a freeze on all land transfers. This freeze created a quandary for TAPS. The Mineral Leasing Act of 1920 granted companies a twenty-five-foot-wide right-of-way on either side of any pipeline they constructed through federal land, but TAPS required more. The land freeze prevented TAPS from easily securing any extra land. Native Alaskan villages also sued to prevent TAPS from using land they desired, which complicated matters further.[77]

Most supporters of the pipeline assumed that the situation would be resolved without delay because the new secretary of the interior, Walter Hickel, was the former governor of Alaska and an outspoken advocate of the pipeline.[78] However, tremendous public interest in the pipeline placed all decisions affecting the project under intense scrutiny. Initially, officials tried to deflect questions associated with potential spills by pointing to the United States National Multiagency Oil and Hazardous Materials Pollution Contingency Plan.[79] The Santa Barbara experience, however, undermined any hope that the plan might be effective in the best of circumstances, much less in remote areas of Alaska.

At a public hearing six months after the initial announcement, state officials expressed irritation with the speed at which federal agencies were resolving the land-related issues and pushed to get the project moving. At that meeting, the governor of Alaska, Keith Miller, asserted that his state had the authority, personnel, knowledge, and management base with which to oversee construction of the highway and pipeline to the North Slope. Officials with the Federal Water Pollution Control Administration disagreed and proposed surveying the stream crossings associated with the proposed pipeline.[80]

The survey turned out to be an expensive proposition, with travel expenses alone hovering around $75,000, not including transportation provided by the army. However, the survey proved useful. Investigators identified a half dozen changes that could be made to protect pristine rivers. For example, along one stretch, they suggested shifting the pipeline to the other side of the highway so as to create a buffer between the pipeline and the river. Along another stretch, they suggested rerouting the pipeline out of the river valley altogether.[81]

Unfortunately for TAPS, investigators also had the opportunity to examine the results of an oil spill that occurred in Prudhoe Bay when a 100,000-gallon bladder tank burst. Approximately a third of the oil reached the Arctic Ocean.[82] Two months later, researchers with the U.S. Geological Survey expressed serious concerns about what would happen if TAPS buried a pipeline carrying hot oil in the unconsolidated ground of the tundra. Interior Secretary Hickel rushed to assure TAPS that "the line will indeed be built," and he praised the industry's cooperation in addressing concerns raised by the various agencies under his control.[83]

However, serious scrutiny of the pipeline had only just begun. Two months later, preservationist and environmental groups—including the Wilderness Society, Friends of the Earth, and the Environmental Defense Fund—sued to delay the pipeline project, pointing to provisions in NEPA that required an environmental assessment for federal actions that could affect the environment in a significant way. They asserted that allowing TAPS to use more federal land than allowed by the Mineral Leasing Act represented such an action. Hence, they argued that the Department of the Interior should comply with NEPA by assessing the pipeline's impact on the environment, analyze alternatives, and submit the assessment for review. The courts agreed and judged the eight-page assessment that the Department of the Interior had already prepared to be insufficient.[84]

In addition to the new legal obstacles, TAPS experienced some internal conflict. A major partner in TAPS, Exxon, appeared to be in no hurry to proceed as it already had easy access to cheaper supplies of crude from Saudi Arabia and Iran. Hence, when Exxon financed an exploratory effort to send a tanker equipped with an ice-breaking prow to Prudhoe Bay and back, which it did in the summer of 1970, some people, including the head of British Petroleum, saw the move as a waste of time. A month later, TAPS dissolved and was replaced by a new company, the Alyeska Pipeline Service Company, in which four additional oil companies also held an interest. Meanwhile, the pipeline remained on hold.[85]

In January 1971, the Interior Department finally released the preliminary draft of its environmental assessment, but critics still judged it as being sorely deficient. Even the Army Corps of Engineers and the Department of Defense, both supporters of the pipeline, were highly critical of the document. Among other things, critics charged that the assessment (1) understated the impact of the project on Valdez, the port at which tankers would load the pipeline oil; (2) failed to demonstrate sufficient knowledge of how the pipeline would affect the permafrost; (3) overestimated the ease with which oil could be removed from the tundra if a spill occurred; (4) did not take migratory patterns of wildlife into full consideration; and (5) did not consider alternatives seriously. Reviewers for the recently established Environmental Protection Agency even asked to examine details that had not yet been determined, such as the protocol for operating shutoff valves. Public hearings on the environmental assessment eventually resulted in 12,000 pages of testimony and evidence.[86]

As the Department of the Interior, aided by Alyeska, revised the environmental impact statement, the pipeline waited, with Alyeska spending millions of dollars just to keep stacks of pipe from rusting and trucks in running order. The Interior Department's next draft, a nine-volume document released in March 1972, discussed, according to one reviewer, "just about every issue raised by environmentalists—from the migratory habits of caribou to the dangers of oil spills." Meanwhile, the Alaska Native Claims Settlement Act removed one of the obstacles to construction. This act provided Native Alaskans with $462 million up front and a 2 percent royalty on all oil produced until an additional $500 million had been paid.[87]

In response to the latest environmental assessment, a group representing fifteen U.S. and Canadian fishing associations charged the Interior Department with glossing over "the difficulties associated with maneuvering in Valdez Arm and Valdez Narrows despite the perilous conditions of hurricane force winds, drift ice . . . and some of the worst fog conditions encountered by mariners anywhere." In addition, the group charged that Alyeska would try to keep the tanker traffic moving even in unsafe weather because if "the line ever stopped pumping hot oil and cooled off, restarting it would be a major problem."[88]

Others came out in favor of an all-pipeline route through Canada, alongside a natural gas pipeline to Chicago that many critics perceived to be inevitable. Although a Canadian pipeline would have been longer, the oil and gas would flow directly from Alaska to the existing Canadian-U.S.

pipeline network with no need for a tanker run. Proponents of the Canadian alternative argued that the environmental assessment actually served as a persuasive argument for their position. Oil companies, critics explained, did not want to be locked into the U.S. market and preferred the option requiring tankers, which gave them easy access to a potentially lucrative Japanese market.[89]

In late 1972, the U.S. District Court of Appeals in Washington, D.C., upheld a ruling that the Department of the Interior had satisfied the requirements of NEPA but ruled that Congress also had to amend the Mineral Leasing Act to allow for a broader right-of-way. The issue then moved from the courts to Congress. Although the vote was close, with the Senate requiring Vice President Spiro Agnew to break the tie, the tide quickly turned in favor of the pipeline. Crude oil prices rose in the summer of 1973, mainly due to actions by the Organization of Petroleum Exporting Countries (OPEC), and this suddenly made the pipeline more attractive than ever. The Yom Kippur War and the subsequent Arab-led embargo of oil sales to the United States provided additional incentives, and President Richard Nixon signed the necessary legislation on November 16, 1973, two weeks after the start of the embargo. On January 23, 1974, the Interior Department issued permits allowing construction to begin on a pipeline system capable of transporting two million barrels of crude per day. By that time, estimates of construction costs had risen to $5.9 billion, about six times the initial estimates.[90]

Given that the original construction plan treated the Trans-Alaska pipeline as little different from a line crossing Oklahoma or Texas, the NEPA-inspired delay proved valuable. Much of the increase in construction costs came from elevating 400 miles of pipeline in areas containing unstable permafrost. Each of the 78,000 vertical support members required to accomplish this task included a refrigerant-filled tube capable of withdrawing heat from the ground. No energy was needed. Heat from the soil caused the refrigerant at the bottom of the support to vaporize and rise up to the top. There, a fin radiated heat, causing the gas to condense into a fluid and drop back down to the bottom of the support. Although the heat transfer mechanism stopped when the temperature of the air rose too high, the device removed enough heat to keep a block of ground around the support member permanently frozen.[91]

In the event that caribou would be afraid to walk under the pipe, which could upset migration patterns, short portions of the elevated line were also buried and refrigerated.[92] Other design changes included insulation sleeves for the pipeline that kept the oil warm enough to pump for twenty-one days after a shutdown. Alyeska also placed ninety-seven electronically operated gate valves, each weighing thirty-five tons, in strategic locations near streams and populated areas. At eighty other points along the line, they also placed check valves that prevented the oil from draining backward into potential breaks, which limited the maximum spill from a single break anywhere in the line to 64,000 barrels. In addition, pipeline designers included a system of cathodic protection to keep the buried portion of the line from corroding. The construction plan also called for inspectors to check every weld on the line with X-rays. Plans for the marine terminal at Valdez were modified to include a facility for treating oily ballast from tankers, a system for recovering vapors from tankers, and a sewer system for capturing drips and accidental spills.[93]

The first crude oil entered the pipeline on June 20, 1977, and reached Valdez forty days later. By that time, actual construction costs reached $8 billion, roughly eight times the original estimate. Within three years, operators in the Prudhoe Bay field had extracted over one billion barrels of oil and were making plans for a large-scale seawater injection project. In those three years, the pipeline experienced one incident of sabotage (16,000 barrels lost) and two leaks due to pipe settling (5,500 barrels).[94]

The tanker route from the port of Valdez to ports in California and Washington was almost as controversial as the pipeline itself, with the main fear being that one of many tankers required to move the oil would eventually be involved in a major accident. Continued increases in the size of tankers only increased that fear. By 1970, the world's fleet of oil tankers contained six ships capable of carrying 300,000 tons of crude, and several capable of carrying over 400,000 tons were on order. New ways to refer to each class of ship also came into use; typical descriptions are shown in table 11.2. Potential spills from such ships could be huge.[95]

Limits on the size of tankers did exist. In the early 1970s, the economics of building tankers capable of transporting more than 400,000 tons of crude were unknown, but estimates suggested that construction costs per ton of capacity leveled off around 500,000 dwt. For example, while con-

Table 11.2. Typical Designations for Various Size Tankers

Description	Size (dead weight tons)
General purpose tankers	25,000 or less
Super tankers	25,000 to 150,000
Very large crude carriers (VLCCs)	150,000 to 300,000
Ultralarge crude carriers (ULCCs)	300,000 and greater

Source: *Elements of Oil-Tanker Transportation* (Tulsa, Okla.: PennWell Books, 1982), 8.

struction costs dropped from $409 per ton for a 50,000 dwt tanker to $199 per ton for a 300,000 dwt tanker, those costs dropped only to $178 per ton for a 400,000 dwt tanker. Operating costs reinforced that pattern. Moving crude from the Middle East to the United States on a 50,000 dwt tanker cost $13 per ton. With a 250,000 dwt tanker, that cost dropped to $5.70 per ton. However, doubling the size of the tanker to 500,000 dwt only dropped the cost of moving crude to $5.15 per ton.[96] Given the added risk of operating a 500,000 dwt tanker, economics suggested that tankers any larger were not likely to be constructed.

In the United States, another important factor limited the size of tankers: the depth of the water at U.S. ports. Tankers capable of carrying 200,000 tons of crude required approximately seventy feet of water. Although fifty ports capable of receiving such ships were scattered throughout the world, no port in the United States had channels deep enough to accommodate such tankers. Tankers capable of carrying 100,000 tons required about fifty feet of water and could dock at several ports in the United States, including: Machiasport, Maine; Long Beach, California; San Pedro, California; Seattle, Washington; and Valdez, Alaska. All other ports in the United States were too shallow for tankers that large. For the run from Valdez to terminals in Puget Sound and around Los Angeles, companies planned to use tankers no larger than 150,000 dwt, which meant that preparing those ports required little dredging. San Francisco, a shallower port, would receive only smaller tankers.[97]

Some oil companies were also interested in using ultralarge crude carriers (ULCCs) and very large crude carriers (VLCCs) to carry crude from foreign ports to the U.S. East Coast. However, by the early 1970s, plans for creating deepwater ports in Maine, Delaware, Maryland, and Florida had been proposed and rejected, in part, on environmental grounds. The plan for the port in Maine—initiated in 1968 by Occidental Petroleum, a company flush with Liberian oil—involved taking advantage of the state's deepwater port by building both a marine terminal and a 300,000-barrel-per-day refinery. Opposition to the plan coalesced around environmental concerns. At a hearing in 1970, Max Blumer of the Oceanographic Institute in Woods Hole, Massachusetts, asserted that the presence of a 300,000-barrel-per-day refinery was incompatible with maintaining an ecologically sound bay. According to his estimates, even a well-run marine terminal would spill .01 per cent of all oil handled or, in this case, about 30 barrels a day. This quantity, he argued, would affect marine life in the area.[98]

Blumer came to that conclusion based on a study he performed after an oil barge spilled 700 tons of fuel oil on the doorstep of the Woods Hole Oceanographic Institute. Following that incident, he studied of the spill's effect on marine life, which showed that the oil had more of an effect on marine organisms than scientists previously thought. He also suggested that the cumulative effects of small but routine spills could be more destructive than one large spill.[99] Although Blumer's estimates and conclusions were challenged, he made a point that people opposed to the project took to heart. Even in the absence of a disaster, some amount of oil would steadily find its way into the water.

In Delaware, a proposal to construct a deep channel was also defeated, with the governor of the state eventually stating, "We in Delaware made a choice, and that was to use our valuable bay and coastal areas for recreation and tourism instead of for another purpose for which they are admirably suited, industrial growth."[100] However, by that time, the urgency of constructing a deepwater port on the East Coast had dissipated. First, a port in New Brunswick, Canada, capable of receiving tankers up to 350,000 dwt was already in operation and officials there planned to construct a set of berths especially designed to transfer crude from VLCCs and ULCCs to tankers small enough to dock at refineries on the East Coast of the United States.[101] Second, the practice of "lightering," which involved the direct transfer of crude from VLCCs and ULCCs to smaller tankers capable of reaching existing marine terminals, made the deepwater ports less critical. Lightened enough, the VLCCs themselves could even reach some ports to unload the rest of their crude.[102]

Another innovation involved moving "ports" to deep water. In Texas, a consortium of oil companies backed a plan to construct an unloading station about thirty miles off the coast of Galveston, which was the closest a 500,000 dwt tanker could get to Galveston without dredging. Even if the consortium decided to dredge, doing so would have been difficult because of the numerous pipelines buried on the sea floor to support offshore operations. No map of those pipelines existed.[103] Instead, the consortium planned to moor tankers to a fixed structure in deep water. Pipelines would carry the crude to refineries onshore.[104]

Another consortium of ten major oil and pipeline companies desired to build a similar deepwater port off the Louisiana coast. With almost five hundred tankers between 200,000 to 400,000 tons in service or on order, port officials in Texas and Louisiana saw the so-called superports as critical. By 1973, the general design of the Texas and Louisiana projects— known, respectively, as Seadock and LOOP (Louisiana Offshore Oil Port) —had been completed. Backers of both projects were ready to apply for permits from federal agencies, a move that would officially trigger environmental assessments of the projects. Aware of how the Alaska pipeline had been delayed for five years and how other deepwater projects had fizzled out in the face of opposition that coalesced around environmental issues, both groups proceeded cautiously.[105] Before releasing any engineering or construction contracts, they pushed for legislation that would allow licensing from a single agency. In 1974, Congress passed a bill making the U.S. Department of Transportation the responsible agency.[106]

Ultimately, energy conservation measures taken during the 1970s kept crude imports from rising, dissipating the urgency associated with increasing the capacity of existing receiving facilities. Eventually, Mobil Oil, a major backer of Seadock, indicated that provisions associated with the Department of Transportation's license were too strict and backed out of the project. Exxon soon followed, and the project, which lingered on paper for another several years, collapsed by the end of the decade. The Louisiana offshore port was eventually completed at a cost of $700 million and started operations in 1982. Initially, it failed to attract the traffic for which it was designed and began receiving crude from smaller tankers, many of which would have otherwise traveled up the Mississippi River to reach refineries there. By the late 1980s, LOOP was operating at full capacity and the practice of lightering in the Gulf of Mexico dwindled.[107]

Environmental Quality as a Political Choice

By the 1970s, the task of extracting, transporting, and refining petroleum had become significantly more complicated than it had been a decade earlier. Engineers and technical managers in the petroleum industry, not accustomed to receiving any real opposition to large-scale construction projects, were disconcerted by challenges to projects such as the Trans-Alaska pipeline and the construction of deepwater ports. Robert O. Anderson, the chief executive of Atlantic Richfield and generally sympathetic to the views of groups seeking to protect the Alaskan wilderness from damage, argued that many who opposed projects such as the Trans-Alaska pipeline were anti-industrialists more intent on reducing the nation's consumption of energy than in actually protecting the environment, and little could be done to satisfy them.[108]

However, many citizens who lobbied for environmental regulations saw reducing the nation's high energy demand and issues of environmental quality as related; various aspects of this perspective were captured in contemporary works such as *The Population Bomb* (Paul Ehrlich, 1968); *The Closing Circle* (Barry Commoner, 1971); *The Limits to Growth* (Donella Meadows et al., 1972); and *Small is Beautiful* (E. F. Schumacher, 1973).[109] In general, these works presented unregulated growth, including industrial growth, as a problem. The various authors focused on different issues, but all agreed on one point: choices had to be made in how people used the shared environment. One could not assume that market forces and economic incentives to use resources as efficiently as possible would somehow produce the best outcome for society in general. Even economists who still put their faith in the market's invisible hand recognized that the market was imperfect and that the costs of many transactions, such as the costs of pollution, were paid by people not involved in the transactions. At the very least, societies had to account for those market imperfections.[110]

The first pictures of Earth from the Moon, broadcast all over the world in the late 1960s, reinforced the notion that limits existed and that industrial societies could not simply proceed full steam ahead without considering the consequences of their development. Reducing pollution and setting environmental standards, more people realized, was not merely a technical task to be performed by experts. That task also involved political choices. Basic questions, though, remained, and addressing those questions would require intense debate.

Redefining Efficiency

WHEN THE *Torrey Canyon* ran aground in 1967 and focused public attention on the problem of oil spills, efforts to establish air and water quality standards in the United States were underway but floundering. In the next five years, dramatic changes in the world of pollution control would occur. Strong federal pollution control legislation related to air and water quality would lay the foundation for a new regulatory system in which industrial facilities would have to monitor and manage their effluents and emissions to an extent never before necessary.

Initially, it appeared as if Congress would simply make fine adjustments to the Water Quality Act of 1965 and the Air Quality Act of 1967, two relatively weak pieces of legislation. However, in 1970, with both political parties seeking to demonstrate their interest in environmental issues, the regulatory flavor of proposed legislation escalated from draft to draft. The final series of federal actions proved far tougher than anybody expected.[1] First, in 1970, the Nixon administration consolidated the pollution control activities of fifteen different agencies into a single organization, the U.S. Environmental Protection Agency (EPA). Congress then empowered this agency by passing two strong pieces of pollution control legislation: the Clean Air Act Amendments of 1970 and the Federal Water Pollution Control Amendments of 1972.[2] As the EPA transformed these laws into enforceable rules and regulations, the economics, as well as the language and rhetoric, of pollution control were gradually transformed.

The full transition to this new regulatory regime promised to be a slow one. The law allowed private groups to sue the EPA for failing to carry out its mission, ensuring that the rule-making process would be deliberate, tedious, and open to intense public scrutiny.[3] Subsequent legislation also gave the EPA authority over other pollution-causing activity, complicating its task even more.[4]

The petroleum industry, partly by choice and partly by chance, emerged as a significant actor in the construction of this new regulatory system. Controversies over automobile emissions, the *Torrey Canyon* spill, the Santa Barbara blowout, and the Trans-Alaska pipeline, all of which received substantial coverage in the popular press, resulted in the petroleum industry being tightly linked to pollution concerns. In addition, as energy producers, oil companies also found themselves challenged on two other fronts: by consumers worried about the rising price of fuel, and by environmental activists who pitted energy consumption and industrial growth against efforts to maintain environmental standards. Finally, on a more practical level, oil companies had to comply with the new pollution control regulations. These regulations would force petroleum companies to integrate environmental objectives into what it meant to operate an "efficient" industrial facility.

Regulatory Change

Initially, many oil companies responded to the growing interest in environmental protection by creating or renaming positions and departments to have the word "environmental" in their titles, such as: the "Environmental Conservation Department" (Shell, 1969); "Coordinator of Environmental Conservation" (Exxon, 1970); "Coordinator, Environmental Conservation" (Sun, 1970); "Director of Environmental Control" (Phillips, 1970); "Department of Environmental Affairs" (Texaco, 1971); "Manager of Environmental Protection" (Atlantic Richfield, 1971); and "Environmental Sciences Department" (Union Oil, 1972). The budget, responsibility, and authority of these new departments and positions varied widely. Some were refinery-level departments or positions; others reported directly to the chief operating officer or to a member of his or her staff. Some involved assigning a new title to a previous position or department. Others represented newly created positions.[5]

None of these titles, of course, reflected an immediate change in the culture of any organization. Those refineries with a history of taking pollution concerns seriously continued to do so. Those with a history of sloppy housekeeping and poor control over their effluents and emissions, often because local regulations were weak or poorly enforced or both, usually continued operating as before.[6] Still, such changes represented one step, however slight, toward establishing a culture in which regulations established to meet environmental objectives were seen as legitimate.

Public relations departments also sought to highlight the "environmental" efforts of their companies. For example, in 1971, when the Petroleum Engineering Publishing Company sponsored twelve awards for "engineering innovation in the field of environmental control," numerous oil companies submitted nominations. However, many projects, including some winners, reflected utilitarian values more consistent with simply using resources more efficiently than with defining and meeting higher standards of environmental quality. (See table 12.1.) Some companies explicitly stressed that their investment in "environmental" projects represented good business practices and were not the result of regulatory pressure— reflecting, to some extent, a continuing belief that such regulations were not necessary.[7]

However, evidence that regulations did make a difference existed. A study of refineries performed by the Council on Economic Priorities indicated that the most important factor in determining the degree to which a refinery managed its emissions and effluents was a local agency with a history of enforcing local pollution control laws. In areas where pollution control laws were enforced, refineries managed their effluents and emissions more carefully than refineries in other areas. The study also found that refineries owned by the same company, but located in areas with different histories of enforcement, varied more in their control of emissions

Table 12.1. Meritorious Awards Program for Environmental Innovation in the Field of Environmental Control, 1971, Petroleum Engineer Publishing Co.

Organization	Category	Project
Mobil Oil	Oil spills (Production sector, land or offshore)	Reduced oil spillage on a set of leases by reducing the number of oil treatment sites from 21 to 1. Was part of a project to automate operation of the lease.
Sun Oil	Soil and water conservation (Production sector)	Laid ninety miles of pipe to obtain fresh water for several waterflood projects, allowing the company to obtain approximately 35 million barrels of oil from mature fields.
Amoco	Oil spills (Transportation sector)	Developed a "brush belt" oil skimmer used in the cleanup of oil spills.
Colonial Pipeline	Soil and water conservation (Transportation sector)	Installed equipment at one of its tank farms to recover oil from water drawn from the bottom of storage tanks.
Gulf Oil	Emissions control (Processing sector)	Developed a hydro-desulfurization process capable of reducing the sulfur content of residual fuel from 4% to 1%.
Phillips Petroleum	Resource utilization (Refining sector)	Added equipment for treating and burning an "off-gas" formerly vented to the atmosphere.
Amoco	Soil and water conservation (Refining sector)	Investigated how to improve the effluent coming from aerated lagoons.
Amoco	Water disposal (Industry wide)	Installed an incinerator to burn spent caustic and oily sludge at one of its refineries.
City of Long Beach, California	Waste utilization (Industry wide)	Began injecting .5 million barrels of brine per day into wells. The brine previously had been discharged into the ocean.
Cities Service	Wildlife conservation (Industry wide)	Allowed public access to 30,000 acres of its land in North Louisiana for hunting, fishing, and other recreational uses.
Shell Oil	Environmental research	Implemented an extensive study of marine oil spills, which include research involving oil spill removal and fingerprinting oils for identifying the source of spills.
Northern Illinois Gas Co.	General environmental improvement	Implemented broad programs to improve aesthetics, including planting grass after repairing pipe, reducing noise at regulator stations, and landscaping right-of-ways.

Source: Compiled from a summary of winning projects, Petroleum Engineering's meritorious awards program, Environment, box 11, Cities Service Oil Company Collection, Western History Collection, University of Oklahoma.

and effluents than refineries owned by different companies in the same area.[8]

For example, the study identified Los Angeles County as having strictly enforced air emission standards. All refineries operating there, regardless of the company, met tough standards. However, the same study identified Los Angeles as having the weakest enforcement of wastewater effluent standards and found that refineries located there—again, regardless of the oil company—did relatively little to treat their waste streams. Just the reverse was true for the Shell and Amoco refineries along the Mississippi in Wood River, Illinois, where local agencies had a history of paying significant attention to water quality but little to air quality. Hence, according to these studies, enforcement mattered more than either corporate culture or the culture of a particular facility.[9]

An important change in the regulation of effluent from industrial facilities such as refineries came in 1970 when the Army Corps of Engineers began requiring all facilities to acquire permits for discharging wastes into the nation's waters. The corps required no new legislation to take on this program because the courts had recently reinterpreted the word "refuse" in the Rivers and Harbors Act of 1899 to include all foreign substances. Two years later, the Federal Water Pollution Control Amendments of 1972 transferred responsibility for the permit program to the recently created EPA.[10]

The members of Congress expressed ambitious goals in this latest piece of water quality legislation: they called for the elimination of pollutants into public bodies of water by 1985. This, however, was just a goal. In terms of actual requirements, legislators mandated that by July 1, 1977, all industrial plants had to reduce their level of pollution-causing discharges to that commensurate with the best practical pollution control technology (BPT). In addition, by July 1, 1983, industrial plants had to reduce their discharges of pollutants to levels associated with the best available control technology (BAT). However, they left the details of making this happen to the EPA and its process for reviewing, creating, and enforcing rules.[11]

In hindsight, Congress gave the EPA a nearly impossible task. Before setting quality standards for effluents released into public waters, the agency first had to identify the "best practical" control technology available and then specify the quality of effluents that could be achieved with that technology. In doing so, analysts with the EPA were expected to take into account the total cost of the pollution control technology in relation to the "effluent reduction benefits," the "age of equipment and facilities involved," and the "processes employed."[12] Hence, to establish meaningful effluent guidelines meant that someone had to examine different categories of industrial plants separately. This task fell to the Effluent Guidelines Division of the EPA, which by 1978 published nearly 1,700 regulations covering over 250 types of industrial processes.[13]

Guidelines for refineries proved difficult to set. The approximately 250 refineries then in operation ranged in size from small specialty plants capable of refining no more than 500 barrels per day to sprawling complexes that refined more than 400,000 barrels of crude each day. In addition, given the possible combination of processes a refinery could use, arriving at meaningful effluent standards for all plants within a single category was no easy task. Furthermore, the EPA discovered that the data it initially planned to use to set standards was inadequate. This data included an EPA-API waste load survey performed in 1972, information from discharge permit applications that refineries submitted, and data collected by monitoring effluents at selected refineries. Although this set of data provided information about the concentration of pollutants in the final discharge from refineries, it did not fully document the type of processes each refinery used and the quantity of water each refinery discharged.[14]

To complete its task of developing effluent limits, the EPA hired consulting firms to collect and analyze more data. Although API technical committees participated in this effort, funding provided by the EPA established sources of expertise outside the API. For example, to more closely correlate effluent levels, waste treatment processes, and production processes, the EPA contracted with the Roy F. Weston Company. That firm, founded by an engineer who previously had been in charge of waste control at a large urban refinery, possessed levels of expertise that matched major oil companies.[15]

The standards that the EPA eventually specified as being obtainable with the best practical technology were based on the performance of refineries that segregated their different waste streams, diverted storm water into holding ponds so as not to overwhelm treatment equipment, maintained two levels of equipment for removing solids and oil, employed filters to remove the most difficult emulsions, and used some method of biological treatment. In addition, the EPA specified its effluent limits for refineries as a fraction of throughput. Therefore, no refinery could meet these standards simply by diluting its waste stream with enough water to lower the concentration of pollutants.[16]

For example, before 1972, the Exxon refinery in Bayway, New Jersey, kept its concentration of waste oil low by diluting the 15 million gallons of process water it generated each day with 186 million gallons of cooling water. Therefore, even though the refinery released an average of 5,000 pounds of oil each day, the concentration of waste oil in its effluent averaged around 3 ppm.[17] By July 1, 1977, however, refineries were expected to discharge less than 6 pounds of oil for every thousand barrels of crude processed. For the Bayway refinery, this meant reducing its oily discharges to approximately 1,000 pounds per day.[18]

The EPA also specified limits for nine other measures of effluent quality, including biochemical oxygen demand, chemical oxygen demand, suspended solids, ammonia, pH, phenols, sulfide, total chromium, and hexavalent chromium. In addition, the effluent limits that refineries were expected to meet by July 1, 1983—those associated with the "best available" technology—assumed that, in addition to the previous waste treatment processes, refineries also used a final stage of treatment with activated carbon. Achieving that level of performance would require most refineries to invest significant capital in their system for collecting and treating wastewater.[19] For the Bayway refinery, the proposed BAT standard meant reducing its oily discharges to approximately 150 pounds per day.[20] To protect specific bodies of water, state-level agencies could also set ambient water quality standards, which could result in even stricter standards for facilities seeking a permit to discharge into those bodies of water.

The simple requirement that industrial facilities apply for discharge permits proved useful to regulators. To complete the application, facilities had to determine what they were releasing as effluent and officially record their results. Such documentation represented an important step toward viewing the disposal of waste in effluents as a transaction that could be monitored.[21] Actual changes in practice, though, occurred slowly. No refiner saw any advantage in being the first to comply with the new standards. If anything, the opposite held. One can certainly imagine the hesitancy with which refiners approached decisions to invest millions of dollars in developing technology that would only increase expectations as to what was practical. This uncertainty also discouraged firms from altering processes or installing new technology ahead of schedule.[22]

As the deadline for compliance approached, the API challenged some of the EPA's standards in court. However, the National Commission on Water Quality, a federally sponsored committee established to analyze the technological, economic, social, and institutional impacts of the Federal Water Pollution Control Amendments of 1972, noted that two refineries with a combined crude capacity of 290,000 barrels per day (representing 2 percent of U.S. refining capacity) already employed technology equivalent to the "best practical" pollution control technology. In addition, seventy-three refineries (representing 54 percent of U.S. capacity) already employed some form of biological treatment, a major component of what the EPA considered to be the "best practical" technology. The commission also concluded that although day-to-day variability in the operation of a refinery would result in some short-term violations of EPA effluent limits, refineries could be expected to satisfy the long-term averages. However, the commission determined that petroleum refineries would have considerable difficulty complying with the 1983 limits. Given this analysis, the courts upheld the limits targeted for July 1, 1977, but ordered the EPA to reconsider the limits targeted for July 1, 1983.[23]

One week before the 1977 deadline, the program director for the National Commission on Water Quality, Joseph G. Moore, Jr., spoke at a forum on the management of refinery wastewater sponsored by the EPA and the API. The title of his talk got right to the point: "What does July 1, 1977, mean now?" What, Moore wondered, would happen when companies not in compliance attempted to obtain or renew discharge permits? Moore suggested that companies not in compliance call their "lawyer's attention to the fact that they should be prepared to respond to litigation shortly after July first." Moore also reminded his audience that even if the EPA decided not to take action against violators, private citizens could institute action both against the violator and against the EPA for failing to execute the provisions of the law.[24]

Indeed, such a lawsuit against the EPA had already complicated the task of setting effluent guidelines. A year earlier, the Natural Resources Defense Council successfully sued to require that the EPA consider sixty-five toxic chemicals when setting effluent limitations for the "best available" technology.[25] For each chemical or class of chemicals having its own effluent limit, the EPA now had to define a testing procedure. Doing so was a difficult task because effluents often contained numerous molecules of a similar chemical structure, some capable of interfering with the detection of others.[26]

Not all refineries required permits. Thirty-four refineries (representing 11 percent of U.S. refining capacity) discharged their effluent into publicly owned waste treatment facilities and, hence, discharged no effluent into any public body of water.[27] However, the central waste treatment facilities into which they discharged effluent did have to obtain a permit. In turn, the managers of those treatment centers set their own constraints on the industrial plants they serviced. One such centralized treatment center, the Joint Water Pollution Control Plant operated by the Los Angeles County Sanitation District, received wastewater from seventeen refineries, one of the largest clusters in the United States.[28]

The permit system, therefore, had an indirect effect even on refineries that discharged to public treatment facilities. For example, after being required to obtain a federal discharge permit in 1974, officials with the Los Angeles County Sanitation District began setting and enforcing stricter standards. Otherwise, the county would have been unable to meet the conditions of its permit. In addition, the county started charging a fee for its services, which amounted to about $250 per million gallons. Previously, industrial plants in the county had been charged a flat property tax for sewer service, which did not cover the full cost of running the treatment plant.[29]

This increased level of control can be seen in the district's rule forbidding plants from discharging storm runoff into the sewer leading to the Joint Water Pollution Control plant. The storm runoff, though infrequent, exceeded the capacity of the centralized plant and overwhelmed its ability to treat wastewater. When one refinery attempted to secure a waiver because its treated effluent ran through the same sewer system as its storm water, the county refused to grant the waiver. In a compromise, county and refinery engineers devised a plan in which the refinery, during storms, automatically diverted its discharge to an experimental activated carbon treatment plant located on the refinery's grounds. Effluent from the activated carbon treatment plant ran directly into a channel leading to the Los Angeles harbor. The plant consisted of a holding reservoir, twelve cells containing 49,000 pounds of activated carbon, and equipment for regenerating the carbon. Because of the expense of regenerating the activated carbon, the refinery could not afford to operate the plant continuously. Still, the experimental plant could be started up on short notice so as to treat bursts of storm water.[30]

In general, then, establishing the technology-based effluent standards mandated by Congress involved a decade of research, discussion, and debate. In the process, standards were attacked from both sides, both for being too lenient and too strict. Amendments to water quality legislation in 1977 acknowledged the difficulties agencies and firms were having in meeting deadlines and goals, and refined the original legislation by setting new deadlines and technological requirements. However, by the mid-1980s, an effective permitting system was in place, with state agencies and the EPA capable of setting and enforcing effluent standards at discharge points. Although the effort to improve the nation's surface water quality fell short of the ambitious goals set by Congress—the complete elimination of pollution-causing discharges into public bodies of water by 1984—the National Pollutant Discharge Elimination System represented an important step toward framing the discharge of effluents as a transaction that could be monitored and managed so as to meet socially constructed measures of environmental quality.

To improve the quality of the nation's air, legislators followed a different strategy than they did in efforts to control water pollution. They charged the EPA with establishing ambient air quality standards. States then had to come up with a way to meet and maintain those standards and to seek EPA's approval for their plans. Although this was the same general approach used by the weak Air Quality Act of 1967, the new legislation gave the EPA the authority to take action if deadlines were not met. In addition, the act singled out several areas subject to special federal authority: automobile emissions, fuel and fuel additives, and performance standards associated with new stationary sources.[31]

This legislation affected the petroleum industry as much as any. The six pollutants for which ambient air quality standards were created—particulates, lead, sulfur dioxide, carbon monoxide, nitrogen oxides, and ozone—were all associated with petroleum products in one way or another.[32] Fuel oil, of course, contained significant amounts of sulfur and nitrogen, all of which ended up as sulfur dioxide and nitrogen oxides when the fuel was burned. Automobiles released emissions high in carbon monoxide and unburned hydrocarbons. Although hydrocarbons did not end up being listed as a criteria pollutant, they were recognized as a precursor to the formation of low-level ozone, a major component of smog. Finally, refiners used tetraethyl lead to boost the octane rating of gasoline, and vehicles that burned leaded gasoline represented the major source of all airborne lead.

The issue of lead in gasoline was also tied to automobile emissions in another way. A provision in the Clean Air Act Amendments of 1970 required a 90 percent reduction in emissions of hydrocarbons and carbon monoxide from automobiles. To meet that requirement, General Motors announced that it would install catalytic converters on its vehicles starting in 1975, a deadline specified in the act.[33] That decision doomed the market for tetraethyl lead. Any lead in the exhaust would render these pollution control devices ineffective.[34]

Ten years earlier, few people would have predicted that General Motors would adopt this technology. For thirty-five years, General Motors and Standard Oil of New Jersey—the world's largest manufacturer of automobiles and the world's largest refiner of gasoline—jointly owned the major manufacturer of tetraethyl lead, the Ethyl Corporation. However, in 1962, these two companies sold the Ethyl Corporation to the Albemarle Paper Manufacturing Company, freeing themselves to develop the sort of catalytic pollution control device that the experts in the automobile industry had been discussing for over a decade.[35] However, another five years passed and no serious plan to use catalytic devices emerged.

In a remarkable use of antitrust law, the U.S. Department of Justice then sued the four largest U.S. automobile makers with conspiring to impede the development of pollution control devices and for having done so since 1953. The suit alleged that the companies agreed to refrain from competition in their own development of pollution control equipment and agreed not to buy patents from outsiders who made advances in pollution control devices. Three years later, in 1970, after the four companies settled with the Justice Department out of court, General Motors made the announcement that it planned to meet emission standards through the use of catalytic converters.[36] This decision set in motion a program to phase out leaded gasoline and resulted in a regulatory innovation that would serve as a model in emissions trading programs.

From the perspective of refineries, the phase-out of leaded gasoline was not a pollution control issue as much as a change in product specification.[37] As automobile manufacturers placed more cars with catalytic converters on the road, refiners had to produce more gasoline that did not need the octane-boosting power of tetraethyl lead. Otherwise, engines burning unleaded gasoline would knock and ping badly, behavior previously suppressed by the tetraethyl lead. To reduce their dependency on tetraethyl lead, refiners needed to install expensive new process equipment: catalytic reformers that converted (re-formed) low-octane, straight gasoline molecules into high-octane, cyclic ones.

Without a strategy to manage the transition, the EPA could only hope that the separate decisions of numerous refiners would result in a supply of unleaded gasoline necessary to meet demand. Firms that designed and constructed refinery equipment, such as the Fluor Corporation, took out full-page advertisements in trade journals, encouraging oil companies to plan ahead. In one advertisement, the firm noted that although talking about the subject of lead reduction "tends to make us look as though we were rubbing our hands gleefully . . . we think the conversion program will hit our industry like a tidal wave."[38]

Some refineries moved quickly in installing the necessary equipment.[39] In 1971, even before the EPA officially set July 1974 as the date by which all major service stations had to start offering unleaded gasoline, the Cities Service refinery in Lake Charles, Louisiana, announced construction plans for a 30,000-barrel-per-day catalytic reformer. Similar announcements from several other oil companies followed.[40] Judging from the speed at which these oil companies moved to increase their capacity for producing unleaded gasoline, they saw being the suppliers of lead-free gasoline as a marketing opportunity.

To ensure that the supply of unleaded gasoline would keep pace with the number of new cars on the road, the EPA also required refiners to phase out their use of tetraethyl lead, with the target for 1978 being .8 grams per gallon (averaged over leaded and unleaded gasolines produced by the refiner), dropping to .5 grams per gallon by the end of 1979. In managing this reduction, though, the EPA faced two dilemmas. The economies of scale associated with catalytic reformers made producing unleaded gasoline more expensive for smaller refiners, resulting in more of a hardship for them to meet the targets. As a result, the EPA gave refineries under a certain size favorable treatment. Second, forcing the remaining refineries to meet the same limits and deadlines was not particularly efficient. Refiners that had already installed reformers could easily meet their targets. Refiners without reformers faced challenges, and they often sought suspensions, which were granted if the refiner showed good faith toward meeting its goals.[41]

In 1982, with favorable treatment to small refineries scheduled to be eliminated in October, few small refineries had invested in the reformers necessary to produce higher-octane gasoline.[42] Expecting those refineries to construct reformers by October simply was not realistic. Yet extending the favorable treatment only rewarded them for not taking action. To manage the phase-out in a way that recognized the distribution of available equipment, yet penalized and rewarded the appropriate parties, the EPA allowed inter-refinery averaging. For the purpose of meeting the lead limits imposed by the EPA, a refinery not capable of meeting its targets could average its production with a refinery that exceeded its targets. As long as the combined average met the target, the EPA considered both refineries in compliance.[43]

In essence, then, a market for lead credits had been created. A refinery that exceeded its target could now sell its credit to any other refinery. With this system of trading in place, the EPA could steadily cut the lead limit without worrying about whether all refineries could meet the limit. Refiners who could exceed their targets efficiently did so, and sold their credits to refiners who found it costly to install the necessary equipment. In this program, significant auditing and enforcement were possible. The EPA checked refiners' reports of tetraethyl lead use against the records of tetraethyl lead manufacturers, cross-referenced reports of trading transactions, and occasionally conducted site audits.[44]

Some system of trading also looked promising for dealing with an even more contentious problem: allowing new sources of emissions in areas not in attainment for a criteria pollutant. Clearly, in areas where the concentration of a criteria pollutant exceeded the ambient air quality standard, pollution control agencies could not allow any net increases in emissions of that pollutant. After all, if an area did not meet the ambient air quality standards for sulfur dioxide, for example, allowing a facility to increase its emissions of sulfur dioxide only made the matter worse. But what should be done if a firm desired to expand its facilities or wanted to construct new facilities in the area? Simply deciding that industry could not expand or move into a nonattainment area was not a politically acceptable alternative. Neither was incrementally reducing the emissions of all other facilities so as to compensate for the new emissions. Even if a mechanism for making such reductions existed, state-level pollution control agencies could not reallocate emissions throughout the region in response to each request for new emissions.

The EPA eventually backed a strategy codified in amendments to the Clean Air Act in 1977. Industrial expansion in a nonattainment area was allowed when the added pollution load was offset by reductions elsewhere in the region.[45] This use of offsets provided the flexibility necessary to allow new sources of emissions. In practice, most firms that installed new sources of emissions secured their offsets by replacing older boilers or shortening the production schedules of older, less efficient processes. In effect, facilities were allowed to offset their own increased emissions with reductions elsewhere.[46]

Firms desiring to construct new facilities in a nonattainment area faced a more difficult challenge. They had to secure offsets—technically, emissions reduction credits—from existing facilities. For example, when the

Hampton Roads Energy Company decided to build a refinery in Hampton Roads, Virginia, the president of the company, Jack Evans, was informed that he would have to secure offsets for emissions of hydrocarbons, a precursor to the criteria pollutant ozone. The planned site of the new refinery was in an area out of attainment for ozone, mainly because Virginia's Dismal Swamp, located ten miles from the site, released over 100 tons of hydrocarbons per day. Officials with the EPA suggested that Evans secure the necessary offsets by paying the owner of a competing refinery to reduce its emissions of hydrocarbons. To Evans, the notion of cleaning up a competitor's refinery seemed absurd, especially since his company's refinery would be cleaner than the other.[47]

The policy of requiring offsets was not absurd, but it did set a precedent. It suggested that existing facilities had a license to emit pollutants and that firms could trade a portion of their license as if it were a commodity. The policy also raised numerous questions as to the exact rules for generating and trading offsets. For example, to save money, could a facility expected to cut emissions from one piece of equipment instead make the necessary cuts at another piece of equipment? In effect, the facility would be putting an imaginary bubble over its plant and treating the facility as one emissions source. Or, could a refinery operating in Southern California generate the emissions reduction credits it needed to offset an increase by putting a service vehicle on the highway to help reduce traffic jams? After all, fewer traffic jams would translate into fewer hydrocarbon emissions.[48] Or, if a firm shut down a facility, should it get credit for the resulting cut in emissions and be able to trade or use that credit sometime in the future? Many of the questions that arose also applied to areas meeting all air quality standards. After all, a provision of the Clean Air Act of 1970 barred the significant degradation of air quality in areas with clean air, which groups such as the Sierra Club sued to have the EPA uphold.[49] In addition, certain requirements, such as new source standards, applied in all areas.

By the mid-1980s, courts had answered many of the trading-related questions that arose, and in 1986 the EPA released an Emissions Trading Policy Statement. The agency encouraged states to use emissions trading as a way to manage the reductions needed to bring polluted areas into compliance. A number of urban areas were still out of attainment for at least one criteria pollutant, and making additional cuts was difficult. Using so-called command and control to specify the reductions that each facility had to make was guaranteed to frustrate all participants. Numerous problems with trading programs also existed. For example, keeping track of what a specific facility was allowed to emit promised to be a nightmare. The permitting system for air emissions had been pieced together, not designed to accommodate trades.[50] At the same time, the fact that such programs were even conceivable reflected the changing culture of pollution control. Increasingly, discharges of contaminants were becoming transactions to be measured and monitored.

Emissions trading was also being discussed as a way to address another pollution concern that came to the fore in the 1980s: acid rain. Sulfur dioxide released by power plants in one airshed could raise the pH of precipitation in distant airsheds. The Reagan administration had resisted action for almost a decade, but as scientists learned more about the dynamics of acid rain, environmental groups pushed for a national strategy aimed at reducing emissions of sulfur dioxide.[51]

General efforts to reduce sulfur dioxide emissions dated back to the late 1960s, with the Department of Health, Education, and Welfare (HEW) recommending that the sulfur content of heavy industrial fuels be limited to .3 percent. The EPA, after its formation in 1970, continued to push that recommendation. However, refineries (and the API) steadfastly opposed any specification that required residual fuel oil to have a sulfur content less than 1 percent.[52] The main problem for refiners was that their residual oil could not be priced much higher than low-sulfur coal. Imposing too low a sulfur limit on residual oil threatened to price it out of the market. Hence, for over a decade, representatives for the API attacked the sulfur specifications on numerous fronts. In addition to questioning the ambient air quality standard that justified the sulfur limit, they also argued that one could burn fuel with more sulfur than .3 percent and still meet that standard. They also argued that the EPA, in recommending a cap on the sulfur content of fuel, violated the intent of the Clean Air Act by explicitly telling states how they should implement the national standards.[53] They also pointed out that the low-sulfur requirement would place pressure on the nation's energy supply.[54] Finally, they framed the problem as a cost-benefit issue, arguing that increases in the cost of the fuel canceled out any benefits gained from burning low-sulfur fuel.[55]

The petroleum industry was not alone in resisting these standards. Producers of high-sulfur coal and the electric power companies that burned this coal argued that even if sulfur limits were justified in polluted urban areas, fuel with a higher sulfur content certainly could be burned in unpolluted areas.[56] Such arguments raised questions about how ambient concentrations were to be measured. What if plants built higher stacks so as to dilute the concentration of sulfur dioxide at ground level? With that point in mind, coal companies took out full-page advertisements that lobbied for measuring ambient concentrations only where people breathed.[57]

The problem of acid rain, however, brought the issue to a head and shifted the tenor of the argument. Oil companies began arguing that if additional cuts in sulfur emissions were necessary, power plants should install scrubbers capable of removing that sulfur from their stacks, a technology that was becoming a practical option. Companies mining high-sulfur coal sided with the API. Electric utilities, on the other hand, strenuously opposed any expectation that they would install scrubbers.[58]

Congress finally addressed this issue with passage of the Clean Air Act of 1990, which, among other things, required the EPA to cut emissions from large power plants by almost 50 percent over twenty years.[59] It also set deadlines that nonattainment areas had to meet in making progress toward meeting all ambient air quality standards. In both cases, innovative uses of emissions trading—similar to that used in the phase-out of leaded gasoline—promised to help regulators manage these two contentious changes without resorting to "command and control" decisions.

In the effort to cut emissions of sulfur dioxide from large power plants, the strategy was to cap the total quantity of sulfur dioxide emissions allowed from a pool of large power plants and to gradually reduce that cap. First, the EPA created enough emission "allowances" to match current emissions, with each facility receiving the allowances it needed. Each allowance represented a license to emit a ton of sulfur dioxide. To achieve the necessary cuts, the EPA then began shrinking the pool of allowances from year to year, without worrying about how to distribute those cuts.

Instead, the operators of each facility decided what was more efficient for them: shifting to low-sulfur fuels, installing scrubbers, securing allowances, or some combination of the three.

Enforcement relied on an system of annual reconciliation. At the end of each year, each power plant had to submit paperwork demonstrating that it possessed enough allowances to match its measured emissions. For example, if a plant was allocated 1,200 emissions allowances at the beginning of the year but actually emitted 1,500 tons of sulfur dioxide, then that facility needed to secure an additional 300 allowances to cover its emissions. In addition to tracking the ownership of allowances, regulators could also audit a power plant to verify that its fuel purchases and record of monitored emissions were consistent with what was reported.

The trading program put into place could also serve as a way to allocate emissions of sulfur dioxide even after target levels were reached. Increases in the demand for electricity could quickly raise the demand for emissions allowances. Without a trading program in place, managing the allocation of those emissions on an ongoing basis would be difficult.

Some states also set up "cap and trade" programs to manage the emissions cuts necessary to bring nonattainment areas into compliance. For example, California's South Coast Air Quality Management District set up a program known as Reclaim that issued allowances for a certain quantity of sulfur dioxide emissions and managed cuts by slowing reducing the number of allowances available.[60] In Illinois, regulators later designed a similar program to reduce emissions of smog-causing hydrocarbons.[61] In both cases, facilities still had to satisfy all operating permits. The need to reconcile emissions with allowances simply provides a way to distribute cuts and to manage a finite pool of emissions on an ongoing basis.

Trading programs, of course, are still in their infancy, and individual programs require significant effort to develop. Effective programs must be designed to achieve specific goals in a way that takes into account the physical problem being managed. Most industrial facilities currently do not participate in any emissions or effluent trading program and probably will not for some time to come. Still, these programs are now a legitimate policy tool and reflect the magnitude of the change that has taken place. A regulatory infrastructure and culture is now in place that requires facilities to secure permits and to respect a combination of technology-based performance standards and ambient standards, making it possible to use trading programs to allocate effluents and emissions.

In the last three decades of the twentieth century, then, the degree to which refineries and other industrial facilities monitor and manage their emissions and effluents has changed dramatically. Emissions and effluents have come to be treated as transactions to be monitored and managed and integrated into the cost of operations so as to satisfy socially constructed measures of environmental quality. Furthermore, in attempting to meet the complex constraints imposed on their emissions and effluents, refiners have gradually integrated environmental performance into the design of industrial processes.

Pollution Control and the Design of Technological Systems

The system of pollution control regulations constructed in the 1970s and 1980s transformed environmental objectives into what the Dutch Committee for Long-Term Environmental Policy has called "boundary conditions."[62] The general assumption is that these boundary conditions, if maintained, initially encourage "end-of-pipe" solutions tacked on to existing processes. However, as firms come to treat these boundary conditions as legitimate constraints faced by all competitors, they invest in alternative technologies that satisfy those boundary conditions as efficiently as possible, resulting in process changes as well as end-of-pipe controls.

To the extent that this pattern holds, political choices designed to accomplish specific environmental objectives can be said to influence the direction of technological development.[63] Over time, as engineers and technical managers make decisions with the need to meet pollution control regulations in mind, the changes they make come to be embedded in the larger technological system. Eventually, distinguishing environmentally driven innovations from other innovations becomes difficult.

In the refining sector of the petroleum industry, one can find evidence of this pattern in the integrated system of reforming, hydro-desulfurization, and sulfur recovery that evolved in the years since officials in California put pressure on refiners to reduce their sulfur dioxide emissions. In this case, pollution control regulations encouraged several interwoven process changes rather than the addition of equipment easy to identify as pollution control equipment. In one of the main changes, refineries began removing sulfur from cuts of petroleum through hydro-desulfurization and delivering the resulting gas to a Claus plant for conversion to elemental sulfur.[64] To what extent can this hydro-desulfurization equipment, the equipment to strip the hydrogen sulfide from the product stream, and the recovery plant be identified as pollution control equipment?

Certainly, few people would consider hydro-desulfurization units to be pollution control equipment. After all, refineries must remove sulfur from many refined products to meet customer expectations. Hydro-desulfurization simply replaced older methods of removing sulfur. However, in adopting this process, refiners gained more control over their emissions of sulfur and eliminated the large quantities of sludges associated with older methods of sulfur removal. Hence, incentives provided by pollution control regulations play some role in any decision to install a hydro-desulfurization unit.

Classifying sulfur recovery plants is also difficult. To some extent, the sulfur that refineries produce became one more product to sell, justifying the installation of this equipment. However, the price of sulfur has fluctuated over the last several decades, in part due to the tremendous amount of sulfur produced by refineries. In 1998, the oil and gas industry in the United States produced 8.2 million tons (70%) of the 11.6 millions tons of elemental sulfur produced domestically. Worldwide, 32.3 million tons (56%) of 57,800 tons came from the recovery of sulfur in oil refineries and gas plants. Given that sulfur production at refineries is dependent on the demand for petroleum products, not sulfur, analysts expect that approximately 90% of all sulfur produced will eventually come from environmentally regulated sources.[65] Even if refineries cannot sell their sulfur at a price high enough to justify its recovery, receiving some amount for whatever sulfur is generated is better than having to pay for its disposal. Managing one's residuals so as to minimize the costs of disposal has become an important goal, and that goal leads to the operation of sulfur recovery plants.[66]

Should the octane-boosting catalytic reformers that refiners installed in response to the phase-out of leaded gasoline be viewed as pollution control equipment? Although octane specifications are driven by performance concerns, not environmental concerns, refiners were prohibited from using tetraethyl lead by air quality legislation. Clearly, some portion of the costs associated with installing and operating catalytic reformers should be associated with meeting environmental objectives, but determining the exact fraction is difficult. Accounting becomes even more complicated when one considers that refiners use the hydrogen generated by reformers as feedstock to their hydro-desulfurization plants.

Some equipment—such as tail gas plants that reprocess the waste gases exiting Claus recovery plants—look more like pollution control equipment. The need for tail gas plants emerged because Claus recovery plants can only convert about ninety-five percent of the incoming hydrogen sulfide to elemental sulfur. Therefore, Claus recovery plants always produce a tail gas consisting of sulfur dioxide, hydrogen sulfide, and whatever impurities may be present in the original feed.[67] This tail gas, if flared, is a significant source of sulfur dioxide. If processing high-sulfur crude, the largest refineries could emit over one hundred tons of sulfur dioxide each day if they flared their tail gases.[68]

In 1972, the EPA recommended that states limit the amount of SO_2 released from Claus recovery plants to .01 pounds for every pound of sulfur produced—meaning that they expected 99 percent of the hydrogen sulfide to be converted into elemental sulfur, not 95 percent.[69] By the mid-1970s, oil companies had developed over a dozen commercial processes for Claus tail gas cleanup, with the Shell Claus Offgas Treating (SCOT) system becoming a widely adopted scheme. In this process, the escaping sulfur dioxide was reconverted to hydrogen sulfide by mixing it with hydrogen and passing the gas over a catalyst. The cost of this process, however, matched the cost of the entire Claus plant, indicating that removing the last 5 percent of sulfur cost as much as removing the first 95 percent.[70] In the absence of pollution control regulations, the investment in tail gas recovery plants certainly could not be justified.[71] Still, even here, some costs can be recovered. In 1980, for example, one refinery installed a process that transformed its captured tail gas directly into a specialized fertilizer, ammonium thiosulfate, which makes that process look less like a piece of pollution control equipment.[72]

Given these complications, anybody estimating the cost of meeting pollution control regulations faces a difficult task. For example, in the early 1970s, the EPA calculated that refiners would spend $900 million to meet new air quality standards, something less than one cent for each gallon of gasoline sold in the United States.[73] However, this figure did not include any investment on capital equipment that would show a return—such as sulfur recovery units—or the cost of operating and maintaining equipment. Finally, the estimate did not include costs associated with meeting new product specifications stemming from pollution control laws, such as the requirements to produce unleaded gasoline and low-sulfur industrial fuels. More recently, the authors of a popular text on petroleum refining estimate that the costs associated with meeting all pollution control regulations raises the cost of refined products by between ten and twenty cents.[74]

Regardless of the precise amount, the capital that companies invest in equipment necessary to satisfy pollution control regulations encourages decision makers in those companies to treat those regulations as legitimate. In the ideal, firms come to see their ability to meet pollution control regulations efficiently as a competitive advantage. Indeed, as the technological system associated with refining continues to evolve under a system of strict pollution control regulations, one can expect innovations adopted in response to environmental constraints to become even more difficult to distinguish from other innovations.

This shift from a guiding ethic rooted solely in the efficient use of resources to one that recognizes the need to comply with regulations based on environmental objectives occurred for many reasons, including changing public values and the emergence of a social movement capable of politically expressing those values. Also critical to the change, though, was the growing recognition on the part of government officials, the public, and even industry itself that a guiding ethic based solely on efforts to increase the efficiency with which an industrial operation processed resources had reached a dead end. Since then, the efficiency-based ethic of self-regulation has gradually been replaced by an ethic that most people, and most firms, now treat as legitimate: reaching consensus on environmental objectives and then regulating pollution-causing discharges to meet those objectives. The notion of efficiency, of course, remains important. Firms still have to be efficient to prosper. However, the notion of efficiency now includes the need to satisfy socially defined environmental objectives expressed in terms of enforceable regulations.

THIRTEEN   **Environmental Objectives**
and the Evolution
of Tankers

IN THE 1970S AND 1980S, as federal and state agencies constructed
a regulatory system capable of controlling emissions and effluents from
industrial facilities, oily discharges from oceangoing tankers remained a
concern. Indeed, in 1970, when the explorer Thor Heyerdahl sailed across
the Atlantic in an Egyptian-style papyrus reed craft, he reported encoun-
tering something that he did not expect to see in open ocean: lots of oil. In
places, oily scum prevented the eight men on Heyerdahl's craft from using
seawater to bathe.[1] In all likelihood, that oil came from tankers flushing
out their empty tanks en route to the Middle East from Europe or heading
south to Venezuela after unloading crude in the United States, two of the
most heavily traveled tanker routes in the world.[2]

The media, and the public, paid much less attention to these intentional discharges than they did to accidental spills, such as those from the *Torrey Canyon* and the blowout at Santa Barbara. The urgency of accidental
spills made for good television news, especially when the spilled oil threat-

ened popular beaches. Images of dying birds covered in oily slime proved to be particularly effective. No similar images existed for intentional discharges of oily water in the middle of the ocean. However, the focus on accidental spills also directed more attention to the equally serious concern of intentional discharges.

Both concerns—intentional discharges and accidental spills—presented policy makers with significant challenges. In attempting to reduce intentional discharges from oil tankers, governments quickly discovered that it was difficult to enforce treaties that specified how much oil a tanker could legally discharge into the sea. As for accidental spills, the industry's voluntary insurance plan spread the cost of each incident over a much larger number of tanker trips, preventing any one firm from having to bear the full risk of a spill.[3] Furthermore, a liability cap kept the total possible damages associated with each incident low enough not to be an incentive for major change.[4]

Given these and other barriers to change, efforts to reduce the amount of oil entering the sea from tankers proceeded slowly. However, in the 1970s and 1980s, shipowners adopted several technological innovations that paved the way for stricter regulations. Because of these changes, the Oil Pollution Act of 1990, which Congress passed in the wake of the *Exxon Valdez* spill, proved far more effective than it might have otherwise.

The amount of information associated with oil spills increased dramatically in the early 1970s. A portion of this increase stemmed from a law that required anybody responsible for an oil spill in U.S. waters to report that spill to the U.S. Coast Guard, regardless of whether the oil came from a tanker, an offshore drilling operation, a pipeline, or an industrial facility.[5] In 1974, the U.S. EPA began publishing a quarterly index of incidents, articles, and reports associated with oil spills, with the first edition running about two hundred pages. In addition, numerous researchers began studying factors associated with the causes, prevention, and containment of oil spills, along with the effects those spills had on marine life.[6]

In their efforts to understand oil pollution concerns, many researchers began by estimating the total amount of oil entering the ocean each year.[7] Although individuals disagreed on the precise amount that came from different sources, most agreed on several points. First, the largest quantity

Table 13.1. Oil Entering the Sea, 1971

Source	Amount (barrels/yr)	Percent (of total)
Spent lubricants	2,190,000	44.7
Tanker operations (oily ballast, slop, bilges, etc.)	1,067,000	21.8
Non-tanker ship operations	600,000	12.3
Tanker and oil barge accidents	320,000	6.5
Refineries and petrochemical plants	300,000	6.1
Non-tanker ship accidents	250,000	5.1
Offshore production	100,000	2.1
Marine terminals (spills, leaks, etc.)	70,000	1.4
Total	4,897,000	100.0%

Source: U.S. Senate, Committee on Commerce, *Hearings on the Navigable Waters Safety and Environmental Quality Act,* 92nd. Cong., 1st. sess., Sept. 22–24, 1971, 298. Data is from Joseph D. Porricelli, Virgil F. Keith, and Richard L. Storch, "Tankers and the Ecology," which was entered into the record of the hearings as pp. 295–328.

of waste oil reached the ocean through municipal sewers and industrial effluents, with the disposal of spent crankcase oil representing a significant problem.[8] Second, oily ballast and slop that ships released in normal operations amounted to more oil than that released during tanker accidents. Third, offshore production contributed some oil to the total, but less than the amount associated with tankers. The estimates used by engineers in the Coast Guard, who were charged with finding ways to improve tanker operations, are summarized in table 13.1.

Clearly, the rate, concentration, and location of each discharge mattered as much as the total quantity. Although the disposal of spent lubricants represented the largest source of oil entering the sea, people discharged relatively small amounts of this oil from cities and towns all over the world and did so steadily throughout the year, making that source of oil a chronic problem. Intentional discharges of oily ballast and oily slop from tankers, distributed throughout shipping lanes, also represented more of an ongoing problem. In contrast, oil reaching the sea from accidents came from a few major incidents, each of which released substantial amounts of oil in a single location over a short period of time. Each source of oil presented regulators with different challenges.

In regards to intentional and accidental discharges of oil from tankers,

Coast Guard engineers raised a number of basic questions. What changes to tankers and tanker practices would reduce the amount of oil entering the sea? What was the immediate cause of most accidents? What was the most practical way to contain and clean up oil spills? What changes in the law might encourage the parties responsible for oil spills to be more careful? What changes in the law would help make enforcement of international treaties more effective? In September 1971, the Coast Guard answered such questions for a Senate committee holding hearings on a bill to protect the environmental quality of ports, waterfront areas, and the navigable waters of the United States.[9]

The Coast Guard engineers estimated that routine discharges associated with flushing and cleaning cargo tanks accounted for approximately 70 percent of all oil released by the world's tanker fleet. Furthermore, they concluded that load-on-top procedures, which the International Maritime Organization (IMO) officially adopted as an international standard in 1969, were not effective. The IMO regulations assumed that tanker crews pumped all oily water and slop left after flushing or cleaning tanks into an empty compartment for settling. As the oil and water separated, crews could discharge up to 60 liters of water every mile while keeping the concentration of oil in the discharged water below 100 parts per million (ppm). In addition, crews could only discharge one fifteen-thousandth of the tanker's capacity on a single voyage. Therefore, empty tankers were expected to arrive in port with about 90 percent of the original residue still on board as dewatered slop. The incoming crude was to be loaded on top of the dewatered slop.[10]

However, the IMO provisions were not practical. Even if all governments and shipowners supported the approach, which they did not, technical problems still prevented the new practices from being effective. For example, high-speed centrifugal pumps used to move fluids from tank to tank inevitably created a highly emulsified mixture difficult to separate using gravity alone. Therefore, even the "clean" water that settled underneath the oily slop usually had a concentration of oil between 300 and 5,000 ppm. In all likelihood, tanker crews, who had no way to determine the concentration of oil in the water they pumped into the sea, discharged water containing significantly more than the 100 ppm allowed. In addition, nobody had any way of determining whether crews discharged a quantity of oil greater than one fifteen-thousandth of the tanker's capacity.[11]

In the end, the Coast Guard engineers concluded that much of the oily residue left in cargo tanks after unloading still reached the sea. Quite simply, attempts to reduce that quantity by specifying how much oil crews could discharge had failed.[12] If the captain of a tanker had reason to discharge oily ballast or slop into the sea, the crew could pump it overboard with little chance of detection. Indeed, tankers still occasionally used darkness as a cover to discharge oil when close to shore. For example, in 1970, when an agent for the Texas Water Quality Board reported finding oil from an unknown source washing ashore, he was told that "it was probably just an oily ballast discharge from a ship that passed in the night," something "seen by old-timers in this area for the past 20 years."[13]

To reduce the amount of oil entering the ocean from routine tanker operations, the Coast Guard engineers recommended the use of segregated ballast tanks, that is, tanks reserved for ballast water. Because no oil would ever enter those compartments, captains could pump ballast water in and out of the tanker without releasing any oil into the sea. The Coast Guard engineers pointed out that segregated ballast tanks would be less controversial when applied to the largest class of tankers because "tankers tend to become weight limited rather than volume limited" as they get larger. In other words, the largest tankers could not fill all their available space with oil and still be seaworthy. Hence, dedicating some of that space for ballast water would not represent as large a concern as on smaller ships. Because compartments that did carry oil would still have to be cleaned from time to time, engineers with the Coast Guard also considered devices such as oil-water separators and methods of burning the emulsified slop to be necessary.[14]

The Coast Guard engineers also suggested ways to reduce accidental discharges of oil. Based on an analysis of 1,416 tanker accidents, they noted that groundings accounted for about 26 percent of all tanker accidents leading to spills, causing them to assert that double bottoms would significantly reduce the probability of spills due to grounding. Not only would a double bottom protect cargo tanks from minor punctures but would also provide extra protection against buckling.[15] In addition, the compartments created by double bottoms could be used as segregated ballast tanks. In that way, one solution solved two problems.

The Coast Guard engineers did not recommend constructing a second hull on the sides of ships. Although ship-to-ship collisions and ship-to-object rammings accounted for 40 percent of all tanker accidents resulting in spills, they argued that double hulls "of a practical depth (1 to 3 meters) would not be effective in preventing spills due to a medium- or high-energy collision." Instead, to reduce the number of collisions, many of which occurred in coastal waters near the entrance to major harbors, the engineers with the Coast Guard recommended a "total systems approach" with the aim of narrowing "the band in which human judgment, and therefore errors in human judgment, could occur."[16]

This total systems approach included better traffic control in harbors, more systematic training for tanker crews, and an improved design for the terminal-tanker interface so that it would be less likely to cause problems when tanker crews and terminal crews did not speak the same language. They also identified design features to improve the maneuverability of tankers, such as: (1) lateral thrusters, which were perpendicular to the main propellers and improved steering at low speeds; (2) controllable pitch propellers, which improved braking by allowing ships to reverse thrust simply by changing the pitch of the blades; and (3) the addition of more sophisticated navigational systems.[17]

Finally, they recommended the use of "flue gas inerting systems," which were also being championed by insurance associations. In the late 1960s and early 1970s, seven tankers, each capable of carrying over 100,000 tons of oil, exploded while empty. Three of the tankers sank. Presumably, oil clinging to the sides of compartments gave off vapors that mixed with oxygen in explosive proportions and were then set off by static electricity or some other source of ignition. No such accidents occurred in similar ships using a flue gas inerting system. In such systems, exhaust gas coming from a ship's flue was pumped into empty cargo tanks, replacing the oxygen-filled air necessary to support explosions.[18]

One year after these hearings, Congress passed the Ports and Waterways Safety Act of 1972, empowering the Coast Guard to control the movement of vessels in the nation's ports, to develop and operate traffic control systems, and to establish and enforce regulations, including design standards, for all vessels carrying polluting substances into U.S. coastal waters. This mandate, though, did not result in any in dramatic changes to tanker practices in the short-term, in part because the regulatory culture of the Coast Guard was weak.[19]

Action at the international level was also necessary. The amount of oil being transported by oceangoing tankers reached 1,366 million tons per year in 1973, a ninefold increase over the amount of oil being shipped when the first international treaty on tanker discharges had been written in 1954.[20] Although participating nations had amended the 1954 treaty several times, most amendments, such as the one calling for load-on-top procedures, proved ineffective or unenforceable or both. Therefore, several nations—this time, including the United States—proposed that the IMO specify the use of segregated ballast tanks as an international standard. The result was the 1973 International Convention for the Prevention of Pollution from Ships (MARPOL). MARPOL required all new tankers to dedicate about 30 percent of their space to segregated ballast tanks so that no ballast water would ever come into contact with oil. If strategically placed in the space created by double hulls, segregated tanks also promised to help reduce the amount of oil released due to groundings.[21]

Segregated ballast tanks, by themselves, could not solve the entire problem of intentional discharges. Cargo tanks still had to be cleaned from time to time. However, in the mid-1970s, a practice known as "crude oil washing" came into wider use after owners of large tankers installed flue gas inerting systems on their ships. Engineers discovered that, with an gas inerting system in place to keep tanks free of oxygen, they could wash down the insides of tanks with crude oil while unloading. As the cargo was being pumped out, crews pumped a portion of the crude oil back through the ship's tank-washing system. Doing so removed about 90 percent of the oil that previously had clung to sides of tanks.[22] Cargo owners liked this system of cleaning tanks because most of the oily residue previously left behind could be unloaded with the rest of the crude. In the largest tankers—those capable of holding 500,000 tons of oil—crude oil washing reduced the amount of oil left in a tanker from about 2,000 tons to approximately 200 tons.[23]

By the late 1970s, any tanker equipped with segregated ballast tanks and a crude oil washing system carried the state of the art in pollution control. In 1977, one ship having those features—the *Tonsina*, a 120,000 dwt tanker purchased by the Standard Oil Company of Ohio for service in Alaska— was touted as being an "ecology class" tanker. The *Tonsina* also came equipped with an oil-water separator capable of handling the smaller volumes of oily water likely to require separation.[24] Tankers such as the *Tonsina* required a minimum of enforcement to ensure that they complied with the discharge levels set by the 1973 MARPOL treaty. After all, tanker captains had no reason not to use crude oil washing if available. In addition, the plumbing on ships with segregated ballast tanks prevented crews from pumping oil into the ballast tanks. Amendments to MARPOL in 1978 recognized the value of crude oil washing systems in reducing oily discharges and added this equipment to its set of international standards for new tankers.[25]

Advances in "fingerprinting" crudes provided tanker owners with another incentive to adopt the latest technology for reducing discharges to the sea. Initially, some regulatory officials advocated adding chemical tracers to crude oil so as to give each load of crude an identifiable signature. In this way, the tanker responsible for an unreported oil slick could be identified and the captain prosecuted.[26] Although no such scheme was put into use, others experimented with more sophisticated ways to identify the "fingerprint" of various crudes and to track down the offending ship. Indeed, by the mid-1970s, the EPA had established routine procedures for taking samples of oil found in the sea and comparing it to samples taken from ships in the vicinity. This enforcement mechanism increased the chances that spilled oil could be linked to the tanker responsible for it.[27] Advances in the use of remote sensing also contributed to the agency's ability to detect spills in open water.[28]

Although the design standards associated with the 1978 amendments to MARPOL were effective in reducing intentional discharges, they could not eliminate accidents. Two oil spills—the wreck of the Liberian-registered *Amoco Cadiz* off the coast of France (1978), and a blowout of the offshore oil well *IXTOC 1* in the Gulf of Mexico (1979)—would soon remind everybody that little had changed since the wreck of the *Torrey Canyon* and the blowout at Santa Barbara a decade earlier.

The *Amoco Cadiz* spilled more oil than any tanker up to that time, around 225,000 tons (1.5 million barrels), almost twice as much as the *Torrey Canyon*. The *IXTOC 1* blowout released more oil into the ocean than any accident before or since, well over 500,000 tons (3.5 million barrels); only the deliberate sabotage of Kuwaiti terminals during the 1991 Gulf War released more oil into the sea.[29] Both the *Amoco Cadiz* incident and the *IXTOC 1* blowout involved poor judgment on the part of those in positions of responsibility. This problem of how to prevent human error from leading to serious consequences raised new questions about the larger technological system in which those errors were embedded.

In chronicling the ebb and flow of daily life on a large tanker, Noël Mostert, a journalist who documented a round-trip he made on a tanker run between the Middle East and Europe in the early 1970s, pointed to boredom as a major problem facing tanker crews. By the time Mostert made his trip, the largest class of tankers were over a quarter-mile long, highly automated, and more like artificial islands than ships. Indeed, at most ports, large tankers did not actually dock but moored offshore and pumped their cargo ashore through pipes. A crew member could make several voyages and never see land up close. In addition, Mostert noted that crews often lost their sense of time, with little marking one day from the next. Hence, when entering a crowded channel, those in charge made what Mostert called a "nightmarish transition from placid, vacuous days to unrelenting pressure, strain and fear."[30]

Research into the causes of collisions supported Mostert's observations. One set of statistics showing that 85 percent of groundings and collisions involved some human error in the course of operations. Improper maintenance, for example, led to equipment failures that preceded some accidents. Language differences between crew members and terminal operators and between equipment manuals and crew members only compounded problems. Crew members sometimes turned valves at the wrong time or failed to turn valves at the right time, resulting in accidents such as tanks overflowing or oil being pumped directly into the sea.[31]

A number of human-related factors came into play in events leading up to the wreck of the *Amoco Cadiz*, which entered the English Channel fully loaded on March 16, 1978. Around 9:30 in the morning, the flange of a pipe carrying pressurized hydraulic fluid to the ship's rudder cracked, spraying the engine room with the fluid. Before the crew could fix the pipe, forces on the rudder snapped mechanical linkages that normally held the rudder in position, rendering it useless even after the pipe was repaired. Meanwhile, the tanker drifted toward shore, which lay less than eight miles away.[32]

The human element entered the picture in full force soon after a 10,000-horsepower tug appeared on the scene. The captain of the drifting tanker informed the tug that he wished to hire the tug to point the tanker toward open ocean. With full power still available, he planned to move away from shore, using the tug to maintain the proper heading. The tug's captain, however, wanted the *Amoco Cadiz* to accept "Lloyd's open form," a salvage agreement that, based on the decision of an arbitration board in London, would award the tug, and all members of the tug's crew, some portion of the property saved. With the tanker drifting toward shore, the two captains, communicating by radio and speaking through translators, haggled over the terms of the agreement. The miscommunication and confusion that ensued ended with the *Amoco Cadiz* being grounded about three miles from the French coast, puncturing a compartment and releasing crude into the sea. Furthermore, the tanker had come to rest at high tide. Therefore, as the tide receded, portions of the hull not sufficiently supported by the water sagged and placed significant strain on structural members. Over the next two weeks, rough seas broke the stranded tanker apart, causing 223,000 tons of crude to be released.[33]

A year later, the blowout of a well in the Gulf of Mexico (the *IXTOC 1*, owned by a Mexican firm) put offshore production in the spotlight. Here too, poor judgment played a significant role. Just before the crew lost control of the well, they encountered a loss of circulation, which meant that the drilling fluid being pumped down the center of the drill stem was not returning through the annular space between the drill stem and the wall of the hole. This situation was not unusual because the weight of a mile-high column of heavy drilling fluid often caused some fluid to flow into porous formations. In most cases, drilling crews simply added material that thickened the drilling fluid and increased its tendency to plug up porous formations.[34]

However, in this case, the crew did not have enough drilling fluid on hand to compensate for the lost material. Instead, they pumped in salt water to make up for the loss. Then, after mixing a new batch of heavy drilling fluid, the driller decided to pull the pipe out of the hole so the crew could change the drill bit. But instead of first circulating the entire column of drilling fluid so as to replace the watered-down portion with heavier material, the crew simply started to lift and disassemble the drill pipe.

Problems occurred when the drill bit had been raised almost level with the sea floor. On the platform, drilling fluid started spraying out of the pipe, indicating that the natural pressure of the formation exceeded the pressure exerted by the column of watered-down drilling fluid. The driller immediately realized that he had to trigger the blowout preventer. However, the drill collar—the heavy pipe located just above the bit—happened to be aligned with the blowout preventer. Policy called for the crew to release the pipe and let it drop back down the hole. Unfortunately, the top segment of pipe, partially unscrewed by this time, had tilted off the vertical and prevented the crew from dropping the pipe back into the hole. Therefore, when the crew triggered the blowout protector, the heavy drill collar was still inside.[35]

Though designed to shear through normal drill pipe, the blowout preventer could not cut through the thick-walled drill collar. Hence, the protector sealed off the space around the drill pipe but could not seal off the space inside, and on July 3, 1979, *IXTOC 1* blew out of control, eventually destroying the rig and spilling between 10,000 and 30,000 barrels of petroleum into the Gulf of Mexico each day. In all, about 500,000 tons (3,500,000 barrels) of crude escaped before crews brought the well under control.[36]

The wreck of the *Amoco Cadiz* and the blowout of *IXTOC 1* both resulted from a chain of events involving equipment failures, design flaws, and poor human judgment. Clearly, such chains of events cannot be easily predicted, monitored, and regulated.[37] Although steps can be taken to reduce the likelihood of such incidents, encouraging those steps through laws and regulations requires significant insight into the problems being solved. Furthermore, the companies responsible for such operations must have a significant interest in preventing such accidents. A company's potential liability must be high enough to justify investing in keeping crews trained, equipment maintained, and systems in place to prevent the escalation of simple problems.[38]

In the early 1970s, a few states challenged the liability cap that the federal government established for oil spills, but those challenges were defeated in court.[39] In the mid-1980s, Congress debated liability-related issues once again, but no legislation came out of those efforts.[40] Then, on March 24, 1989, the tanker *Exxon Valdez* ran aground on a reef in Prince William Sound off the coast of Alaska. The resulting spill raised new questions about spill-related costs for which shipowners were liable and focused public attention on ongoing calls for legislative action.[41]

In terms of sheer magnitude, the *Exxon Valdez* spill was not huge. Its release of 11 million gallons ranks about thirtieth on the list of large spills, about seven times smaller than the largest tanker spill on record (table 13.2). In terms of volume, the wreck of the *Torrey Canyon* (1967) and the breakup of the *Amoco Cadiz* (1978) both rank higher than the *Exxon Valdez* incident.[42] That the *Exxon Valdez* spill occurred in a relatively fragile ecosystem located in pristine waters certainly magnified its significance. But the *Exxon Valdez* spill was also troubling because it happened in a place—the port serving the Trans-Alaska pipeline—where major oil companies had promised that they would do everything possible, technologically and organizationally, to prevent such events from occurring and to take quick action when they did. Clearly, more action was needed if spills were to be prevented.

A year later, Congress passed the Oil Pollution Act of 1990, which eliminated federal liability caps in the event of "the violation of an applicable Federal safety, construction, or operating regulation" or "gross negligence or willful misconduct." In addition, the act allowed the recovery of additional costs—including damages to natural resources, lost revenues, and lost economic potential—substantially raising the potential cost of a spill.[43] The act also required vessels operating in U.S. waters to demonstrate proof of financial responsibility, preventing independent tanker owners with few assets from taking unwarranted risks. Furthermore, the law phased out the use of single-hull tankers in U.S. waters over a twenty-

Table 13.2. Largest Tanker Spills

Rank	Tanker	Year of spill	Location of spill	Gallons (millions)
1	*Atlantic Empress*	1979	W. Indies	76
2	*ABT Summer*	1991	Angola	72
3	*Castillo De Bellver*	1983	S. Africa	70
4	*Amoco Cadiz*	1978	France	65
5	*Haven*	1991	Italy	42
6	*Odyssey*	1988	Atlantic	39
7	*Torrey Canyon*	1967	England	36
8	*Seq Star*	1972	Oman	36
9	*Hawaiian Patriot*	1977	Hawaii	30
10	*Independenta*	1979	Turkey	28
32	Exxon Valdez	1989	Alaska	11
	Iraqi release	1991	Kuwait	240
	IXTOC blowout	1979	Gulf of Mexico	140

Source: National Research Council, *Tanker Spills: Prevention By Design* (Washington, D.C.: National Academy Press, 1991), 16. Updated with additional data from the International Tanker Owners Pollution Federation, "Past Spills—Statistics—1974–1999" (http://www.itopf.com/stats.html), 2000.

five-year period.[44] Two years later, the International Maritime Organization amended its international treaty on marine oil pollution (MARPOL 73/78) and approved a plan to retire single-hull tankers throughout the world by 2023.[45]

Performance versus Design Standards

In encouraging the development of technological systems that help achieve environmental objectives, is it more desirable to set and enforce performance standards or to set design standards? Clearly, performance standards provide industrial firms with more flexibility in how to respond to a specific regulation. However, in efforts to reduce oily discharges from tankers, design standards—such as requiring tankers to be equipped with flue gas inerting systems, crude oil washing systems, segregated ballast tanks, and double hulls—have proven more effective. Indeed, in the 1960s and 1970s, performance standards in the form of limits on intentional discharges failed as a regulatory mechanism. They were unenforceable and resulted in no significant changes to industry practice.

Liability, of course, is also an incentive to technological change. The increased exposure to liability that the Oil Pollution Act of 1990 imposed upon tanker owners certainly was a powerful incentive for the industry to take more aggressive action in preventing spills. But barriers to change remained. The phase out of single-hull tankers coordinated a change that some firms would have resisted, even in the face of increased liability.

First, a spurt of tanker construction in the early 1970s along with the reduced demand for petroleum due to conservation efforts in the second half of 1970s had skewed the age distribution of the world's fleet.[46] Many tankers would be approaching the end of their useful economic life toward the end of the century. What would happen when companies began replacing those ships? In the absence of any requirement to phase out single hulls, shipowners ordering new tankers might have continued using single hulls. Each new single-hull ship ordered would have represented a missed opportunity to secure an important incremental improvement for a quarter century.

Second, determining whether the cost of a particular change is justified by the accompanying reduction in risk is difficult. The incremental protection that double hulls provided was not so great as to have justified immediately scrapping or retrofitting single-hull tankers. Tanker transport is a commodity bought and sold in a competitive market, with tankers competing not only with each other but also with other modes of transport such as pipelines.[47] Even a small difference in the per-gallon transportation cost translates into a significant amount. The owner of a fleet that unilaterally decided to operate only double-hull tankers would still have to compete with rates offered by the operators of single-hull tankers.

Third, double hulls are not a technological silver bullet capable of addressing all spill-related concerns. In the 1980s, tankers spilled about one-five-hundredth of 1 percent of all oil they transported through U.S. waters.[48] While double hulls promised to reduce that spill rate by three-quarters, they could not reduce it to zero.[49] Even with double hulls, large spills can still occur. In low-energy collisions, double hulls make a difference by preventing any oil from spilling in cases where a single-hull ship would have experienced a significant release. The authors of one assessment study estimated that, on average, a double-hull tanker is about five times less likely than a single-hull tanker to discharge oil in a collision or grounding.[50] However, if a rock pierces the outer hull far enough, it will also pierce the inner hull. When a structure the size of a horizontal skyscraper hits a rock at twelve knots, whether it has one or two hulls makes little difference. A fully loaded 200,000 dwt tanker moving at 12 knots possesses more kinetic energy than eight Boeing 747s coming in for a landing, and much of that energy is dissipated at points of contact.[51] Indeed, double hulls probably would not have prevented the *Exxon Valdez* from releasing oil.[52]

Double hulls also create opportunities for failures not possible with single hulls. For example, water flowing into the airspace between the hulls could result in a double-hull tanker becoming unstable. Oil vapors in the same air space increase the chance of explosion. Corrosion, too, presents more of a challenge because crews must pump salt water into the air space when ballast is needed. Therefore, the outer hull gets attacked by salt water from both sides, increasing long term risks associated with poor maintenance. Such problems, if anticipated, can be prevented, but no guarantee exists that a particular double-hull design will provide the same level of protection to the marine environment as another. Indeed, some double-hull designs show problems with stability and, in some circumstances, could release more oil than single-hull tankers in the same circumstances.[53]

Fourth, other hull-related changes to improve a tanker's spill performance were also possible. Table 13.3 compares the performance of double hulls with some of these other design alternatives. The numbers represent each alternative's ability to reduce the risk of spillage and are relative to MARPOL tankers, which refers to tankers that met all international standards in place before the *Exxon Valdez* spill. For example, one alternative makes use of the fact that petroleum and petroleum products are lighter than water. If the hydrostatic head created by the height of the cargo compartment is less than the hydrostatic head associated with the sea, water will flow into a punctured single hull, not out. Therefore, by installing a watertight deck that cuts horizontally through the middle of all compartments, naval architects could significantly reduce the internal hydrostatic head on the cargo.

Table 13.3. Oil Spill Performance of Various Tanker Designs

	Outflow relative to a MARPOL fleet	
Design Alternative	Small tanker (40,000 dwt)	Large tanker (200,000 dwt)
Double bottom	66%	40%
Double sides	133%	109%
Double hull	56%	29%
Hydrostatic control	43%	51%
Smaller tanks	72%	64%
Oil-tight deck with double sides	47%	28%
Hydrostatic control with double sides	57%	27%
Hydrostatic control with double hull	50%	26%

Source: National Research Council, *Tanker Spills: Prevention By Design* (Washington, D.C.: National Academy Press, 1991), 166.

The table suggests that double hulls represent only one of many hull-related design changes possible to reduce the risk of oil spills. In addition, the size of a tanker, the ability of a structure to withstand the stresses and strain of salvage operations, and any changes affecting maneuverability and stopping ability also affect the size of potential spills. Therefore, the safest tanker would, in addition to having a double hull, be kept below a specific size and have numerous small compartments, a mid-deck, and an auxiliary propulsion system. Other factors not related to tanker design—such as the quality of navigational aids, communication between crews and the Coast Guard, systematic training and inspection, tug escorts in critical areas, and improved navigational systems—also affect the spill rate. Clearly, choices had to be made.

At the same time, there are good reasons to assume that tanker owners would have eventually moved to double hulls in response to liability changes, though not necessarily in time to prevent another generation of single-hull tankers from being built. First, double hulls meshed well with efforts to reduce intentional discharges, especially provisions that tankers reserve about 30 percent of their volume for ballast space. After all, the gap between hulls can be used to hold segregated ballast water. Double hulls also complemented crude oil washing systems. In single-hull tankers, numerous structural components along the hull blocked the spray of wash oil from reaching surfaces to which oil clung. With double hulls, many of these structural components lie outside the cargo compartments. Indeed, approximately 4 percent of the world fleet (by tonnage) was already protected by double hulls before 1990, which suggests that there were good reasons to make the change.

Second, in terms of its effect on consumers, the cost of constructing tankers with double hulls was not a major barrier to change. A tanker with a double hull costs approximately 10 percent more to construct than the equivalent MARPOL-approved single-hull structure. If the added capital costs are distributed over the life of a tanker and factored in with operational costs, including insurance, the added transportation costs hovered around one and a half cents per gallon. The precise amount, of course, varied with the size of the tanker and the length of the voyage. Still, requiring tankers to be equipped with double hulls did not dramatically affect the price of refined products.[54]

Therefore, with the liability cap removed, firms were more receptive to government-enforced standards that ensured all firms took some steps to reduce the risk of spills. After all, firms that failed to take aggressive action penalized the entire industry through higher insurance rates and poorer public relations. As much as anything, then, the phasing out of single hulls by the Oil Pollution Act of 1990 can be viewed as coordinating a technological change rather than as imposing a technological solution. Changes in liability represented the more important legislative action. Furthermore, in the long term, other factors—such as better navigational aids, better communication between crews and the Coast Guard, systematic training and inspection, tug escorts in critical areas, and improved subsystems—play as important a role in reducing spillage as design changes.[55]

Today, new tankers discharge less oil into the sea—both in terms of intentional and accidental discharges—than tankers in the early 1970s. With segregated ballast tanks and crude oil washing systems, the drop in intentional discharges of oily wastes has been significant, approaching a reduction of an order of magnitude. Spills due to accidents have also decreased, though not as dramatically or as systematically as intentional discharges. When averaged over the period 1970–74, the amount spilled each year was approximately 234,000 tons. The highest five-year running average occurred in the period 1975–79, when approximately 402,000 tons were spilled each year. By 1994–98, that running average had dropped to approximately 54,000 tons of oil spilled each year. With better navigational systems in place and the phaseout of single-hull tankers continuing, we can expect this quantity to drop even further.

Though significantly reduced, the risk of large spills has not been eliminated. As long as tankers transport oil, spills and accidents will occur. Furthermore, additional ways to reduce the risk of a major spill are always possible. For example, some states have unsuccessfully attempted to require that tugs be used to guide large tankers through sensitive areas. However, in this case, the federal government has resisted efforts by states to set provisions that are stricter than federal standards.[56] In general, though, the technological evolution of oil tankers over the last three decades has been significantly influenced by environmental objectives expressed through regulatory mechanisms such as treaties, national laws and regulations, court-determined liability, and insurance premiums.

FOURTEEN

Closing the Loop

EFFORTS TO REDUCE ONE TYPE of pollution have often increased other types of pollution, with the problem simply being shifted from one form to another. For example, in the 1930s, when some refineries starting burning their acid sludge to reduce the amount of acid released into streams, they ended up releasing more sulfur dioxide into the air. Even in the 1970s, engineers managing the effluent at one refinery joked that they should just "airstrip" a particularly troublesome contaminant and let the "air pollution people" take care of it.[1] In the words of historian Joel Tarr, controlling pollution has too often been a "search for the ultimate sink."[2]

In this search for the ultimate sink, the nation's groundwater has often ended up the loser. For example, when firms were pressured to control their emissions and effluents in the years after World War II, many responded by burying more wastes. In the process, they created disposal sites that would leak contaminants into groundwater for years to come.[3] Even after passage of strong pollution control laws in the early 1970s, groundwater initially remained unprotected. The lack of attention paid to groundwater is reflected in EPA's initial listing of thirty-eight refineries as "zero-discharge" plants. Those refineries, representing 4 percent of the total U.S. refining capacity, ran their wastewater into on-site ponds or injected it into deep wells. Because these refineries did not discharge any effluent into public bodies of water, the EPA did not require them to obtain discharge permits. Of course, wastewater seeping from lagoons could and did reach groundwater. Poorly maintained injection systems could provide an even more direct path to groundwater contamination.[4] In the absence of additional legislation, however, injection wells and lagoons simply fell outside federal law.

Wastes sent to landfills also remained uncontrolled. Congress had passed a Solid Waste Disposal Act in 1965, but this piece of legislation, reminiscent of earlier air and water pollution legislation, focused on research, education, and demonstration projects. The Resource Recovery Act of 1970 focused on the reclamation of energy and material from solid wastes, requiring the EPA to submit annual reports on progress made in recycling, but it set no enforceable objectives. Industrial firms could still dispose of most wastes in landfills, regardless of what hazards they posed.

Creating and enforcing legislation aimed at protecting groundwater has consumed many of the resources devoted to pollution control since the mid-1970s. First, the Safe Drinking Water Act of 1974 mandated that the EPA regulate the use of injection wells. Soon after, the Resource Conservation and Recovery Act (RCRA) of 1976 charged the agency with tracking hazardous wastes from "cradle to grave," providing a mechanism for preventing hazardous wastes from disposal in an uncontrolled manner. Tracking hazardous industrial wastes, though, did not eliminate problems caused by past disposal practices. To deal with that concern, Congress eventually passed the Comprehensive Environmental Response, Compensation, and Liability Act (CERCLA) of 1980, also known as Superfund. Together with weaker legislation that focused on toxic products such as pesticides, these laws attempted to close the loop left open by the passage of air and water pollution regulations alone.

Closing the Loop on Waste, Leaks, and Spills

The complexity of protecting the quality of the nation's groundwater can be seen by examining the implementation of soil- and groundwater-related regulations in the petroleum industry. These regulations include those associated with tracking hazardous wastes, preventing leaks from tanks and pipelines, and cleaning up sites contaminated by leaks and wastes disposed of in the past. In implementing these regulations, regulators and oil companies alike have faced significant challenges.

Early on, the effort to implement RCRA raised numerous questions as to what wastes should be considered hazardous and how those wastes should be disposed.[5] For example, soon after the passage of RCRA, the keynote speaker at a conference on refinery wastewater called for more studies associated with the disposal of sludge, noting that "nobody yet knows how much sludge we are going to generate" from "all these water pollution efforts."[6] Some of this sludge came from biological treatment processes that converted emulsified organics into biological mass, a technology the EPA included in the "best practical" wastewater treatment available to refiners. Other sludges came from oil-water separators, clarifiers, and tanks for breaking emulsions.

Refiners had to dispose of all this sludge in one way or another, and many did, at least initially, by spreading it on a convenient tract of land. In a practice known as landfarming, disposal crews spread the waste sludge on the ground in four- to six-inch layers and allowed it to dry for about one week. Then, after adding nutrients, they mixed the sludge into the soil. The land could be used to dump more sludge in about a month.[7] However, as regulators set about the task of determining which wastes should be listed as hazardous, they raised concerns about these sludges. If impurities such as metals were present, they could not be biologically digested and could accumulate in the soil. Potentially, rain trickling through a sludge farm could then carry these impurities to aquifers or surface bodies of water.[8] Wastes in other sectors of the industry, such as drilling fluids, also came under question.[9]

Eventually, the EPA did list some oily sludges as hazardous waste, a decision that resulted in refineries having to track that material until disposed of in a manner considered safe. This need to track some of their sludges from "cradle to grave" have made refiners accountable for contaminants in such wastes to an extent previously impossible. Operators at refineries and other industrial facilities now must fill out a manifest when transporting for disposal any wastes listed as hazardous. The transporter then sends a copy of the manifest back to the refinery and to the appropriate regulatory agency to verify that a licensed disposal site received the waste. To be licensed, a disposal facility must be able to dispose of the listed material in an appropriate manner. For example, if the disposal site is a landfill, it must be constructed so that water trickling through the soil can be collected and treated. If it is an incinerator, it must be designed to meet specific combustion standards.[10]

Soil contaminated from past leaks and spills also presented challenges. Refineries inevitably contaminated some soil due to past operations. Huge quantities of petroleum pass through a refinery's pipes and tanks, and leaks and spills accumulate over time. In addition, engineers often buried special-purpose storage tanks and pipes used to move petroleum products around a facility, making it difficult to detect any leaks that occurred. Where leaks and spills occurred, petroleum products accumulated in the soil and sometimes ended up reaching groundwater. The extent to which this contamination presented an immediate problem depended, of course, on the local geology and the location of any aquifers used to supply water for drinking or industrial uses. At some refineries, regulators—often backed by court order—have pushed for the installation of extensive groundwater monitoring systems and systems for preventing the escape of contaminated groundwater.[11]

Past disposal practices also came back to haunt refiners. If, in the past, workers dumped sludge removed from storage tanks holding leaded gasoline, then soil in the disposal area inevitably contained high levels of lead. Or, if an old refinery, especially one constructed sometime before strict wastewater laws went into effect, discharged its effluent to a small stream, then contaminants inevitably settled out and gradually accumulated in the stream's sediment. When contaminants are discovered in the sediment, should they be removed and disposed of as hazardous material? If so, who should pay for the effort and what standards of "cleanliness" should be used? How much sediment should be removed from how much of the stream? Studying, debating, and litigating such questions—whether under RCRA, CERCLA, or state-level programs—can consume millions of dollars, and many refineries have had to confront such problems. The liability for such cleanups has even become an item on the balance sheets of firms that operate refineries.[12]

The complexity of protecting the quality of the nation's groundwater can also be seen in regulations specifically designed to prevent future leaks from underground storage tanks. In 1984, an amendment to RCRA required the EPA to address concerns associated with underground storage tanks, such as those installed at gasoline filling stations. Many of these tanks were steel vessels having no corrosion protection. After a certain age, they were sure to leak. Coordinating fifty state-level programs in which ten of thousands of individual station owners—many operating marginal businesses—were expected to upgrade or replace their tanks required both a major educational program and a detailed schedule of compliance.[13]

Evidence confirming that the program to upgrade underground storage tanks was important came when the additive methyl tertiary butyl ether (MtBE) began showing up in drinking water supplies. When added to gasoline, this chemical improves the fuel's octane rating and results in exhaust containing fewer unburned hydrocarbons. Initially employed by some oil companies to improve the octane rating of their gasoline, its use increased after the EPA began enforcing a provision in the Clean Air Act of 1990 that required the sale of oxygenated fuels in areas with an ozone problem. Additives such as MtBE helped refiners meet that requirement. Unfortunately, in encouraging the use of this chemical to address an air quality problem, the EPA put groundwater sources all over the United States at risk. MtBE, which adds a turpentine-like odor to water even in small quantities, is highly soluble and difficult to remove. Leaks and spills of gasoline containing MtBE inevitably occurred and introduced this chemical into groundwater systems throughout the United States.[14]

Although debate exists over the level of MtBE's toxicity, the chemical's presence in groundwater is clearly undesirable. Critics charged that, in failing to consider the larger ecology of industrial operations, officials responsible for the EPA's air quality programs encouraged wide dissemination of a chemical that posed a serious water quality threat. In March 2000, the EPA banned the additive, using its authority under the Toxic Substances Control Act (TSCA). Regardless of how one views the decisions to encourage and later ban MtBE, the debate over use of this additive illustrates the interconnectedness of pollution concerns and the complexity of addressing one problem without creating another.[15]

Problems associated with long-distance pipeline systems leaking or spilling their contents also demanded some attention. In the early 1970s, approximately 15 million gallons of petroleum and petroleum products spilled or leaked from oil pipelines each year, equivalent to the capacity of a small tanker. Corrosion was responsible for some problems, but the use of corrosion prevention technologies kept most stretches of line in good condition. Operators of heavy construction equipment represented a larger problem. For example, in 1974, of the 1,477 gas pipeline failures that occurred, 1,030 (70%) of them were excavation-related accidents. A typical accident report contained explanations such as "the road grader dented and gouged the pipe which then fractured and ripped apart." In that same report, the investigator also noted that one of the operators "was not familiar with the locations of pipelines in the area and did not recognize pipeline markers as warning devices." The investigator also noted that the other operator "was aware of the pipeline right-of-way and saw pipeline markers but did not . . . take any special precautions as he crossed the pipeline with the road grader because he believed that all pipelines were buried 5 feet or more."[16]

To prevent such accidents, the National Transportation Board recommended that all states develop "one-call" systems: a single telephone number that contractors could call for information about any pipelines or electrical lines buried in work areas.[17] Over the next two decades, federal, state, and local agencies and the numerous companies that maintained buried pipelines and cables coordinated the development of one-call services operated by private firms. The effectiveness of one-call systems now in place varies widely, with problems arising from excavators not bothering to call, poor information provided by companies that own underground facilities, and the difficulty of maintaining a quality one-call service while making a profit. In 1999, the Office of Pipeline Safety issued a report that documented these difficulties, but also noted that new technologies such as global position systems (GPS), geographic information systems (GIS) systems, and various tools for tracing lines were making a difference.[18]

Some progress has been made in reducing the quantity of oil spilled from pipelines. By the late 1990s, the amount of oil lost from long-distance pipeline systems dropped to about 5,880,000 gallons per year, approximately a third of the annual losses from the 1970s.[19] Still, the industry has a long way to go before eliminating pipeline leaks as a concern. Even where one-call systems are effective, pipeline leaks, small and large, still occur. In the period 1993 to 1998, excavation accidents still accounted for about 25 percent of all crude oil pipeline breaks. In addition, despite significant advances in corrosion protection, small portions of a pipeline can still deteriorate, sometimes resulting in leaks that go undetected for long periods before the line ruptures. Corrosion was responsible for about 25 percent of all incidents that occurred in the period 1993 to 1998. Storm-related breaks, failed pipe, failed welds, equipment malfunction, operator errors, and vandalism accounted for the remaining 50 percent.[20]

In the mid-1990s, regulators also started taking a closer look at pipeline operations. In 1992, Texas alone experienced forty-two large (over 10,000 gallons) pipeline spills, resulting in 2,300,000 gallons being spilled.[21] Then, in 1993, a major pipeline spill led to calls for stricter regulations. On March 28, a high-pressure oil pipeline ruptured in Fairfax County, Virginia, sending a 100-foot plume of fuel oil into the air. The line, operated by the Colonial Pipeline Company, released over 400,000 gallons of oil before it could be shut down and fully drained. That incident set in motion discussions as to what could be done to prevent such events from happening and how to reduce the amount of damage when they occurred.[22]

Clearly, additional steps can always be taken to reduce both the risk of failures and the amount of damage caused by those failures. Even a century ago, engineers could have run one pipe inside another to provide two layers of protection.[23] To detect small leaks, they also could have installed some form of detection between the fuel-carrying pipe and casing. Even today, though, costs discourage firms from using such systems in all but the most sensitive areas. In Japan, for example, engineers included such a system when designing a twenty-nine-mile pipeline carrying jet fuel along a populated corridor. The detection system continually transmits signals through a special coaxial cable attached to the bottom of the fuel-carrying pipe. The impedance of this particular cable is designed to change where saturated with fuel, allowing operators not only to detect leaks but also to pinpoint the location of the leak by analyzing changes in the propagation of reflected waves.[24]

Ways to detect corrosion problems inside a pipeline are also available. One of the best methods involves running inspection devices through the inside of a pipe to collect information about the pipe's condition. However, in the early 1990s, no regulations required such inspections and no standards existed as to what actions should be taken as a result of such inspections. Regulations associated with the placement of flow-monitoring devices and shutoff valves were also lax. No regulations specified a minimum distance between shutoff valves, and regulations that required companies to place shutoff values on either side of a stream crossing allowed companies to install manual valves, which made quick responses impossible.[25]

Firms, of course, require time and incentives to integrate these technologies and practices into their systems. However, as in the tanker industry twenty years ago, incentives associated with reducing the risk of a major accident may not be strong enough to encourage change. Pipelines move huge quantities of oil long distances, and insurance spreads the costs of spills over much of that quantity. For example, in 1986, damage from spills amounted to $300 million. When spread over all oil transported through pipelines in the United States, those damages raised the cost of transporting a barrel of oil one thousand miles by about five cents, or about one-eighth of a cent per gallon.[26] In the absence of regulations requiring all pipeline companies to implement the same level of protection, few companies are likely to take aggressive action on their own.

Although the Colonial spill did not result in strong legislation, Congress did pass the Accountable Pipeline Safety and Partnership Act of 1996. This act charged the Office of Pipeline Safety with conducting risk management demonstration projects with pipeline companies, so as to test the effectiveness of various strategies companies might employ to reduce spills and limit damage. These projects are likely to lay the foundation for future regulatory requirements associated with the design, inspection, and repair of pipelines. A focusing event, such as the rupture of a high-profile pipeline, may be necessary before Congress enacts legislation mandating such regulations, but consensus as to the specific form of regulations may be easier to reach after these demonstration projects are complete.[27]

Brine Concerns Revisited

In the effort to protect groundwater, injection wells also came under more scrutiny. Even before Congress passed the 1972 amendments to the Federal Water Pollution Control Act, a deputy chief at the EPA, Stanley M. Greenfield, noted that the agency was "watching deepwell injection very closely."[28] By then, approximately a hundred firms, including refineries and petrochemical plants, had started using injection wells for the disposal of chemical wastes. For example, in Texas, facilities operated by companies such as Armco Steel, Ethyl Corporation, DuPont, Celanese Chemical, Rohm and Haas, and Sinclair Petro-Chemicals all used deep wells to dispose of chemical wastes.[29] Among other things, Greenfield expected to see more "collisions" between these wells and efforts to protect freshwater aquifers.

Injection wells in oil fields also caught Greenfield's attention. Indeed, he pointed to brine injection wells as a major concern, especially in areas where the number of people dependent on groundwater had increased.[30] When the EPA examined the industry's use of injection wells more closely, researchers documented groundwater contamination from oil field brines in seventeen states.[31] This finding did not surprise anybody familiar with the survey conducted by the Interstate Oil Compact Commission fifteen years earlier. The newer studies simply reinforced what everybody already knew: oil producers lifted a lot of salt water and had a limited number of ways to dispose of that water, with injection wells being the most attractive option. That finding, though, gave weight to a section in the Safe Drinking Water Act of 1974 that explicitly ordered the U.S. EPA to protect drinking water supplies from possible contamination by injection wells.[32]

From the perspective of protecting groundwater supplies, other methods of brine disposal, such as allowing the salt water to seep into the ground, still represented the greater concern. Hence, as the EPA turned its attention to brine disposal, these other methods also received new attention. As it happened, oil-producing states tended not to enforce prohibitions against these older methods too strictly. Doing so would have forced some oil producers to abandon their operations, something that state agencies such as the Texas Railroad Commission did not wish to see happen. To keep marginal oil fields producing, the agency occasionally granted exemptions to state environmental regulations associated with the disposal of brine.[33]

For example, in the late 1960s, residents in Corpus Christi started complaining about a system of disposal ditches and creeks that carried salt water from nearby oil fields to Cayo del Oso, a saltwater bay. This drainage system, operated since 1939 by a salt water disposal corporation, had originally been constructed to solve a pollution problem. However, some residents noted that the oil field brine could affect the ecology of an estuary due to the presence of minerals in proportions not normally found in seawater.[34] Some brines, for example, contained high concentrations of lead and barium.[35] Therefore, discharging oil field brine directly into a saltwater body such as the Gulf of Mexico or the Pacific Ocean, once an accepted method of disposal, now raised concern.

Injection systems, of course, were the obvious alternative. In Corpus Christi, though, injection systems were ruled out because numerous old wells in the area might serve as conduits to a local aquifer. Brine injected underground could easily find its way up to the aquifer through leaky casing and poorly plugged holes. Where the cost of meeting regulations would put local producers out of business, the Railroad Commission often ruled in the favor of oil producers.[36] Such decisions prompted a frustrated spokesperson for the Texas Parks and Wildlife Department to complain that he personally knew "of 100 [brine disposal] cases in this past year where we were overruled" by the Railroad Commission.[37]

After two years of study, officials with the EPA concluded that existing state-level regulations did not sufficiently protect groundwater. Specifically, they decided that operators of injection wells should monitor the integrity of their injection wells to be sure that they were properly sealed and did not allow injected fluids to travel back up to aquifers through a leaky or poorly sealed casing. Toward this end, the EPA proposed having the operators of injection wells routinely monitor the pressure in the annular space between the casing and injection pipe and to have their wells inspected for mechanical integrity every five years. They also proposed that operators putting new injection wells into service be required to search for and fix any leaky well within one-quarter mile of the new injection well.[38]

Representatives from various state agencies, the API, and the Interstate Oil Compact Commission strongly protested the EPA's initial set of regulations. The provision for routine pressure tests, if approved, would force many operators to install additional fittings and valves. In addition, the proposed regulations required that operators make an inventory of all other wells in the area before drilling a new injection well, with the purpose being to determine which of those wells might require remedial work. The regulations, many argued, represented useless paperwork.[39] As evidence, a spokesperson for the API pointed to East Texas. In that field, he asserted, the East Texas Salt Water Disposal Corporation had been injecting 750,000 barrels of salt water each day for years, and none of the ten municipalities drawing fresh water from wells in the area "ever had an incident of contamination."[40]

The cost of complying with the proposed regulations quickly became the subject of debate. A study sponsored by the API concluded that, if approved, the new regulations would cost producers about $5 billion dollars in the first five years of implementation. In addition, the study indicated that approximately six billion fewer barrels of oil would be produced over the next twenty years because producers operating in marginal oil fields would be forced to abandon their wells. A study sponsored by the EPA placed those numbers at less than one-tenth of the API-sponsored study.[41] Officials with the Interstate Oil Compact Commission also lobbied against the need for new regulations, arguing that the existing regulations were sufficient and that most of the brine pollution identified by the EPA came from past practices.[42] However, in 1980, the regulations proposed by the EPA (with revisions) went into effect, affecting 160,000 class II injection wells in thirty-one states. In most cases, state agencies were given the responsibility of designing programs to meet the criteria established by the EPA.

In 1987, after eight years of implementation, a study performed by the General Accounting Office concluded that some state agencies were lax in enforcing the new regulations. In that study, investigators checked the records that state agencies maintained for a random sample of injection wells in Texas, Oklahoma, Kansas, and New Mexico, and discovered that officials in the first three states were issuing permits without requiring evidence that operators had conducted the necessary pressure tests. In addition, the study pointed to twenty-three instances in the United States of brine injection wells contaminating groundwater.[43]

But even if a regulatory agency had the will to aggressively protect fresh water from potential contamination by injection wells, other problems— none of them new—made the task difficult. Unless individuals or towns operated freshwater wells in an area and noticed a change, no easy way of detecting problems existed. Some researchers specifically discouraged the digging of monitoring wells because such wells could contribute to any problem that existed. Even if inspectors could sample the groundwater in the area around oil fields, they could not learn much unless they studied changes in the water over long periods of time—an expensive proposition.[44] Finally, even if someone discovered a change in water quality, pinning down the source of contamination usually proved difficult.[45]

In Texas, the Railroad Commission did not even have enough people to routinely inspect the 50,000 injection wells over which it had authority.[46] Investigators typically responded only to specific complaints. For example, in the period 1982 to 1985, the commission received approximately 6,500 specific complaints having something to do with injection wells, abandoned wells, or pollution of some kind. In about one hundred of those instances, inspectors found that salt associated with an oil or gas operation had contaminated potable groundwater. Of these, most could be traced either to abandoned wells that had not been plugged properly or to old saltwater disposal pits. Based on this data, the authors of a report commissioned by the Texas Department of Agriculture concluded that "most of the problems of groundwater contamination resulting from oil and gas production activities stem from unsound past practices that are no longer permitted: abandoned and improperly plugged wells that are frequently not a part of the public record, saltwater disposal pits, and salt water disposal down the casing of an injection well without a tube-and-packer arrangement."[47] Of course, such problems, even though caused by past practices, were still concerns.

Problems also existed in smaller oil producing states, even in the older fields of New York and Pennsylvania.[48] To reduce the need to inspect injection wells in all these areas, the EPA proposed new design standards for injection wells. In many injection wells, a single leak could allow injected fluid to escape, which meant only one layer of protection existed. The new standards put three layers of protection in place. Producers could continue to operate existing wells with one and two layers of protection, but they had to test those wells every year and every three years, respectively.[49] Not surprisingly, many in the oil industry argued that the regulations were too strict. In particular, the editor of the *Oil and Gas Journal* noted that "The U.S. has 170,000 active injection wells and 2.2 million abandoned or inactive wells. Against those numbers," he pointed out, "23 cases of aquifer damage look like isolated problems."[50]

Although the API and other interest groups criticized aspects of the strict new regulations proposed by federal regulators, they did not question the legitimacy of regulation. They had come to accept compliance with well-defined regulations as a logical way to maintain the level of environmental quality desired by the broader society. Indeed, the ability of the federal government to impose strict regulations on the operation of brine injection wells depended on this acceptance. In the case of regulations affecting brine injection wells, the largest challenge involved getting small producers operating marginal wells in older fields to invest a significant percentage of their revenue in complying with regulations.

New extraction technology also raised new issues relevant to the protection of groundwater in oil-producing areas. In the 1960s, oil producers began experimenting with enhanced methods of recovering petroleum from mature oil fields, with some of the more promising methods involving the injection of water containing solvents and detergents. These chemicals helped to strip oil away from the microscopic surfaces to which it clung.[51] Oil producers also experimented with recovery operations involving the injection of steam. Higher oil prices in the 1970s made such projects more attractive, increasing the intensity of secondary and tertiary recovery efforts in oil-producing regions.[52]

By the late 1970s, one aspect of these recovery operations—the withdrawal of fresh groundwater for use in steam injection projects—became a public issue. In such projects, using fresh water to generate steam was far less expensive and troublesome than using brine for the same purpose. However, farmers in Texas and Oklahoma had come to depend on huge quantities of water and realized that sources of groundwater were not inexhaustible. Indeed, groundwater levels in aquifers such as the Ogallala were dropping from year to year, causing some concern that the entire supply would be exhausted by 2020.[53]

At issue was the question of whether oil producers had the right to remove fresh water from the hydrological cycle. State agencies experimenting with ways to recharge aquifers in parts of Texas, Oklahoma, and New Mexico hoped to return more of the water withdrawn for agricultural and municipal use back into the aquifers from which that water came.[54] However, when oil companies withdrew large amounts of fresh water from those same aquifers to inject into oil-bearing formations, that water could not be returned to an aquifer. Hence, in 1977, when Amoco began purchasing sewage water from the city of Hobbs, New Mexico, to use in a major waterflood, the good press it received for making use of wastewater did not impress those concerned about future water supplies.[55] After all, water leaving a sanitation plant could still be returned to an aquifer. Water injected deep underground into oil-bearing porous formations saturated with brine could not. In short, any fresh water used in an enhanced oil recovery project was, for all practical purposes, removed from the hydrological cycle.

In a more controversial case, the Oklahoma Water Resources Board granted an oil producer the right to pump up to sixteen billion gallons of fresh water from the Ogallala for use in recovering oil. Farmers opposed the project, claiming that the oil producer was wasting a valuable resource. They noted that oil producers could use salt water for enhanced recovery projects without too great an additional expense.[56] Residents fighting to conserve supplies of groundwater, including some leaseholders who received royalties from oil and gas production, saw the expense of using brine instead of fresh water as a cost of doing business. Some opponents of the board's action went further and framed the problem in terms of pollution. They maintained that the company was polluting the groundwater by mixing it with salt water and chemicals. The fact that the groundwater was being moved from its original formation before being contaminated did not make the pollution any more palatable. Closing the loop, in this case, meant respecting the hydrological cycle.[57]

A New Industrial Ecology

Perhaps more than any other issue, global warming illustrates both the importance of thinking in ecological terms and the complexity of doing so. In this case, carbon dioxide, a gas previously thought of as a natural by-product of combustion and an integral part of the carbon cycle, has come to be seen as a form of pollution with the potential to alter the world's climate. And again, the petroleum industry, as a producer of fossil fuels, emerges as a central actor in debates over this issue.

If governments eventually reach consensus on the need to cap emissions of carbon dioxide, the challenges will be immense. Constraints on emissions of carbon dioxide conflict with a fundamental assumption made by many people in industrialized societies: that industrial development should be encouraged in all countries and that the results will be good for everybody. However, consuming energy—typically by burning fossil fuels and generating carbon dioxide—is a key component of that development. Regulating emissions of carbon dioxide while encouraging steady increases in energy consumption is obviously problematic.

Because of the tight link between industrial development and energy consumption, regulating carbon dioxide emissions will be far more complex than regulating other types of pollution.[58] Theoretically, if limits on emissions of carbon dioxide appear necessary, trading schemes, such as the one used to allocate emissions of sulfur dioxide from power plants, could be used to allocate the pool of available emissions. However, at the international level, numerous issues that were not controversial at the national level suddenly become very controversial. Even the initial allocation of emissions allowances becomes far more complex. Should the status quo—the current profile of emissions—be used in determining the initial allocation? Or should governments receive allocations based on the population they represent or the area of land they control? Serious enforcement of any constraints on emissions of carbon dioxide will also require sovereign nations to recognize the authority of an enforcing agency, which represents still another level of complexity.

Efforts to address the issue of global warming by limiting emissions of carbon dioxide may also create new environmental concerns. For example, nuclear power, which does not contribute to the problem of global warming, may suddenly become more attractive. Is encouraging the spread of nuclear facilities wise? How does one integrate concerns related to nuclear power into decisions associated with preventing global warming? New questions about the use of injection wells will also emerge. Over the last century, industrialized societies have withdrawn huge amounts of carbon (in the form of oil and coal) from below the earth's surface and have released that carbon into the atmosphere through combustion. Given that the carbon originally came from the ground, some may argue that emissions of carbon dioxide should be injected back into the ground. For example, to satisfy emissions limits, power plants burning fossil fuels might inject their stack gases deep underground, dissolving those gases in the brine-saturated porous formations that exist throughout the world. Regardless of whether that alternative is either feasible or wise, the issue will be raised.

Despite the obvious difficulties of managing the global carbon cycle by regulating human activity that affects that cycle, the mere idea that such a regulatory system might be possible demonstrates the dramatic changes that have taken place in the last century. In the early twentieth century, it would have been difficult to even imagine a system in which emissions and effluents and other forms of industrial and urban discharges could be monitored and managed so as to maintain specific measures of environmental quality. A few people might have seen such a system as being desirable, but anybody could have pointed to numerous political, technological, and economic reasons why such a system was not possible.

Today, however, most people can visualize a regulatory system in which material flows, including the residuals of industrial operations, are integrated in a larger industrial ecology shaped not only by economic and technological constraints, but also by environmental regulations and the limits of natural ecosystems. Although many issues still have to be resolved, arguments in favor of setting and achieving environmental objectives are no longer undermined by an efficiency-minded ethic of industrial self-regulation. Engineers and technical managers who focus on making industrial operations more efficient now take for granted the need to comply with pollution control regulations. Although debate continues on the details of regulation, efforts to meet the regulations that exist have redefined the notion of industrial efficiency to include environmental objectives.

Conclusion

IN *The Closing Circle: Nature, Man, and Technology,* a popular and influential book written in the early 1970s, biologist-author Barry Commoner described his first law of ecology succinctly: everything is connected to everything else. In an ecosystem, he explains, interconnected parts act on one another in complex ways. Technology, Commoner emphasized, should be considered as one of those parts. Indeed, he concludes the book by asserting that the ecological stability of the natural world depends on our ability to integrate technological systems into the larger ecology of the natural world.

In the United States, the system of pollution control regulations that has been constructed over the last three decades represents an important step in this direction. Pollution control regulations do more than "control" pollution from industrial and municipal facilities. They also "regulate" human uses of the physical environment to sustain those uses valued by society. These regulations link the flow of industrial and municipal wastes to a larger web of human values, resource flows, and natural limits. Not only do they place constraints on disposal practices, they also integrate the costs of environmental consequences into economic decisions.

Indeed, they restructure the economics of waste disposal to reflect environmental objectives.

The existing regulatory system for controlling pollution is neither elegant nor uniformly effective. Technological systems, social institutions, and the ecology of the natural world are complex, and so must be any system that links the three. For example, air quality legislation and water quality legislation differ in approach not only because of their different histories but also because the human uses of air and water differ, as do the ways in which these components of the physical environment interact with ecosystems. Regulations to protect groundwater have likewise developed in complex ways, often evolving to fit the loose edges of other regulations and closing a loop that, if left open, would undermine efforts to manage industrial wastes adequately.

As the history of pollution control in the petroleum industry shows, political and technical leaders were already wrestling with the issue of industrial pollution in the first two decades of the twentieth century. Indeed, the debates leading up to the Oil Pollution Act of 1924, which temporarily expanded into a general debate over water pollution, revealed both the extent of concern and the lack of consensus on any particular way to address the problem of pollution. Debates over air pollution generally took place at the local level and focused on the burning of coal, though proposals to add tetraethyl lead to gasoline did trigger some debate at the national level.

In both of these early debates, people who argued in favor of government-enforced regulations were at a disadvantage. Leading engineers and technical managers suggested that decisions involving pollution should be based on the efficient use of resources. This line of argument shifted the focus away from environmental objectives. In the debate over water pollution, industrial leaders argued that engineers were already reducing releases of pollution-causing discharges and argued that what industry needed was time and assistance, not penalizing regulations. And in the case of leaded gasoline, Charles Kettering of General Motors argued that tetraethyl lead, which allowed refiners to use a greater percentage of the nation's crude oil as motor fuel, was as good an example of "scientific conservation" as could be imaged. At a time when people were concerned about the adequacy of the nation's oil reserves, Kettering's argument proved more compelling than the argument against leaded gasoline made by health professionals.

In the petroleum industry, this ethic of utilitarian conservation was soon reinforced by efforts in which increasing the efficiency of industrial operations overlapped with efforts to address pollution concerns. For example, conservation-minded leaders justified regulating the extraction of oil using the logic of efficiency. As it happened, these regulations on the rule of capture also allowed petroleum engineers to solve serious pollution concerns associated with the disposal of oil field brines, and to do so in a way that increased the efficiency of extraction even further. Hence, regulations justified not by environmental objectives but by the efficient use of resources facilitated the ability of oil producers to meet various pollution concerns.

Engineers in other sectors of the petroleum industry also validated the assumption that efforts to increase the efficiency of industrial operations would also reduce the discharge of pollution-causing wastes. In the 1930s and 1940s, pollution-causing discharges per barrel of oil transported and processed also declined as engineers made pipeline systems and refineries more efficient. Furthermore, to demonstrate the industry's ability to regulate itself, the American Petroleum Institute (API) sponsored a manual on the disposal of refinery wastes that summarized the state of the art in waste disposal. In such a context, discussions of environmental objectives and efforts to put mechanisms in place to achieve those objectives failed to gather momentum.

Only in the 1950s did efficiency-based arguments against the need for strong pollution control legislation fail. By that time, the pollution problems that remained were often the most difficult to solve, yet the economic incentives to address those concerns were dwindling. For example, in the absence of regulations, refiners had little incentive to prevent fugitive leaks from the thousands of valves they maintained. Not only were the economic incentives associated with the recovery of lost material insufficient, legal incentives were also minimal. After all, nobody was likely to win a nuisance or damage suit based on those emissions. Those leaks did not even interest pseudoregulators such as fire insurance companies. Yet, as investigators in California determined, those leaks could still contribute to a city's smog problem.

In addition, the scale of industrial production had grown dramatically since the 1920s, overwhelming many of the "per unit of production" reductions that engineers had secured. Therefore, in the 1950s, the nation's airsheds and bodies of water were still getting dirtier, even as the expectations of a growing middle class were rising. Advocates of systematic pollution control regulations could no longer be easily dismissed.

Any transition from one set of rules to another is bound to meet with resistance. Such changes complicate strategic decisions, frustrating people who see decisions they made under one set of assumptions rendered obsolete by a shift in the rules. Examining the response of decision makers in the petroleum industry to the change from a efficiency-justified system of self-regulation to a government-enforced system of pollution control reveals the complexity of this change.

Initially, federal legislators passed laws that funded research and educational programs related to pollution control. Then, in the mid-1960s, with the need for stronger action becoming more apparent, Congress passed a wave of legislation aimed at defining air and water quality standards. In the late 1960s, after it became clear that the first wave of legislation would have little effect on industrial practices without stronger enforcement mechanisms, Congress opened the door to a regulatory system that fully embraced a new guiding ethic—that society had to first reach consensus on environmental objectives and then put rules in place to achieve those objectives.

Even if everybody accepts the notion that regulations are needed to achieve environmental objectives, numerous factors can still affect the pace at which a specific regulation comes to be accepted as legitimate. Based on the process by which regulations affecting the petroleum industry were transformed into enforceable rules, these factors include the following:

the level of consensus on the environmental objective a regulation is designed to achieve;

the availability of technology to measure and monitor the specific activity being regulated, a clear connection between this activity and the accompanying environmental objective, and the degree to which consistent enforcement is possible;

the cost of compliance, the degree to which competitors face the same costs, and the degree to which past investments are affected;

the level of flexibility that firms have in how they comply and the degree to which firms can gain a competitive advantage by complying efficiently; and

the degree to which technological changes related to pollution control come to be integrated into the larger technological system.

With the shift from an efficiency-justified ethic of self-regulation to an ethic rooted in the definition and pursuit of environmental objectives, industry also lost control over the production of scientific and technological knowledge related to the effects of industrial wastes on ecological and geophysical systems. In the first half of the century, industrial firms and trade organizations exerted a monopoly over the production of much of that knowledge. As long as the effort to address pollution concerns was framed in terms of using resources efficiently, few outside groups challenged this control. The few that did, such as the U.S. Bureau of Mines, generally framed problems in the same way as industry leaders. In any case, neither the bureau nor industrial groups initiated pollution-related research unless pressed to do so in response to conflicts that reached the courts or the ears of legislators.

In the history of petroleum-related pollution control, independent centers of pollution-related expertise first emerged in response to the Los Angeles smog problem. Federal research funds kept these centers of expertise independent of industry throughout the 1960s, allowing them to pursue research questions that decision makers in industry might have avoided. The creation of agencies such as the Federal Water Pollution Control Agency and, later, the U.S. EPA, both of which controlled substantial research funds, further extended the network of university researchers and consulting firms capable of producing and evaluating complex technical information.

In the 1970s, industry-sponsored researchers still influenced technical debates on environmental issues, but industry no longer controlled the agenda. The starting point for debate had also changed. The central issue was no longer whether government-enforced regulatory action was needed to prevent pollution-related conflicts. Instead, the main questions revolved around the measures of environment quality that a society should seek to sustain and the environmental objectives that should be set. Occasionally, opponents of specific regulations questioned the extent to which the accompanying objective represented an efficient use of resources, but such arguments no longer carried the same power. Today, technical experts can legitimately question whether a specific regulatory approach is the most efficient way to achieve an objective, but decisions about the objective itself are ultimately political decisions.

By the 1990s, after two decades of debate, a regulatory regime had been constructed in the United States that defined how industrial facilities could and could not use the shared environment as a sink for their residual wastes. This regime included systems for (1) monitoring and setting limits on the concentration of various contaminants in effluents; (2) monitoring and setting limits on various types of emissions; (3) tracking wastes listed as hazardous from "cradle to grave"; (4) regulating fluids injected into underground formations; and (5) cleaning up contamination caused by past industrial activity. In addition, special provisions to reduce various types of accidental discharges—such as oil spills, pipeline leaks, and leaks from underground storage tanks—also apply to particular types of industrial operations.

As this regulatory system has come to be accepted as legitimate, what it means to operate an efficient industrial process gradually has been redefined to include environmental objectives. Engineers, of course, still think in terms of thermodynamic efficiency when designing equipment, but the ability of a facility to meet pollution control regulations is now an important measure of performance. To some extent, the release of contaminants into the environment has become a transaction that people expect to be measured, monitored, and paid for. By integrating human activity and human values into the larger ecology of the natural world, the system of regulation that has been created represents one of the major accomplishments of the twentieth century.

Numerous challenges remain in the effort to manage the discharge of pollution-causing contaminates into the environment. For example, attempts to integrate nonpoint discharges into the larger ecology of industrial processes, human activity, and natural ecosystems represent a major challenge. In part, though, such concerns now receive the attention they do because industrial and municipal discharges have been brought under greater control. Global warming and the reframing of carbon dioxide as a contaminant emitted to the environment also represents a major new area of debate. Although challenges remain, the most difficult phase of this transition to a new regulatory system—recognizing the need to set and manage environmental objectives—lies in the past.

Notes

Introduction

1. References to many studies that examine pollution-related conflicts associated with industrial change can be found in the bibliographic essay by Jeffrey K. Stine and Joel A. Tarr, "At the Intersection of Histories: Technology and the Environment," *Technology and Culture* 39 (October 1998): 601–40.

2. For concerns associated with domestic sanitation, see the articles on wastewater and sewers in Joel A. Tarr, *The Search for the Ultimate Sink: Urban Pollution in Historical Perspective* (Akron, Ohio: The University of Akron Press, 1996). For issues associated with the changing nature of work and new forms of social control, see Steven J. Ross, *Workers on the Edge: Work, Leisure, and Politics in Industrializing Cincinnati, 1788–1890* (New York: Columbia University Press, 1985).

3. David Stradling, *Smokestacks and Progressives: Environmentalists, Engineers, and Air Quality in America, 1881–1951* (Baltimore: Johns Hopkins University Press, 1999), 21–36.

4. For an example of an early water pollution conflict, see John T. Cumbler, "Conflict, Accommodation, and Compromise: Connecticut's Attempt to Control Industrial Wastes in the Progressive Era," *Environmental History* 5 (July 2000), 314–34. For a study that examines the response to industrial pollution from the perspective of industrial hygiene, see Christopher Sellers, "Factory as Environment: Industrial Hygiene, Professional Rivalries and the Modern Sciences of Pollution," *Environmental History Review* 18 (spring 1994): 55–83.

5. To get a sense of the faith that various people put in the effort to increase efficiency and eliminate waste, see Stuart Chase, *The Tragedy of Waste* (New York: Macmillan Company, 1927). Although Chase does not discuss pollution-related concerns, the approach he takes is certainly consistent with efforts to fight pollution. Also see Robert B. Westbrook, "Tribune of the Technostructure: The Popular Economics of Stuart Chase," *American Quarterly* 32 (1980): 387–408.

6. Legislation that gave the existing regulatory system definite shape includes the National Environmental Policy Act of 1969; the Clean Air Act of 1970; the Federal Water Pollution Control Amendments of 1972; the Federal Insecticide, Fungicides, and Rodenticides Act of 1972; the Safe Drinking Water Act of 1974; the Toxic Substances Control Act of 1976; the Resource Conservation and Recovery Act (RCRA) of 1976; and the Comprehensive Environmental Response, Compensation, and Liability Act (CERCLA) of 1980.

7. Much, of course, has been written about the "gospel of efficiency," with the classic being Samuel P. Hays, *Conservation and the Gospel of Efficiency: The Progressive Conservation Movement, 1890–1920* (Cambridge, Mass.: Harvard University Press, 1959).

8. For the argument that the rising expectations of a growing middle class played a central role in the rise of the environmental movement, see Samuel P. Hays *Beauty, Health, and Permanence: Environmental Politics in the United States, 1955–1985* (New York: Cambridge University Press, 1987).

9. This book does not focus on the social and political conflicts that emerge as oil fields are developed. For studies that do, see Paul Sabin, "Searching for Middle Ground: Native Communities and Oil Extraction in the Northern and Central Ecuadorian Amazon, 1967–1993, *Environmental History* 3 (April 1998): 144–68; Myrna Santiago, "Rejecting Progress in Paradise: Huastecs, the Environment, and the Oil Industry in Veracruz, Mexico, 1900–1935," *Environmental History* 3 (April 1998): 169–88; and Nancy Quam-Wickham, "Cities Sacrificed on the Altar of Oil: Popular Opposition to Oil Development in 1920s Los Angeles," *Environmental History* 3 (April 1998): 189–209.

10. For an examination of the key role that "system builders" play in the development of complex technological systems, see Thomas P. Hughes, *Networks of Power: Electrification in Western Society, 1880–1930* (Baltimore: Johns Hopkins University Press, 1983).

11. For the notion of that disasters can serve as focusing events capable of triggering political action, see Thomas A. Birkland, *After Disaster: Agenda Setting, Public Policy, and Focusing Events* (Washington, D.C.: Georgetown University Press, 1997).

Chapter 1. Pollution Concerns Articulated

1. For more on the practices of the Standard Oil Trust, see Ida M. Tarbell, *The History of the Standard Oil Company* (New York: McClure, Phillips and Co., 1904); Allan Nevins, *Study in Power: John D. Rockefeller, Industrialist and Philanthropist,* 2 vols. (New York: Charles Scribner's Sons, 1953); Ralph W. Hidy and Muriel E. Hidy, *Pioneering in Big Business, 1882–1911: History of the Standard Oil Company (New Jersey)* (New York: Harper and Brothers, 1955); Jerome T. Bentley, "The Effect of Standard Oil's Vertical Integration into Transportation on the Structure and Performance of the American Petroleum Industry, 1872–1884" (Ph.D. diss., University of Pittsburgh, 1976); Roger M. Olien and Diana Davids Olien, *Oil and Ideology: The Cultural Creation of the American Petroleum Industry* (Chapel Hill: University of North Carolina Press, 2000).

2. Andrew Hurley, "Creating Ecological Wastelands: Oil Pollution in New York City, 1870–1900," *Journal of Urban History* 20 (May 1994): 340–64.

3. Federal Oil Conservation Board, *Complete Record of Public Hearings, February 10 and 11* (Washington D.C., Government Printing Office, 1926), address of G. S. Davison (President, Gulf Refining), 45–48.

4. J. G. Burr to House Rivers and Harbors Committee, Jan. 19, 1921, Pollution of Waters, Commerce Papers, Herbert Hoover Presidential Library (HHPL).

5. J. S. Gutsell, *Danger to Fisheries from Oil and Tar Pollution of Waters,* Bureau of Fisheries document no. 910, appendix 7 to *Report of the U.S. Commissioner of Fisheries* (Washington, D.C.: Government Printing Office, 1921).

6. "Against Water Pollution," *New York Times,* July 10, 1921, 24.

7. Report by Herman M. Biggs, House Committee on Rivers and Harbors, *Hearings on Oil Pollution of Navigable Waters,* 67th Cong., 2d sess., Oct. 25, 1921, 63–64.

8. "Protecting the Beaches," *New York Times,* January 16, 1921, sec. 8, 1.

9. U.S. Public Health Service, *Oil Pollution at Bathing Beaches,* public health report (Washington, D.C.: Government Printing Office, 1924).

10. Statement of George W. Booth, House Committee on Foreign Affairs, *Hearing on Oil Pollution of Navigable Waters,* 67th Cong., 2d sess., Feb. 15–18, 1922, 21–25.

11. Report by Biggs, House, *Pollution of Navigable Waters,* Oct. 25, 1921, 63–75.

12. Report by Theodore Horton, House, *Pollution of Navigable Waters,* Oct. 25, 1921, 12–15.

13. "Dock Company Indicted," *New York Times,* Feb. 2, 1921, 24.

14. Report by Horton, House, *Pollution of Navigable Waters,* Oct. 25, 1921, 14.

15. Albert Cowdrey, "Pioneering Environmental Law: The Army Corps of Engineers and the Refuse Act," *Pacific Historical Review* 44 (August 1975): 331–49.

16. W. W. Moore to A. E. Amerman, Mayor, City of Houston, Dec. 11, 1920, J. S. Cullinan Papers, folder 4, box 77, J. S. Cullinan Collection, Houston Metropolitan Research Center, Houston Public Library.

17. A. E. Amerman, Mayor, City of Houston to W. W. Moore, Dec. 13, 1920, folder 4, box 77, J. S. Cullinan Collection.

18. Colonel Benjamin Casey Allin III, *Reaching For The Sea* (Boston: Meador Publishing, 1956), 102.

19. "River Oil Burns the *Granite State*," *New York Times*, May 24, 1921, 1; statement of Booth, House, *Pollution of Navigable Waters*, Feb. 15–18, 1922, 24.

20. H.R. 7369, 67th Cong., 1st sess., 1921. The previous bill (66th Cong., H.R. 16022) received little attention.

21. Most earlier debates involving water pollution revolved around local issues associated with urban sanitation; see Joel A. Tarr, *The Search For The Ultimate Sink: Urban Pollution in Historical Perspective* (Akron, Ohio: University of Akron Press, 1996).

22. Statement of Clay Stone Briggs, House, *Pollution of Navigable Waters*, Oct. 25, 1921, 41.

23. H.R. 8783, 67th Cong., 1st sess., 1921.

24. Statement of David Neuberger, House, *Pollution of Navigable Waters*, Oct. 25, 1921, 6–9.

25. Statement of James William Marshall, House, *Pollution of Navigable Waters*, Oct. 25, 1921, 17–18.

26. Statements of Van H. Manning and J. H. Hayes, House, *Pollution of Navigable Waters*, Oct. 25, 1921, 24–27.

27. Statement of Samuel A. Taylor, House, *Pollution of Navigable Waters*, Oct. 25, 1921, 38–39.

28. Statement of John H. Buck, House, *Pollution of Navigable Waters*, Oct. 25, 1921, 36–37.

29. Statement of F. S. McIlheny, House, *Pollution of Navigable Waters*, Oct. 25, 1921, 27–28.

30. Statement of J. I. Tierney, House, *Pollution of Navigable Waters*, Oct. 25, 1921, 51–52.

31. Statement of Francis Taylor, House, *Pollution of Navigable Waters*, Oct. 25, 1921, 18–19.

32. Statement of William Gibbs, House, *Pollution of Navigable Waters*, Oct. 25, 1921, 31–32.

33. Report by Thurlow Nelson, "Some Aspects of Pollution as Affecting Oyster Propagation," House, *Pollution of Navigable Waters*, Oct. 25, 1921, 55–57.

34. Neuberger, House, *Pollution of Navigable Waters*, Oct. 25, 1921, 6–9.

35. "Demand Action on Oil Pollution Bill," *New York Times*, Aug. 12, 1922, 4.

36. Ellis W. Hawley, "Herbert Hoover, the Commerce Secretariat, and the Vision of an 'Associative State,' 1921–1928" *Journal of American History* 62 (1974): 116–40. Also see Guy Alchon, *The Visible Hand of Planning: Capitalism, Social Science, and the State in the 1920s* (Princeton, N. J.: Princeton University Press, 1985).

37. Statement of Herbert Hoover, House Committee on Rivers and Harbors, *Hearings on the Pollution of Navigable Waters*, part two, Dec. 7–8, 1921, 67th Cong., 2d sess., 92–93.

38. Statement of Allen T. Treadway, House, *Pollution of Navigable Waters*, Dec. 7–8, 89–90.

39. Statement of John Q. Tilson, House, *Pollution of Navigable Waters*, Dec. 7–8, 90–91.

40. Statement of Benjamin L. Rosenbloom, House, *Pollution of Navigable Waters*, Dec. 7–8, 94–99.

41. Comment by Benjamin L. Rosenbloom, House, *Pollution of Navigable Waters*, Dec. 7–8, 111.

42. Comment by Dr. Patterson, House, *Pollution of Navigable Waters*, Dec. 7–8, 106.

43. Statement of Charles Dorrance, House, *Pollution of Navigable Waters*, Dec. 7–8, 122–124.

44. Statement of Wells Goodykoontz, House, *Pollution of Navigable Waters*, Dec. 7–8, 141.

45. Statement of J. T. Travers, House, *Pollution of Navigable Waters*, Dec. 7–8, 112–21.

46. Statement of H. F. Moore, House, *Oil Pollution of Navigable Waters*, Feb. 15–18, 1922, 11.

47. In correspondence with Hoover just before the next set of hearings, representatives of the American Petroleum Institute specifically used this language; R. L. Welch, American Petroleum Institute, to Herbert Hoover, Commerce Secretary, Feb. 1, Pollution of Waters, Commerce Papers, HHPL.

48. Statement of Herbert C. Hoover, House, *Oil Pollution of Navigable Waters*, Feb. 15–18, 1922, 4–6.

49. Statement of Eugene F. Moran, House, *Oil Pollution of Navigable Waters*, Feb. 15–18, 1922, 17–21.

50. Statement of H. Bowles, House, *Oil Pollution of Navigable Waters*, Feb. 15–18, 1922, 45–51.

51. Joint Resolution (Pub. Res. 65), 67th Congress, approved July 1, 1922.

52. Stanley Hornbeck, State Department, to Christian Herter, Assistant to Commerce Secretary Hoover, Nov. 21, 1923, Pollution of Waters, Commerce Papers, HHPL.

53. U.S. Bureau of Mines, *Pollution By Oil of the Coast Waters of the United States*, preliminary report, Sept., 1923.

54. This collection of bills included H.R. 51, H.R. 203, H.R. 612, H.R. 3319, S. 42, S. 936, and S. 1388, 68th Cong., 1st sess.

55. Comments of David Neuberger, Senate Committee on Commerce, *Hearings on the Pollution of Navigable Waters*, Jan. 9, 1924, 68th Cong., 1st sess., 108.

56. Statement of Herbert C. Hoover, House Committee on Rivers and Harbors, *Hearings on the Pollution of Navigable Waters*, Jan. 23–25 and 29–30, 1924, 68th Cong., 1st sess., 9–11.

57. Fred W. Powell, *The Bureau of Mines: Its History, Activities, and Organizations*, service monographs of the United States government, no. 3 (New York: D. Appleton and Co., 1922), 4–5; A. Hunter Dupree, *Science in the Federal Government: A History of Politics and Activities to 1940* (Cambridge, Mass.: Belknap Press, 1957), 280–83.

58. Act of Feb. 25, 1913; 37 Stat. 681.

59. Editor's introduction, *Chemical and Metallurgical Engineering* 25 (August 1921): 353.

60. Rodney P. Carlisle and August W. Giebelhaus, *Bartlesville Energy Center: The Federal Government in Petroleum Research, 1918–1983* (Washington, D. C: U.S. Dept. of Energy, 1984).

61. Leonard M. Fanning, *The Story of the American Petroleum Institute: A Study and Report with Personal Reminiscences* (New York: World Petroleum Policies, 1959), 17–25.

62. E. W. Esmay, "The American Petroleum Institute: An Account of Its History, Functional Activities, and Organization Structure," 1942, 1–5, folder "1942 API," box 45, Bernard Majewski Collection, American Heritage Center, University of Wyoming.

63. Fanning, *American Petroleum Institute*, 1–25.

64. Van H. Manning, Chairman of Committee on Improvements and Methods, "Proposed Organization of the Division of Research and Statistics of the American Petroleum Institute," Sept. 1919, curator's exhibit files, National Museum of American History, Smithsonian.

65. "API Report on Fundamental Research in Petroleum," box 37, series 3, Sun Company Collection, Hagley Museum and Library.

66. Letter concerning A.R.A. Rules of Interchange, W. A. Boyd, Jr., to the Petroleum Industry, May 5, 1925, box 20, series 1D, Sun Company Collection, Hagley Museum and Library. For another example of the API's aggressive response to potential changes in fire and safety regulations, see letter on ordinances affecting tank trucks delivering gasoline in New York City, W. R. Boyd, Jr., to Hon. Thomas J. Brennan (Fire Commissioner, New York City), June 23, 1925.

67. B. W. Dunn (Chief Inspector, Bureau of Explosives) to R. P. Anders, July 28, 1924, box 16, series 1D, Sun Company Collection, Hagley Museum and Library; B. W. Dunn to L. I. Doyle (Explosives Agent, Interstate Commerce Commission), July 29, 1924; Doyle to Dunn, July 30, 1924; W. R. Boyd, Jr., to the Petroleum Industry, July 31, 1924.

Chapter 2. Concerns in the Oil Fields

1. Gerald D. Nash, *United States Oil Policy, 1890–1964: Business and Government in the Twentieth Century* (Pittsburgh, Pa.: University of Pittsburgh Press, 1968), 42–48.

2. Roger M. Olien and Diana D. Olien, *Wildcatters: Texas Independent Oilmen* (Austin, Tex.: Texas Monthly Press, 1984), 13–17.

3. Harold F. Williamson, Ralph L. Andreano, Arnold R. Daum, and Gilbert C. Klose, *The American Petroleum Industry: The Age of Energy, 1899–1959* (Evanston, Ill.: Northwestern University Press, 1963), 43, 299–300.

4. By the time Drake drilled his well in Pennsylvania, oil for kerosene was already being produced from seeps and hand-dug wells in other regions, most notably the Oil Springs Region of Ontario, Canada. For works documenting Pennsylvania as the center of the early oil industry, see S. Eaton, *Petroleum: A History of the Oil Region of Venango County, Pennsylvania* (Philadelphia: J. P. Skelly and Co., 1866); Andrew Cone and Walter Johns, *Petrolia: A Brief History of the Pennsylvania Petroleum Region From 1859 to 1869* (New York: D. Appleton and Co., 1870); J. T. Henry, *Early and Later History of Petroleum* (Philadelphia: Jas. B. Rodgers, 1873); John J. McLaurin, *Sketches in Crude Oil* (Harrisburg, Pa.: published by the author, 1896); Paul H. Giddens, *The Birth of the Oil Industry* (New York: Macmillan, 1938). Also see Harold F. Williamson and Arnold Daum, *The American Petroleum Industry: The Age of Illumination, 1859–1899* (Evanston, Ill.: Northwestern University Press, 1959), 63–310.

5. To place the petroleum industry of the early twentieth century in an international context, see Daniel Yergin, *The Prize: The Epic Quest for Oil, Money, and Power* (New York: Simon and Schuster, 1991), 19–34.

6. For a chronicle of the discovery of these and other fields, see Carl Coke Rister, *Oil! Titan of the Southwest* (Norman: University of Oklahoma Press, 1949).

7. David Leven, *Petroleum Encyclopedia: Done in Oil* (New York: Ranger Press, 1942), 94–95.

8. H. W. Bell and J. B. Kerr, *The El Dorado, Arkansas Oil and Gas Field* (Little Rock, Ark.: U.S. Bureau of Mines, 1922), 8. Also see also Kenny A. Franks and Paul F. Lambert, *Early Louisiana and Arkansas Oil: A Photographic History, 1901–1946* (College Station, Tex.: Texas A&M University Press, 1982), 107–12.

9. For photographs of these craters, see Franks and Lambert, *Early Louisiana and Arkansas Oil*, 108.

10. *National Petroleum News* 13 (January 19, 1921): 19.

11. "History of the Arkansas Fuel Oil Corporation," January 1961, 1–4, box 28, Cities Service Oil Company Collection, Western History Collection, University of Oklahoma. After the discovery of the Caddo oil and gas field in northern Louisiana, M. L. Benedum and J. L. Trees of Pittsburgh constructed a natural gas pipeline to carry gas from Caddo to Little Rock, Arkansas.

12. Trade periodicals such as *Oil Weekly, Oil and Gas Journal,* and *National Petroleum News* kept the industry informed about all drilling activity and reported discoveries.

13. Franks and Lambert, *Early Louisiana and Arkansas Oil*, 107.

14. For a description of the tools of the scientific prospector, see Louis S. Panyity, *Prospecting for Oil and Gas* (New York, John Wiley and Sons, 1920). For essays that capture this transition from the perspective of oil field workers, see Paul F. Lambert and Kenny A. Franks, eds., *Voices from the Oil Field* (Norman: University of Oklahoma Press, 1959). For a study of one group of scientific and technical entrepreneurs encouraging the transition, see Geoffrey C. Bowker, *Science on the Run: Information Management and Industrial Geophysics at Schlumberger, 1920–1940* (Cambridge, Mass.: MIT Press, 1994).

15. A classic example of overdevelopment occurred in the Cherry Grove field of Pennsylvania. After an encouraging show of oil from several wells, thousands of people descended on the area and invested in leases and production equipment. However, production in the field plummeted after several months, forcing most operators to abandon their investment. See Michael W. Caplinger and Philip W. Ross, *The Historic Petroleum Industry in the Allegheny National Forest,* prepared for the USDA Forest Service (Morgantown: West Virginia University, 1993), 37–45.

16. See J. C. Yancey, *Why and Where Oil is Found* (New York: published by the author, 1919); Thomas P. Comverse, *Oil and Where to Find It* (Amarillo: Tex.: Russell and Cockrell, 1920); G. H. Cox, C. L. Dake, and G. A. Muilenburg, *Field Methods in Petroleum Geology* (New York: McGraw Hill, 1921); Alfred Peterson, *Oil and Gas: Be Your Own Geologist* (Kansas City, Mo.: Franklin Hudson, 1921).

17. Lester C. Uren, *Petroleum Production Engineering* (New York: McGraw Hill, 1924), 8.

18. For the proceedings of a conference in which petroleum geologists and engineers met to discuss various questions involving the origins of oil, see *AIMME Transactions* 66 (1921). Also see

William Harvey Emmons, *Geology of Petroleum* (New York: McGraw-Hill, 1921); John M. Mac-Farlane, *Fishes: The Source of Petroleum* (New York: Macmillan, 1923).

19. William Vlachos and C. A. Vlachos, *The Fire and Explosion Hazards of Commercial Oils* (Philadelphia, Pa.: Vlachos and Company, 1921), 131.

20. S. F. Peckham, *Report on the Production, Technology, and Uses of Petroleum and its Byproducts* (Washington, D.C.: U.S. Bureau of Census, 1885), 59–74; Cone and Johns, *Petrolia*, 174. Also see T. S. Hunt, *Notes on the History of Petroleum or Rock Oil* (Washington, D.C.: Smithsonian Institution, 1862).

21. J. E. Hackford, "Nature of Coal," *AIMME Transactions* 66 (1921): 217–28. Hackford, a scientist from England, presented chemical evidence that petroleum could not have come from coal, new evidence for a conclusion that U.S. geologists had long since reached.

22. Chester W. Washburne, "Oil Field Brines," *AIMME Transactions* 66 (1921): 269–94. See also U.S. Geological Survey, *Chemical Relations of the Oil-Field Waters in San Joaquin Valley, California*, prepared by G. Sherburne Rogers (Washington, D.C.: U.S. Geological Survey, 1917), 25–26.

23. David White, "Genetic Problems Affecting the Search for New Oil Regions," *AIMME Transactions* 66 (1921): 176–98.

24. "Arkansas Lease Prices Higher than at Homer; Buyers Get Close In," *National Petroleum News* 13 (January 19, 1921): 17; "Daily Production at El Dorado Increased 3000 Barrels," *Oil Weekly* 26 (April 29, 1922): 46.

25. "El Dorado Strives to Handle Crowds," *National Petroleum News* 13 (January 19, 1921): 22; John Ise, *The United States Oil Policy* (New Haven: Yale University Press, 1926), 71–72; Charles E. Bowles, *The Petroleum Industry* (Kansas City, Mo.: Schooley Printing, 1921), 28–29. One entrepreneur who came to El Dorado hoping to put together some land deals was billionaire-to-be H. L. Hunt; see Stanley H. Brown, *H. L. Hunt* (Chicago: Playboy Press, 1976), 48–83.

26. Not everybody, of course, benefited from the development of oil fields, and numerous people often opposed development; for a broader picture, see Nancy Quam-Wickham, "Cities Sacrificed on the Altar of Oil: Popular Opposition to Oil Development in 1920s Los Angeles," *Environmental History* 3 (April 1998): 189–209.

27. George Hill (Vice President, Houston Oil) to Edward Whitaker (Executive Committee, Houston Oil), January 25, 1921, folder 6, box 9, George A. Hill, Jr., Collection, Houston Metropolitan Research Center.

28. Joe Simmons (Superintendent, El Dorado) to E. H. Buckner (Vice President, Houston Oil), June 27, 1921, folder 6, box 9, George A. Hill, Jr., Collection, Houston Metropolitan Research Center.

29. Bell and Kerr, *El Dorado*, 22.

30. Ibid., 7–12. The Arkansas Conservation Commission had been formed in 1917 to monitor gas production elsewhere in the state, but was not equipped to deal with a field such as El Dorado.

31. See C. P. Bowie, *Extinguishing and Preventing Oil and Gas Fires*, Bureau of Mines, Bulletin no. 170 (Washington, D.C.: Government Printing Office, 1920).

32. Bell and Kerr, *El Dorado*, 10. For a general account of conditions in U.S. oil fields in the period 1921 to 1923, see Ise, *Oil Policy*, 105–22.

33. Until 1918, over 90 percent of all oil wells were drilled using this method; see R. E. Collom, *Prospecting and Testing for Oil and Gas*, Bureau of Mines, Bulletin no. 201 (Washington, D.C.: Government Printing Office, 1916), 115.

34. By 1921, when oil crews in Kansas, Texas, and Oklahoma were routinely drilling to depths of 4,000 feet, fields with oil less than 1,500 feet below the surface were considered to be shallow fields (Bowles, *Petroleum Industry*, 57).

35. For examples of such incidents, see Mody C. Boatright and William A. Owens, *Tales from the Derrick Floor: A People's History of the Oil Industry* (Lincoln: University of Nebraska, 1970.) For a description and history of cable tool drilling, see J. E. Brantly, "Percussion-Drilling Sys-

tem," in American Petroleum Institute, *History of Petroleum Engineering*, (Dallas, Tex.: API, 1961), 133–269.

36. For a description and history of rotary drilling, see J. E. Brantly, "Hydraulic Rotary-Drilling System," in API, *History of Petroleum Engineering*, 273–444.

37. Bell and Kerr, *El Dorado*, 22.

38. Before the Cushing oil field was discovered in 1912, most petroleum geologists were employed by the U.S. Geological Survey or state agencies charged with making an inventory of mineral resources. Oil companies paid more attention to geologists after they played an important role in the development of the Cushing field. See "Historical Notes," box 21, Cities Service Oil Company Collection, Western History Collection, University of Oklahoma. Also see Ellen Sue Blakey, *Oil on their Shoes: Petroleum Geology to 1918* (Tulsa, Okla.: American Association of Petroleum Geologists, 1985).

39. John R. Suman, "Evolution by Companies," in API, *History of Petroleum Engineering*, 68.

40. Eugene G. Leonardon, "Logging" in API, *History of Petroleum Engineering*, 502; Sullivan Machinery Company, *Diamond Drilling for Oil* (Chicago: Sullivan Machinery Company, 1925).

41. Uren, *Petroleum Production*, 71.

42. Bell and Kerr, *El Dorado*, 25, 79.

43. For more on oil field slang, see Lalia Phipps Boone, *The Petroleum Dictionary* (Norman: University of Oklahoma Press, 1952); Jerome B. Robertson, *Oil Slanguage: An Explanation of Terms and Slang of Oil Fields From Pennsylvania to California, Texas to Montana—and Around the World* (Evansville, Ind.: Petroleum Publishers, 1954); Arthur T. King, *Oil Refinery Terms in Oklahoma*, American Dialect Society Publication no. 9 (Greensboro, N.C.: Woman's College of the University of North Carolina, 1948).

44. Bell and Kerr, *El Dorado*, 23.

45. Ibid., 12.

46. Charles Van Ormer Millikan, *Oil-Well Cementing Practices in the United States* (New York: API, 1959), 1.

47. Ibid., 3. A method for using hydraulic cement to seal off water in oil wells had been patented in 1871 (John R. Hill, "Improved Mode of Closing the Water Courses Encountered in Drilling Oil Wells," U.S. Patent no. 112,596) but the practice had not come into wide use. Also see Paul R. Waddell and Robert F. Niven, *Sign of the 76: The Fabulous Life and Times of the Union Oil Company of California* (Los Angeles: Union Oil Company, 1976), 137–38.

48. Suman, "Evolution by Companies," in API, *History of Petroleum Engineering*, 65–66.

49. Charles A. Warner, "Sources of Men," in API, *History of Petroleum Engineering*, 43.

50. Millikan, *Oil Well Cementing*, 6; Bowles, *Petroleum Industry*, 109.

51. F. B. Tough, *Methods of Shutting off Water in Oil and Gas Fields*, Bureau of Mines, Bulletin no. 163 (Washington, D.C.: Government Printing Office, 1918); Millikan, *Oil Well Cementing*, 8–10; transcript, interview of geologist George Abell by Samuel Myres, Permian Basin Petroleum Museum, Midland, Texas; George Livermore, "Drilling and Completion Practices," Carl Coke Rister folder, Permian Basin Petroleum Museum; Ruth H. McCoy, "Halliburton Oil Well Cementing Company" (M.A. thesis: Oklahoma Agricultural and Mechanical College, 1958).

52. Oil well drillers in the area used cement earlier than 1920 but not for sealing casing. For example, see Franks and Lambert, *Early Louisiana and Arkansas Oil*, 38–39; the authors point to the use of cement by Benedum and Trees in 1908 to control a gas well.

53. For comments on the conscious effort to replace the word "gusher" with "blowout," see Petroleum Extension Service, *Prevention and Control of Blowouts* (Austin: University of Texas, 1958). Also see Joseph A. Pratt, *The Growth of a Refining Region* (Greenwich, Conn.: JAI Press, 1980), 227.

54. Bell and Kerr, *El Dorado*, 12.

55. Bowles, *Petroleum Industry*, 62–67

56. U.S. Geological Survey, *Oil-Field Waters*, 14.

57. E. DeGolyer, "Concepts on Occurrence of Oil and Gas," in API, *History of Petroleum Engineering*, 28.

58. Bell and Kerr, *El Dorado*, 12.

59. The classic paper on the importance of solution gas was not published until 1926: C. E. Beecher and I. P. Parkhurst, "Effect of Dissolved Gas upon the Viscosity and Surface Tension of Crude Oil," in *Petroleum Technology and Development in 1926* (New York: American Institute of Mining and Metallurgical Engineers, 1927).

60. In 1917, 785 compression plants and 102 absorption plants were already in operation in U.S. oil fields. Casinghead gasoline, which contained large amounts of volatile butane (C_4H_{10}), received a bad reputation as a wild, explosive gasoline because many operators sold it as motor fuel without further processing. See Vlachos and Vlachos, *Fire and Explosion Hazards*, 126–27.

61. Bell and Kerr, *El Dorado*, 62–65.

62. Ibid., 55–62; includes diagrams, photos, and description of bleeder pipes and dehydration equipment,

63. Ibid., 60–62.

64. James H. Wescott, *Oil: Its Conservation and Waste* (New York: Beacon, 1930), 195.

65. Stanley C. Herold, *Oil Well Drainage* (Stanford, Calif.: Stanford University Press, 1941), 325.

66. Theodore E. Swigart, "Handling Oil and Gas in the Field," in API, *History of Petroleum Engineering*, 931.

67. Ogden S. Jones, *Fresh Water Protection from Pollution Arising in the Oil Fields* (Lawrence: University of Kansas Publications, 1950), 8. Some oil field brine contained as much as 248,000 parts per million of dissolved solids, or about 70 pounds of dissolved solids per barrel.

68. Wescott, *Oil: Its Conservation and Waste*, 207.

69. *Pulaski Oil Co. v. Edwards*, no. 11334, July 31, 1923, *Oklahoma Reports: Cases Determined in the Supreme Court of the State of Oklahoma*, vol. 91, pp. 56–58.

70. Federal Oil Conservation Board, *Complete Record of Public Hearings, February 10 and 11* (Washington D.C: Government Printing Office, 1926), 48–65.

71. A. W. Ambrose, "Analysis of Oil-field Water Problems," *AIMME Transactions* 66 (1921): 245–68.

72. Bowles, *Petroleum Industry*, 65.

73. Williamson et al., *Age of Energy*, 389.

74. Ibid., 321–29.

Chapter 3. Keeping Oil in the Pipelines

1. Harold F. Williamson, Ralph L. Andreano, Arnold R. Daum, and Gilbert C. Klose, *The American Petroleum Industry: The Age of Energy, 1899–1959* (Evanston, Ill.: Northwestern University Press, 1963), 339–62.

2. H. W. Bell and J. B. Kerr, *The El Dorado, Arkansas Oil and Gas Field* (Little Rock, Ark.: U.S. Bureau of Mines, 1922), 11; "Two Wells in New El Dorado Field Reported Flowing Pipe Line Oil," *Oil Weekly* 21 (March 19, 1921): 34; "El Dorado Producing 40,000 Barrels—Pipe Lines Nearly Ready," *Oil Weekly* 22 (June 18, 1921): 34; *Doherty News* (April 22, 1921): 23, old publications, box 26, Cities Service Oil Company Collection, Western History Collection, University of Oklahoma.

3. "El Dorado Pool of South Arkansas Only 36 Miles Away From Pipelines," *National Petroleum News* 13 (January 19, 1921): 21; "El Dorado Producing 40,000 Barrels—Pipe Lines Nearly Ready," *Oil Weekly* 22 (June 18, 1921): 46.

4. "Keen Demand for El Dorado Crude," *Oil Weekly* 23 (August 20, 1921): 36; John L. Loos, *Oil on Stream! A History of the Interstate Oil Pipe Line Company, 1909–1959* (Baton Rouge: Louisiana State University Press, 1959), 44–45.

5. Arthur M. Johnson, *The Development of American Petroleum Pipelines, 1862–1906* (Ithaca: Cornell University Press, 1956), 11.

6. See the correspondence concerning El Dorado run tickets between George Hill (Vice President of the Houston Oil Company) and Joe Simmons (superintendent of El Dorado leases), folder 6, box 9, George A. Hill, Jr., Collection, Houston Metropolitan Research Center.

7. "Twenty-Three Producers Listed as Completions in El Dorado," *Oil Weekly* 22 (June 11, 1921).

8. Loos, *Oil on Stream,* 45.

9. The typical barrel could hold about forty-two gallons. For the history of using the barrel as a unit of volume in the petroleum industry; see Robert Hardwicke, *The Oilmen's Barrel* (Norman: University of Oklahoma, 1958).

10. Andrew Cone and Walter Johns, *Petrolia: A Brief History of the Pennsylvania Petroleum Region From 1859 to 1869* (New York: D. Appleton and Co., 1870), 96–107. Artificial freshets were originally used to float timber.

11. Williamson et al., *Age of Energy,* 164–68; J. T. Henry, *Early and Later History of Petroleum* (Philadelphia: Jas. B. Rodgers, 1873), 289; Paul H. Giddens, *The Birth of the Oil Industry* (New York: Macmillan, 1938), 101–3. Henry notes that one teamster made nineteen hundred dollars for nine weeks of service. Giddens notes that 6,000 horse teams were engaged in the trade prior to 1862.

12. Cone and Johns, *Petrolia,* 106.

13. Ibid., 101.

14. Giddens, *Oil Industry,* 143. Also see Johnson, *American Petroleum Pipelines,* 4–9; Harold F. Williamson and Arnold Daum, *The American Petroleum Industry: The Age of Illumination, 1859–1899* (Evanston, Ill.: Northwestern University Press, 1959), 164–68.

15. Giddens, *Oil Industry,* 142–44.

16. "Petroleum Panorama: Commemorating 100 Years of Petroleum Progress," *Oil and Gas Journal* 57 (January 28, 1959): F3; *Brief for Defendants on the Facts: Historical Statement—Pipelines—Competitors,* vol. 1, submitted to the Circuit Court of the U.S. for the Eastern Division of the Eastern Judicial District of Missouri, *U.S. v Standard Oil,* 16 (copy in API Reference Library); Williamson and Daum, *Age of Illumination,* 228–29.

17. In 1865, the largest refinery in Western Pennsylvania required two hundred people to produce 1,800 barrels of kerosene per week: William Wright, *The Oil Regions of Pennsylvania* (New York: Harper and Brother, 1865), 200.

18. For an insightful account of Rockefeller and Standard Oil that debunks the stereotype of Standard Oil as being all-powerful and ruthlessly monopolistic, see Roger M. Olien and Diana Davids Olien, *Oil and Ideology: The Cultural Creation of the American Petroleum Industry* (Chapel Hill: University of North Carolina Press, 2000). For other perspectives, see Ida M. Tarbell, *The History of the Standard Oil Company* (New York: McClure, Phillips and Co., 1904); Allan Nevins, *Study in Power: John D. Rockefeller, Industrialist and Philanthropist,* 2 vols. (New York: Charles Scribner's Sons, 1953); Ralph W. Hidy and Muriel E. Hidy, *Pioneering in Big Business, 1882–1911: History of the Standard Oil Company (New Jersey)* (New York: Harper and Brothers, 1955). For a concise summary of Rockefeller's strategy see Daniel Yergin, *The Prize: The Epic Quest for Oil, Money, and Power* (New York: Simon and Schuster, 1991), 35–55.

19. Johnson, *American Petroleum Pipelines,* 1–25. Standard Oil eventually quashed efforts of railroads to expand into the refining business. For a more general analysis of the strategy followed by railroads, see Alfred D. Chandler, Jr., *The Visible Hand: The Managerial Revolution in American Business* (Cambridge, Mass.: Belknap Press, 1977).

20. Arthur M. Johnson, *Petroleum Pipelines and Public Policy, 1906–1959* (Cambridge, Mass.: Harvard University Press, 1967), 5. For an account that captures the spirit with which independent producers viewed Standard, see Charles A. Whiteshot, *The Oil-Well Driller: A History of the World's Greatest Enterprise, The Oil Industry* (Mannington, W. Va.: published by the author, 1905).

21. For a history of this pipeline, see Robert D. Benson and William S. Benson, *History of the Tide Water Companies* (New York: published by the authors, 1913); Williamson and Daum, *Age of Illumination,* 405–7.

22. Pennsylvania legislators did not pass a Free Pipe Line Bill until 1883; this bill gave pipeline companies the right to cross railroad rights-of-way (Johnson, *American Petroleum Pipelines,* 121–22).

23. For a photograph of a Columbia Conduit truck crossing the railroad, see Robert O. Anderson, *Fundamentals of the Petroleum Industry* (Norman: University of Oklahoma Press, 1984), 18.

24. Charles Morrow Wilson, *Oil Across the World: The American Saga of Pipelines* (New York: Longmans, Green, and Co., 1946), 48–49.

25. Benson and Benson, *Tide Water Companies,* 9.

26. Wilson, *Oil Across the World,* 49.

27. Benson and Benson, *Tide Water Companies,* 9.

28. Johnson, *American Petroleum Pipelines,* 121.

29. Hidy and Hidy, *Pioneering in Big Business,* 725 n. The Pennsylvania Railroad failed to prevent the pipeline.

30. See "Affidavits of suffering land owners from devastations caused by pipe line companies," a collection of statements by farmers of Union County, Pennsylvania, against the National Pipeline Company, 1883, Pamphlet Collection, Hagley Museum and Library.

31. Hidy and Hidy, *Pioneering in Big Business,* 81.

32. Johnson, *American Petroleum Pipelines,* 249.

33. Ibid., 4.

34. Loos, *Oil on Stream,* 10.

35. Johnson, *Pipelines and Public Policy,* 79–81.

36. Ibid., 85–87.

37. T. Harry Williams, *Huey Long* (New York: Knopf, 1970), 118. During World War I, Long had been offered $12,000 by Standard Oil of Louisiana for Long's share of a lease that had been obtained by Long for $1,000. Long, holding out for more, lost his opportunity when World War I ended and Standard withdrew its offer.

38. Loos, *Oil on Stream,* 50–51, 82–97. After becoming an interstate carrier, Standard of Louisiana also purchased the segment of the main trunk that crossed Arkansas. Also see Williams, *Huey Long,* 173–80.

39. Charles E. Bowles, *The Petroleum Industry* (Kansas City, Mo.: Schooley Printing, 1921), 77; "Transportation of Oil by Pipeline," in *Sketches of the Midwest,* no date, 51–54, box 23, Cities Service Oil Company Collection, Western History Collection, University of Oklahoma. The reason for keeping each station isolated from the other stemmed more from problems in getting pumps to operate in synchronization than in any desire for that isolation.

40. George S. Wolpert, Jr., *U.S. Oil Pipelines* (Washington, D.C.: American Petroleum Institute, 1979), 14.

41. U.S. Steel, *The Making, Shaping, and Treating of Steel,* 7th ed. (Pittsburgh: U.S. Steel, 1957), 739.

42. Bowles, *Petroleum Industry,* 76

43. Loos, *Oil on Stream,* 39–40. For natural gas lines, companies could use pipe greater than twelve inches in diameter because pumping pressures did not have to be so high on gas lines.

44. Bell and Kerr, *El Dorado,* 11.

45. Lester Charles Uren, *Petroleum Production Engineering* (New York: McGraw Hill, 1924), 570.

46. Williamson and Daum, *Age of Illumination,* 603–4.

47. For photographs of digging and pipe-screwing machinery available in the early 1920s, see Uren, *Petroleum Production Engineering,* 567–69.

48. Ibid., 565. Using heavy collars at the joints increased the allowable line pressure for a given pipe.

49. *Oil and Gas Journal* 57 (January 28, 1959): F30.

50. "Evolution of Natural Gas Transportation Systems Makes a Long and Colorful Story," *The Empire* (August 1939), historical material, box 21, Cities Service Oil Company Collection, Western History Collection, University of Oklahoma. Also see Darwin Payne, *Initiative in Energy: The Story of Dresser Industries, 1880–1978* (New York: Simon and Schuster, 1979), 105. Dresser invented and manufactured the rubber-packed couplings. Field tests by the Bureau of Mines

showed that leakage rates of natural gas from threaded pipe "varied from 17,000 to 15,000,000 cubic feet per year per mile of equivalent three-inch main;" M. J. Kirwan, "Activities of the Petroleum Research Station," paper presented at the International Petroleum and Exposition and Congress, October 8, 1924, Tulsa, Oklahoma, Organization of Bartlesville, box 17, Records of the Petroleum Experiment Center, Bureau of Mines, RG 70-2, National Archives, Southwest Region, Fort Worth.

51. Loos, *Oil on Stream,* 27.

52. "Recollections of Ben Fisher," November 13, 1959, Pipeline History, box 29, Cities Service Oil Company Collection.

53. "Transportation of Oil by Pipeline," in *Sketches of the Midwest,* 51–54, box 23, Cities Service Oil Company Collection.

54. James H. Wescott, *Oil: Its Conservation and Waste* (New York: Beacon, 1930), 195.

55. Magnolia right-of-way lease, box 31, series 7, Sun Company Collection, Hagley Museum and Library.

56. Louisiana Standard had a "telegraphone" system, which provided telegraph and crude telephone service, running along one pipeline in 1909: Loos, *Oil on Stream,* 29. Texaco still had some telegraph lines in service in 1930; Texas Company, *Texaco* (Houston: The Texas Company, 1931), 12.

57. R. C. McWane, *Pipe and the Public Welfare* (New York: Stirling Press, 1917), 65–67.

58. Loos, *Oil on Stream,* 40; Kenny A. Franks and Paul F. Lambert, *Early Louisiana and Arkansas Oil: A Photographic History, 1901–1946* (College Station, Tex.: Texas A&M University Press, 1982), 85.

59. For mention of practices in Louisiana in 1920, see Erick Larson, *Pipeline Corrosion and Coatings* (New York: American Gas Journal, 1938), 4. For quantity of material, see M. B. Sweeney to J. N. Pew, August 3, 1920, box 31, series 7, Sun Company Collection, Hagley Museum and Library.

60. This worst-case estimate assumes that the station would continue to pump oil into the severed line for several hours and that half of the oil remaining in the line after pumping stopped would drain through the break.

61. Interview of Burt E. Hull by W. A. Owens, August 24, 1953, tape 129, Oral History of the Oil Industry, Center for American History, University of Texas, Austin.

62. Johnson, *American Petroleum Pipelines,* 115.

63. *West's Texas Digest, 1840-Date,* Mines and Minerals to Mortgages, (St. Paul, Minn.: West Publishing Company, 1957), section 104. In *Texas Company v Giddings,* 148 S. W. 112 (1912), the court held that the owner of a petroleum pipeline was liable for damages if crude oil leaking from a pipeline contaminated "percolating waters" under the land of a property owner.

64. Interoffice correspondence, J. N. Pew, Jr., to John Pew, box 8, series 1F, Sun Company Collection, Hagley Museum and Library.

65. "Report of the Work Done by the Petroleum Experiment Station," November 1921, box 2, 017.41, Records of the Petroleum Experiment Center, Bureau of Mines, RG 70-2, National Archives, Southwest Region, Fort Worth.

66. Theodore E. Swigart, "Handling Gas and Oil in the Field" in American Petroleum Institute, *History of Petroleum Engineering* (New York: API, 1961), 941. Others before Wiggins in the Bureau of Mines had written about evaporation losses but not as systematically.

67. Glenn Marston, ed., *Principles and Ideas for Doherty Men: Papers, Addresses, and Letters by Henry L. Doherty,* vol. 5, printed for the use of Doherty organizations, 1923, p. 242, box 19, Cities Service Oil Company Collection, Western History Collection, University of Oklahoma.

68. J. H. Wiggins, "Evaporation Loss May Be Cut Third to Half By Shutting Out Air," *National Petroleum News* 13 (April 20, 1921): 77–85.

69. Ludwig Schmidt, "Evaporation Losses of Petroleum and Gasoline," report submitted to U.S. House of Representatives, Committee of Interstate and Foreign Commerce, November 1934, box 106, Records of Research Groups, Bartlesville Petroleum Research Center, Bureau of Mines, RG 70-4, National Archives, Southwest Region, Fort Worth.

70. U.S. Bureau of Mines, *Evaporation Losses of Petroleum in the Mid-Continent Field,* Bulletin no. 200, prepared by J. H. Wiggins (Washington, D.C.: Government Printing Office, 1921); U.S. Bureau of Mines, *Methods of Decreasing Evaporation Losses of Petroleum,* Bureau of Mines, Technical Paper no. 318, prepared by J. H. Wiggins (Washington, D.C.: Bureau of Mines, 1923); "Conservation Saves Millions Says Wiggins," *Petroleum Age* 17 (March 15, 1926): 42, 49.

71. "Tank Roofs to Stand One Pound Pressure Cut Evaporation Loss to .21 Per Cent," *National Petroleum News* 18 (July 21, 1926): 27.

72. For example, as a way to reduce vapor losses from oil exposed to air, Jersey Standard marketed Sealite, a foamy covering that eventually proved ineffective: Henrietta M. Larson and Kenneth W. Porter, *History of Humble Oil and Refining: A Study in Industrial Growth* (New York: Harper and Brothers, 1959), 125. Goodyear pushed rubber breathing bags: "Rubber Bags Seen as Boon to Jobbers," *Petroleum Age* 17 (April 1, 1926): 20; Goodyear to Sun Oil, July 24, 1926, box 27, series 1D, Sun Company Collection, Hagley Museum and Library.

73. National Fire Protection Association, "Laws and Ordinances Regulating the Use, Handling, Storage, and Sale of Flammable Liquids," advance publication, 1925, box 20, series 1D, Sun Company Collection, Hagley Museum and Library.

74. API, "Report on the Records of Oil Tank Fires in the United States, 1915–1925," April 1925, 5, box 20, series 1D, Sun Company Collection, Hagley Museum and Library.

75. Series of correspondence, 1925, "M" folder, box 22, series 1D, Sun Company Collection, Hagley Museum and Library.

76. API, "Report on the Records of Oil Tank Fires in the United States, 1915–1925," April 1925, 19, box 20, series 1D, Sun Company Collection; "API to Keep Up Records on Oil Industry Fires," *National Petroleum News* 18 (August 4, 1926): 95.

77. J. Howard Pew to W. C. L. Eglin, June 13, 1924, box 18, series 1D, Sun Company Collection, Hagley Museum and Library.

78. W. C. L. Eglin to J. Howard Pew, June 24, 1924, box 18, series 1D, Sun Company Collection, Hagley Museum and Library.

79. W. R. Boyd, Jr. (Assistant General Secretary, API) to J. Howard Pew (President, Sun Oil), May 22, 1925, box 20, series 1D, Sun Company Collection, Hagley Museum and Library.

80. *Petroleum Age* 17 (April 15, 1926): 15; clipping and handwritten notes, box 27, series 1D, box 27, Sun Company Collection, Hagley Museum and Library; Paul R. Waddell and Robert F. Niven, *Sign of the 76: The Fabulous Life and Times of the Union Oil Company of California* (Los Angeles: Union Oil Co., 1976): 189–91.

81. "100 Million Barrels of California Oil Stored Without Insurance," *National Petroleum News* 18 (August 25, 1926): 64. Only Standard of California, which self-insured its stores of crude oil, was unaffected.

82. "Lightening an Ever Present Menace to Tanks," *Petroleum Age* 17 (February 1, 1926): 24–25; "Tank Fires Due to Lightening and Other Causes—Methods of Prevention," *National Petroleum News* 18 (August 25, 1926): 49–50.

83. Ludwig Schmidt, "Report of Inspection Trip Through West Texas Fields," October 10, 1928, box 106, Records of Research Groups, Bartlesville Petroleum Research Center, Bureau of Mines, RG 70-4, National Archives, Southwest Region, Fort Worth.

84. For pictures of open storage in the Smackover field (Arkansas) during the 1930s, see Franks and Lambert, *Early Louisiana and Arkansas Oil,* 140–41.

Chapter 4. Refineries, Pollution Concerns, and Technological Change

1. Charles E. Bowles, *The Petroleum Industry* (Kansas City, Mo.: Schooley Printing, 1921), 96; H. G. James, *Refining Industry of the United States* (Oil City, Pa.: Derrick Publishing, 1916), 7–11; Harold F. Williamson, Ralph L. Andreano, Arnold R. Daum, and Gilbert C. Klose, *The American Petroleum Industry: The Age of Energy, 1899–1959* (Evanston, Ill.: Northwestern University Press, 1963), 110–66.

2. H. W. Bell and J. B. Kerr, *The El Dorado, Arkansas Oil and Gas Field* (Little Rock, Ark.: U.S. Bureau of Mines, 1922), 11.

3. Andrew Cone and Walter R. Johns, *Petrolia: A Brief History of the Pennsylvania Petroleum Region* (New York: D. Appleton and Co., 1870), 50–95. The authors include a report prepared in 1855 by Prof. B. Silliman of Yale, who was commissioned by backers of the Drake well to see if Pennsylvania "rock oil" could be used for illumination purposes. Silliman indicated that the oil could be distilled and that some fractions could be used as an illuminant, establishing the basic purpose and processes of a refinery four years before the first commercial oil well. Others before Silliman had pioneered the distillation and refining of illuminants. For example, in 1850, five years earlier than Silliman's report, a chemist in England named James Young patented a process for refining a thick liquid found oozing out of a coal mines. Also see Harold F. Williamson and Arnold Daum, *The American Petroleum Industry: The Age of Illumination, 1859–1899* (Evanston, Ill.: Northwestern University Press, 1959), 26–60, 202–32; Williamson et al., *Age of Energy,* 110–12.

4. John L. Enos, *Petroleum Progress and Profits: A History of Process Innovations* (Cambridge, Mass.: MIT Press, 1962), 6–8. Enos notes the appointment in 1896 of William Burton, a chemist, to the position of refinery superintendent at Standard Oil's Whiting Refinery—an unusually high position for a chemist at a time when most refinery workers had no technical education—and points to Burton, a pioneer in thermal cracking, as being one of the first in the wave of chemists to enter the industry. For more on the story of Lima crude, see Ralph W. Hidy and Muriel E. Hidy, *Pioneering in Big Business, 1882–1911: History of the Standard Oil Company (New Jersey),* (New York: Harper and Brothers, 1955), 155–68.

5. For an example of such diagrams in an introductory text written for insurance underwriters, see William Vlachos and C. A. Vlachos, *The Fire and Explosion Hazards of Commercial Oils* (Philadelphia, Pa.: Vlachos and Company, 1921), 135. For examples in an introductory text aimed at mechanical engineers, see Harold Moore, *Liquid Fuels for Internal Combustion Engines: A Practical Guide for Engineers* (New York: D. Van Nostrand, 1917), 5.

6. Kendall Beaton, *Enterprise in Oil: A History of Shell in the United States* (New York: Appleton-Century-Crofts, 1957), 509.

7. Robert O. Anderson, *Fundamentals of the Petroleum Industry* (Norman: University of Oklahoma Press, 1984), 77.

8. This description of batch refining is based on Bowles, *Petroleum Industry,* 87–109; J. W. Coast, Jr., "Refining Petroleum," *Doherty News* (October 1919), box 26, Cities Service Oil Company Collection, Western History Collection, University of Oklahoma.

9. Frederick A. Talbot, *The Oil Conquest of the World* (Philadelphia: J. B. Lippincott, 1914), 140.

10. Hidy and Hidy, *Pioneering in Big Business,* 298.

11. Water gas was made by passing steam over incandescent coke, which produced a mixture of hydrogen and carbon monoxide that, when enriched with oil gas, was sold as an illuminating gas: Emil R. Riegel, *Industrial Chemistry* (New York: Reinhold Publishing, 1942), 271–75.

12. Coast, "Refining Petroleum," 18.

13. Bowles, *Petroleum Industry,* 91.

14. For a detailed history of these innovations, see: Enos, *Petroleum Progress;* Williamson et al., *Age of Energy,* 110–66; F. H. Garner, "Distillation," in *Petroleum: Twenty-Five Years Retrospect, 1910–1935,* ed. Institution of Petroleum Technologists (London: Aldine House, 1935), 82–85.

15. For a description of Standard's experience with the Livingston continuous distillation process, see Hidy and Hidy, *Pioneering in Big Business,* 421–28.

16. Williamson et al., *Age of Energy,* 126; Hidy and Hidy, *Pioneering in Big Business,* 421–28; Garner, "Distillation," 82–83.

17. In 1855, William Silliman, Jr., the chemist who analyzed Pennsylvania oil for the backers of the Drake well, noticed that intense heat would decompose heavy oil into more volatile components. Ernest Raymond Lilley, *The Oil Industry: Production, Transportation, Resources, Refining, and Marketing* (New York: D. Van Nostrand, 1925), 406; Enos, *Petroleum Progress,* 1; Williamson and Daum, *Age of Illumination,* 216.

18. J. H. Adams in James H. Wescott, *Oil: Its Conservation and Waste* (New York: Beacon, 1930), 108–09.

19. Enos, *Petroleum Progress*, 4, 18–27.

20. Because of overlapping patents, some of those controlling cracking technology entered into a patent pool, a move which eventually resulted in charges that they had violated antitrust laws. One suit over thermal cracking patents was still going on in 1930. Enos, *Petroleum Progress*, 114–21; Wescott, *Conservation and Waste*, 137–46.

21. Wescott, *Conservation and Waste*, 111.

22. Enos, *Petroleum Progress*, 144–45; August W. Giebelhaus, *Business and Government in Industry: A Case Study of Sun Oil, 1876–1945* (Greenwich, Conn.: JAI Press, 1980).

23. Carleton Ellis and Joseph V. Meigs, *Gasoline and Other Motor Fuels* (New York: D. Van Nostrand, 1921), 105.

24. R. R. Sayers et al., *Investigation of Toxic Gasses from Mexican Crude and Other High-Sulfur Petroleum Products*, Bureau of Mines, Bulletin no. 231 (Washington, D.C.: Government Printing Office, 1925); "Safety Engineers are Told of Hydrogen Sulfide Gas," *National Petroleum News* 18 (September 19, 1926): 77; Interview of A. J. Thaman and J. H. Anderson by Maude Ross, January 29, 1960, Tape. No. 216, Oral History of the Oil Industry, Center for American History, University of Texas at Austin. Also see Joseph A. Pratt, "Letting the Grandchildren Do It: Environmental Planning During the Ascent of Oil as a Major Energy Source," *Public Historian* 2 (1980), 52–59.

25. Coast, "*Refining Petroleum*," 18.

26. Cone and Johns, *Petrolia*, 576–77. Refiners had been using sulfuric acid to strip impurities from kerosene since the beginning of the industry.

27. Harold S. Bell, *American Petroleum Refining* (New York: D. Van Nostrand, 1923), 192; W. L. Nelson, *Petroleum Refinery Engineering* (New York: McGraw Hill, 1941), 577; Coast, "Refining Petroleum," 19.

28. David T. Day, ed., *A Handbook of the Petroleum Industry* (New York: Wiley and Sons, 1922), 334–37.

29. Interview of W. B. Hamilton by Mody Boatright, August 16, 1952, Tape 57, Oral History of the Oil Industry.

30. A. D. Smith, "Refining," in Day, *A Handbook of Petroleum Industry*, 1–2.

31. Bell, *American Petroleum Refining*, 77.

32. Joel A. Tarr, "Industrial Wastes and Public Health: Some Historical Notes," *American Journal Public Health* 75 (September 1985): 1059–67.

33. "Cities Service Company: Past, Present, and Future," speech to Columbian Division Sales Meeting, Lake Charles, Louisiana, February 27, 1974, box 21, Cities Service Oil Company Collection, Western History Collection, University of Oklahoma.

34. Ibid.

35. Address of G. S. Davison, Federal Oil Conservation Board, *Complete Record of Public Hearings, February 10 and 11* (Washington D.C., Government Printing Office, 1926), 45–48.

36. Estimate by the Bureau of Mines reported in Wescott, *Conservation and Waste*, 175; Wescott gives losses for each year in the period 1918 to 1929. For estimates of losses at specific refineries operating in the years after WWI, see Coast, "Refining Petroleum," 17; Bell, *American Petroleum Refining*, 72; Day, *A Handbook of the Petroleum Industry*, 331.

37. Wescott, *Conservation and Waste*, 154–72. Also see J. H. Wiggins, "Evaporation Loss May Be Cut Third to Half By Shutting Out Air," *National Petroleum News* 13 (April 20, 1921), 77–85.

38. Lilley, *Oil Industry*, 409.

39. Wescott, *Conservation and Waste*, 100–1.

40. Bell, *American Petroleum Refining*, 76; U.S. Bureau of Mines, *Pollution By Oil of the Coast Waters of the United States*, preliminary report (Washington D.C.: Bureau of Mines, 1923), 38–39.

41. Bell, *American Petroleum Refining*, 412.

42. Ibid.

43. Statement of C. A. Holmquist, Chief Engineer, New York Department of Health, Senate Committee on Commerce, *Pollution of Navigable Waters*, 68th Cong., 1st sess., January 9, 1924, 110.

44. R. J. Hansen to Herbert Hoover, Dec. 8, 1921, pollution of waters, Commerce Papers, Herbert Hoover Presidential Library (HHPL).

45. Enos, *Petroleum Progress*, 143.

46. Wescott, *Conservation and Waste*, 100. Estimates of acid sludge generated assume that one to five pounds of acid were used for each barrel of distillate, that 1 to 4 percent of the distillate ended up in the acid sludge, and that 50 percent of all crude was washed with acid.

47. Hidy and Hidy, *Pioneering in Big Business*, 190–92, 287; Raymond Foss Bacon and William Allen Hamor, *The American Petroleum Industry* (New York: McGraw-Hill, 1916), 584–92.

48. H. E. Michner to J. H. Pew, March 12, 1924, folder A, box 16, series 1D, Sun Company Collection, Hagley Museum and Library.

49. Norman Hurd Ricker, unpublished autobiography, 45–46, box 23, Norman Hurd Ricker Papers, Woodson Research Center, Rice University. When hired by Humble in 1923, Ricker's first job was to figure out how to recover fuller's earth without damaging its filtering capacity.

50. See Christine M. Rosen, "Noisome, Noxious, and Offensive Vapors, Fumes, and Stenches in American Towns and Cities, 1840–1865," *Historical Geography* 25 (1997): 49 82; Joel A. Tarr, "Searching for a Sink for an Industrial Waste," in *The Search for the Ultimate Sink: Urban Pollution in Historical Perspective* (Akron, Ohio: The University of Akron Press, 1996), 385–411; David Stradling, *Smokestacks and Progressives: Environmentalists, Engineers, and Air Quality in America, 1881–1951* (Baltimore: Johns Hopkins University Press, 1999).

51. Sedley Hopkins Phinney to Herbert Hoover, Nov. 5, 1923, Pollution of Water, Commerce Papers, HHPL.

52. Ibid.

53. Statement of J. H. Hayes, Senate Committee on Commerce, *Pollution of Navigable Waters*, January 9, 1924, 83.

54. Statement of Van H. Manning, Senate Committee on Commerce, *Pollution of Navigable Waters*, January 9, 1924, 93–94.

55. Statement of Francis S. McIlheny, Senate Committee on Commerce, *Pollution of Navigable Waters*, January 9, 1924, 105–6.

56. Ibid., 105–6.

57. Statement of C. A. Holmquist, Senate Committee on Commerce, *Pollution of Navigable Waters*, January 9, 1924, 109–13.

58. Herbert Hoover to S. Wallace Demsey, June 17, 1922, Pollution of Water, Commerce Papers, HHPL; House Committee on Rivers and Harbors, *Hearings on the Pollution of Navigable Waters*, 68th Cong., 1st sess., January 23–30, 1924, 9–10; Herbert Hoover to Secretary of War, February 14, 1924, Pollution of Water, Commerce Papers, HHPL.

59. J. S. Frelinghuysen to Herbert Hoover, January 30, 1924, Pollution of Waters, Commerce Papers, HHPL; Hoover to Frelinghuysen, January 30, 1924. For mention of personal attacks on Hoover's character by those seeking stronger legislation, see undated note, Pollution of Waters, Commerce Papers, HHPL; Hoover to J. S. Frelinghuysen, February 6, 1924, Pollution of Waters.

60. Francis S. McIlheny to J. Howard Pew, June 17, 1924, box 18, series 1D, Sun Company Collection, Hagley Museum and Library.

Chapter 5. What to Do with Tankers?

1. For example, see House Committee on Foreign Affairs, *Hearing on Oil Pollution of Navigable Waters*, 67th Cong., 2nd sess., Feb. 15–18, 1922, 5.

2. Harold F. Williamson, Ralph L. Andreano, Arnold R. Daum, and Gilbert C. Klose, *The American Petroleum Industry: The Age of Energy, 1899–1959* (Evanston, Ill.: Northwestern University Press, 1963), 75–78.

3. J. D. Henry, *Thirty Five Years of Oil Transport: Evolution of the Tank Steamer* (London: Bradbury, Agnew, and Co., 1907), 13.

4. Daniel Yergin, *The Prize: The Epic Quest for Oil, Money, and Power* (New York: Simon and Schuster, 1991), 65–70.

5. Henry, *Oil Transport*, 11; L. A. Sawyer and W. H. Mitchell, *Tankers* (New York: Doubleday,

1967), 7; C. R. H. Bonn, *The Oil Tanker* (Glasgow: Association of Engineering and Shipbuilding Draughtsmen, 1922), 20.

6. Henry, *Oil Transport,* 9–13; Laurence Dunn, *The World's Tankers* (New York: John De Graff, 1956), 12; Sir Basil Kemball Cook, "Ocean Transport," in *Petroleum: Twenty-Five Years Retrospect, 1910–1935,* ed. Institution of Petroleum Technologists (London: Aldine House, 1935), 122; and Harold F. Williamson and Arnold Daum, *The American Petroleum Industry: The Age of Illumination, 1859–1899* (Evanston, Ill.: Northwestern University Press, 1959), 642–43. Other ships had previously incorporated one or another of the *Gluckauf*'s features, but most authors recognize the *Gluckauf* as the prototype of the dedicated modern tanker.

7. John L. Enos, *Petroleum Progress and Profits: a History of Process Innovations* (Cambridge, Mass.: MIT Press, 1962), 142–43.

8. The Texas Company, *Texaco* (Houston: Texas Company, 1931), 55; interview of Torkild Rieber by William A. Owens, January 25, 1954, tape 147c, Oral History of the Oil Industry, American History Center, University of Texas at Austin.

9. Craig Thompson, *Gulf: A Human Story of Gulf's First Half Century* (Pittsburgh. Pa.: Gulf Oil Company, 1951), 14–24.

10. The Mexican Eagle Oil Company was a British-owned company that received preferential treatment from the Mexican government as a way to keep the power of U.S.-owned companies in check; see Ludwell Denny, *We Fight For Oil* (New York: Alfred A. Knopf, 1928), 45.

11. Sawyer and Mitchell, *Tankers,* 14.

12. Dunn, *The World's Tankers,* 43. Also see Ernest R. Lilley, *The Oil Industry: Production, Transportation Resources, Refining, and Marketing* (New York: D. Van Nostrand, 1925), 304–05.

13. Herbert John White, *Oil Tank Steamers: Their Working and Pumping Arrangement Thoroughly Explained* (Glasgow: James Brown and Sons, 1920), 31.

14. White, *Oil Tank Steamers,* 31–39. White did not recommend this procedure if the tanker was in a confined harbor.

15. Comments of J. H. Hayes, Senate Committee on Commerce, *Hearings on the Pollution of Navigable Waters,* 68th Cong., 1st sess., Jan. 9, 1924, 83–88; statement of Robert F. Hand, 79–81, 96–97.

16. Robert C. Tuttle (Atlantic Refining) to J. H. Pew (Sun Oil), June 6, 1925, box 20, series 1D, Sun Company Collection, Hagley Museum and Library, Wilmington, Delaware.

17. Interview of Torkild Rieber by William A. Owens, January 25, 1954, tape 147g, Oral History of the Oil Industry, American History Center, University of Texas at Austin.

18. Ibid., tape 147f. For a general work on Lloyd's of London and its influence, see D. E. W. Gibb, *Lloyd's of London* (London: Macmillan, 1957). Plimsoll load lines, required by law on English ships, did not become required by law on U.S. ships until the mid-1920s.

19. For example, see Société anonyme d'armement, d'industrie, et de commerce to Sun Oil Company, June 30, 1924, box 19, series 1D, Sun Company Collection, Hagley Museum and Library. An officer on the ship S.S. *Sunoil* erroneously opened a valve at the wrong time and contaminated 800 tons of "New York Red Oil." The incident was covered by insurance.

20. White, *Oil Tank Steamers,* 45.

21. Bonn, *Oil Tanker,* 33.

22. Elijah Baker III, *Introduction to Steel Shipbuilding* (New York: McGraw Hill, 1943), 37.

23. U.S. Bureau of Mines, *Pollution By Oil of the Coast Waters of the United States,* preliminary report, Sept. 1923, p. 109.

24. Interview of Rieber by Owens, tape 147g.

25. Williamson et al., *Age of Energy,* 275.

26. "Memorandum on Behalf of the American Steamship Owner's Association," April 17, 1924, box 23, series 1D, Sun Company Collection, Hagley Museum and Library.

27. Ibid.

28. Mather and Co. to J. H. Pew, May 7, 1924, box 81, series 1D, Sun Company Collection, Hagley Museum and Library.

29. "Steps Taken to End Harbor Oil Menace," *New York Times,* August 24, 1921, 28; "Urges Oil Salvage to Bar Dock Fires," *New York Times,* September 9, 1921, 8.

30. Statement of Murray Hulbert, House Committee on Rivers and Harbors, *Hearings on the Pollution of Navigable Waters,* 67th Cong., 2d sess., December 7 and 8, 1921, 108–12.

31. Statement of Robert F. Hand, Senate, *Pollution of Navigable Waters,* Jan. 9, 1924, 79–81, 96–97.

32. Statement of Commander R. D. Gatewood, House Committee on Foreign Affairs, *Hearing on Oil Pollution of Navigable Waters,* 67th Cong., 2d sess., February 15–18, 1922, 52–63.

33. Statement of George W. Edmunds, House, *Hearings on Oil Pollution,* February 15–18, 1922, 6–20.

34. Interdepartmental Committee on Pollution of Navigable Waters, "Survey of Proceedings and Findings to March 31, 1923," 12, Pollution of Water, Commerce Papers, Herbert Hoover Presidential Library (HHPL).

35. Statement of Mr. Bowles, House, *Hearings on Oil Pollution,* February 15–18, 1922, 45–50; newspaper clipping, "Device May Rid Ocean of Pollution from Oil," n. d., box 472, Pollution of Waters, Commerce Papers, HHPL; H. E. Michner to J. Howard Pew, April 23, 1924, box 16, series 1D, Sun Company Collection, Hagley Museum and Library; Supervising Inspector General of Steamboat-Inspection Service to C. Herter, December 12, 1923, box 472, Pollution of Waters, Commerce Papers, HHPL; minutes of the sixth meeting of the Interdepartmental Committee on Oil Pollution, April 16, 1923, box 472, Pollution of Waters, Commerce Papers, HHPL.

36. Statement of H. Bowles, House, *Hearings on Oil Pollution,* February 15–18, 1922, 45–50;

37. Statement of Fred K. Nielson, House, *Hearings on Oil Pollution,* February 15–18, 1922, 37–44.

38. U.S. Bureau of Mines, *Pollution By Oil,* 17–20.

39. Oliver Duke to J. Howard Pew, April 23, 1924, box 110, series 1F, Sun Company Collection, Hagley Museum and Library.

40. Christian A. Herter to Judge Davis, November 27, 1923, Pollution of Waters, Commerce Papers, HHPL.

41. Statement of J. H. Hayes, Senate, *Pollution of Navigable Waters,* Jan. 9, 1924, 83–91.

42. *Preliminary Conference on Oil Pollution of Navigable Waters,* June 8–16 (Washington, D.C.:, Government Printing Office, 1926), 436.

43. G. A. B. King, *Tanker Practice: The Construction, Operation, and Maintenance of Tankers* (London: Maritime Press, 1969), 113.

Chapter 6. Validating a Guiding Ethic

1. Joseph C. Robert, *Ethyl: A History of the Corporation and the People Who Made It* (Charlottesville: University Press of Virginia, 1983), 93–127. See also Stuart Leslie, *Boss Kettering: Wizard of General Motors* (New York: Columbia University Press, 1983), 38–60.

2. David Rosner and Gerald Markowitz, "'A Gift of God'?: The Public Health Controversy over Leaded Gasoline During the 1920s," in *Dying for Work: Workers' Safety and Health in Twentieth-Century America,* ed. David Rosner and Gerald Markowitz (Bloomington: Indiana University Press, 1987), 121–39; Joseph A. Pratt, "Letting the Grandchildren Do It: Environmental Planning During the Ascent of Oil as a Major Energy Source," *Public Historian* 2 (1980), 42–45. Also see Joseph C. Aub, Lawrence T. Fairhall, A. S. Minot, Paul Reznikoff, and Alice Hamilton, *Lead Poisoning* (Baltimore: Williams and Wilkins Co., 1926).

3. Robert, *Ethyl,* 113–19.

4. Rosner and Markowitz, "'A Gift of God'?" 129–30.

5. David A. Hounshell and John K. Smith, Jr., *Science and Corporate Strategy: Du Pont R&D, 1902–1980* (New York: Cambridge University Press, 1988), 152.

6. Rosner and Markowitz, "'A Gift of God'?," 122–23; Leslie, *Kettering,* 165; Robert, *Ethyl,* 122; Pratt, "Letting the Grandchildren Do It," 43–45.

7. Leslie, *Kettering,* 165; Hounshell and Smith, *Du Pont,* 154.

8. Robert, *Ethyl,* 122.

9. U.S. Bureau of Mines, *Experimental Studies on the Effects of Ethyl Gasoline and Its Combustion Products: Report of the Bureau of Mines to the General Motors Research Corporation and the Ethyl Gasoline Corporation,* prepared by R. R. Sayers et al. (Washington, D.C.: Government Printing Office, 1927).

10. *The Use of Tetraethyl Lead and Its Relation to Public Health,* Public Health Service, bulletin no. 163, 1926.

11. "Widening Sale of Ethyl Gasoline Carries Challenge to Refiners," *National Petroleum News* 18 (September 22, 1926): 21–23.

12. Rosner and Markowitz, "'A Gift of God'?" 121. Also see Christopher Sellers, "The Public Health Service's Office of Industrial Hygiene and the Transformation of Industrial Medicine," *Bulletin of the History of Medicine* 65 (1992): 42–73.

13. James Whorton, *Before Silent Spring: Pesticides and Public Health in Pre-DDT America* (Princeton University Press: Princeton, New Jersey, 1974).

14. Alice Hamilton, *Industrial Poisons in the United States* (New York: Macmillan, 1925) and *Exploring the Dangerous Trades: The Autobiography of Alice Hamilton, M.D.* (Boston: Little, Brown, and Co., 1943).

15. William Graebner, "Hegemony Through Science: Information Engineering and Lead Toxicology, 1925–1965," in *Dying For Work,* ed. Rosner and Markowitz, 140–59.

16. M. H. Leister to J. Howard Pew, March 11, 1932, box 148, series 1, Sun Company Collection, Hagley Museum and Library. Also see August W. Giebelhaus, *Business and Government in Industry: A Case Study of Sun Oil, 1876–1945* (Greenwich, Conn.: JAI Press, 1980), 75–79.

17. "Report of Joint Educational Committee to the Board of Directors of the American Petroleum Institute," final subcommittee draft, box 16, series 1D, Sun Company Collection, Hagley Museum and Library.

18. J. Edgar Pew to J. Howard Pew, June 3, 1924, box 16, series 1D, Sun Company Collection, Hagley Museum and Library; "Report of Joint Educational Committee."

19. "Report of Joint Educational Committee," statements of Courtland Smith (motion pictures), E. K. Hall (AT&T), and George Baker (publicity agent).

20. Ibid., statement of Bruce Barton. Also see Stephen Fox, *The Mirror Makers: A History of American Advertising and Its Creators* (New York: William Morrow and Co., 1984), 101–14.

21. C. C. Scharpenberg, "Standardizing of Oil-Field Equipment," in American Petroleum Institute, *History of Petroleum Engineering* (New York: API, 1961), 1197.

22. David A. Hounshell, *From the American System to Mass Production, 1800–1932: The Development of Manufacturing Technology in the United States* (Baltimore: Johns Hopkins University Press, 1984), 32–51.

23. Larry R. Lagerstrom, "Constructing Uniformity: The Standardization of International Electromagnetic Measures, 1860–1912" (Ph.D. diss., University of California, Berkeley, 1992), 8.

24. Steven Usselman, "Running the Machine: Technological Change in the Railroad Industry, 1865–1910" (Ph.D. diss., University of Delaware, 1986), 240–48.

25. Bruce Sinclair, *Centennial History of the American Society of Mechanical Engineers, 1880–1980* (Toronto: University of Toronto Press, 1980). Also see Monty Calvert, *The Mechanical Engineer, 1830–1910: Professional Cultures in Conflict* (Baltimore: Johns Hopkins University Press, 1967); Edwin Layton, *The Revolt of the Engineers: Social Responsibility in the American Engineering Profession* (Cleveland, Ohio: Case Western University Press, 1971).

26. "A National Need," *Scientific American* 82 (1900): 307.

27. A. Hunter Dupree, *Science in the Federal Government: A History of Policies and Activities* (Baltimore: Johns Hopkins University Press, 1986), 271–77. Also see David Cahan, *An Institute for an Empire: The Physikalische-Technische Reichsanstalt, 1871–1918* (Cambridge: Cambridge University Press, 1989).

28. Rexmond Cochrane, *Measures for Progress* (Washington, D.C.: National Bureau of Standards, 1966), 90–91.

29. Bruce Seely, *Building the American Highway System: Engineers as Policy Makers* (Philadelphia: Temple University Press, 1987), 127.

30. P. G. Agnew, "Twenty Years of Standardization," *Industrial Standardization* 9 (October 1938): 232; Clifford B. Le Page, "Twenty-Five Years—the American Standards Association, Origins," *Industrial Standardization* 14 (December 1943): 318.

31. Herbert Hoover, *The Cabinet and the Presidency, 1920–1933*, vol. 2 of *The Memoirs of Herbert Hoover* (New York: Macmillan, 1951), 66.

32. "Paul Gough Agnew, 1881–1954," *Magazine of Standards* 25 (February 1954): 47. Agnew also had a hand in the formation of the International Standards Organization; *New York Times*, January 10, 1954, 86.

33. *Engineering News Record* 82 (April 10, 1919): 700.

34. API, Report on Activities, 1932, box 65, series 1D, Sun Company Collection, Hagley Museum and Library; API, Report on the Committee of Technical Relationships, July 6, 1927, box 38, series 7, Sun Company Collection; Minutes of the Second Conference on Standardization in the Mechanical Industries, June 28, 1928, and related correspondence, API Files, box 38, series 7, Sun Company Collection.

35. W. R. Trelford to R. B Anderson, July 24, 1928, API, box 35, series 1D, Sun Company Collection, Hagley Museum and Library; J. Howard Pew to F. J. Huffman, April 25, 1924, box 18, series 1D, Sun Company Collection. Compare to level of standardization evident by the early 1930s; see API, "Code for Measuring, Sampling, and Testing Crude Oil," box 64, series 1D, Sun Company Collection.

36. W. R. Boyd to J. E. Pew, October 22, 1928, box 37, series 7, Sun Company Collection, Hagley Museum and Library; Pew to Boyd, October 26, 1928, box 37, series 7, Sun Company Collection.

37. "Refiners Frankly Describe Processes as Plant Superintendents of W.P.R.A. Meet," *National Petroleum News* 18 (July 21, 1926): 32; the "frank" talk mainly had to do with corrosion and safety.

38. "Boards and Subcommittees of the API," September 1932, box 65, series 1D, Sun Company Collection, Hagley Museum and Library.

39. "Twenty Years of Standardization," *Industrial Standardization* 9 (October 1938): 232, R. M. Hudson (Chief, Division of Simplified Practice) to Commerce Secretary Herbert Hoover, "The Future of Standardization," April 6, 1927, box 569, Standardization, Commerce Papers, Herbert Hoover Presidential Library (HHPL).

40. Charles A. Whiteshot, *The Oil-Well Driller: A History of the World's Greatest Enterprise, The Oil Industry* (Mannington, W. Va.: published by the author, 1905), 179.

41. Ralph W. Hidy and Muriel E. Hidy, *Pioneering in Big Business, 1882–1911: History of the Standard Oil Company (New Jersey)*, (New York: Harper and Brothers, 1955), 27. The company, Astral Oil, was affiliated with Standard Oil.

42. Harold F. Williamson and Arnold Daum, *The American Petroleum Industry: The Age of Illumination, 1859–1899* (Evanston, Ill.: Northwestern University Press, 1959), 313–17, 523; Andrew Cone and Walter R. Johns, *Petrolia: A Brief History of the Pennsylvania Petroleum Region from 1859 to 1869* (New York: D. Appleton and Co., 1870), 578–79. In 1880, seventeen of thirty-eight states still had no special legislation regulating kerosene.

43. Carleton Ellis and Joseph V. Meigs, *Gasoline and Other Motor Fuels* (New York: D. Van Nostrand, 1921), 63–71; Warren Platt, "Standardization—A Present Problem," *National Petroleum News* 18 (July 21, 1926): 63.

44. "Reciprocal Insurance as Applied to Oil Industry Risks," *National Petroleum News* 13 (April 6, 1921): 41–47. For the origin of mutual insurance pools for factories, see Betsy W. Bahr, "New England Mill Engineering: Rationalizing and Reform in Textile Mill Design, 1790–1920" (Ph.D. diss., University of Delaware, 1987).

45. "Eastern Refiners Plainly Told Faults Connection with Fire Hazard," *National Petroleum News* 18 (September 22, 1926): 58.

46. Lalia Phipps Boone, *The Petroleum Dictionary* (Norman: University of Oklahoma Press, 1952); "Strictly API," *Tomorrow's Tools Today* 22 (3rd quarter 1956): 30–33, Lane Wells, box 92,

Records of Research Groups, Bartlesville Petroleum Research Center, Bureau of Mines, RG 70-4, National Archives, Southwest Region, Fort Worth.

47. U.S. Corps of Engineers, Report of the Secretary of War to the House of Representatives, *Pollution Affecting Navigation or Commerce on Navigable Waters,* 69th Cong., 1st sess., 1926, H. Doc. 417, 10.

48. Ibid., 24.

49. Ibid., 10.

50. H.R. 10903, introduced March 31, 1926, 69th Cong., 1st sess. For mention of other bills, see "Discriminatory Legislation Against Oil," *National Petroleum News* 18 (September 22, 1926): 56–57.

51. API, "Survey of Oil Pollution Conditions in the United States," 1927, 1–2, box 35, series 1D, Sun Company Collection, Hagley Museum and Library. Emphasis is in the original.

52. Ibid., 3–6.

53. J. H. Pew to W. R. Boyd, Jr., June 21, 1928, box 35, series 1D, Sun Company Collection, Hagley Museum and Library.

54. Joseph A. Pratt, *The Growth of a Refining Region* (Greenwich, Conn.: JAI Press, 1980), 226.

55. House Committee on Rivers and Harbors, *Hearings on a Bill to Amend the Oil Pollution Act of 1924,* 71st Cong., 2d sess., May 2, 3, and 26, 1930, 22.

56. Federal Oil Conservation Board, *Complete Record of Public Hearings, February 10 and 11* (Washington D.C: Government Printing Office, 1926), 48–65. Also see R. D. Bush, "Conservation of Gas in the Production of Oil," *Oil Bulletin* (September 1927); notes of meeting, J. S. Desmond, June 11, 1929, box 128, Records of Research Groups, Bartlesville Petroleum Research Center, Bureau of Mines, RG 70-4, National Archives, Southwest Region, Fort Worth; A. W. Ambrose, "Only 20% of the Oil is Recovered," *Oil and Gas Journal* (Oct. 10, 1919), 62, 66.

57. John Ise, *The United States Oil Policy* (New Haven: Yale University Press, 1926), 162–63, 281; Leonard M. Logan, Jr., *Stabilization of the Petroleum Industry* (Norman: University of Oklahoma Press, 1930), 6, 137–50, 153; Giebelhaus, *Business and Government,* 124–26. For an extensive discussion of all three men, see Roger M. Olien and Diana Davids Olien, *Oil and Ideology: The Cultural Creation of the American Petroleum Industry* (Chapel Hill: University of North Carolina Press, 2000).

58. David D. Leven, *Petroleum Encyclopedia: Done in Oil* (New York: Ranger Press, 1942), 81–82; U.S. Bureau of Mines, *Methods for Increasing the Recovery of Oil From Sands,* Bulletin no. 148, prepared by J. O. Lewis (Washington: Government Printing Office, 1917): 37.

59. Henry L. Doherty, "Suggestions for Conservation of Petroleum By Control of Production," address at meeting of the American Institute of Mining and Metallurgical Engineers, February 18, 1925, Unit Operations and Conservation, box 28, Cities Service Oil Company Collection, Western History Collection, University of Oklahoma; James J. Delamy, "Chronology of Mr. Doherty's work on Oil Conservation," 1933, Unit Operations and Conservation, box 28, Cities Service Oil Company Collection.

60. "Conservation," an address by Mark Requa to the API, November 19, 1920, box 454, Oil Pamphlets, Oil, Commerce Papers, Herbert Hoover Presidential Library (HHPL).

61. Mid-Continent Oil and Gas Association, "Investigation of Imports of Petroleum and Their Effect on Domestic Markets," April 1921, box 452, Oil, Commerce Papers, HHPL.

62. Ira Jewell Williams (representing Standard Oil) to Mark L. Requa, May 12, 1921, box 452, Oil, Commerce Papers, HHPL. Also see Daniel Yergin, *The Prize: The Epic Quest for Oil, Money, and Power* (New York: Simon and Schuster, 1991), 196–97; Ise, *Oil Policy,* 451–86.

63. Although it would take a full decade for events surrounding the Teapot Dome scandal to unfold, the most damaging revelations had already emerged; see Burl Noggle, *Teapot Dome: Oil and Politics in the 1920s* (New York: W. W. Norton and Co., 1962). Later, geologists would find that the oil reserve was not an independent reservoir.

64. Speech, George Otis Smith to International Petroleum Congress, October 7, 1924, box 542, Oil, Conservation, Commerce Papers, HHPL.

65. George Otis Smith, "Where the World Gets Oil and Where Will Our Children Get it When American Wells Cease to Flow?" *National Geographic* 37 (February 1920): 202.

66. The suit revolved around the legality of a "patent pool" organized by companies that owned patents associated with thermal cracking. Forty-four of the companies were listed as secondary defendants. See Harold F. Williamson, Ralph L. Andreano, Arnold R. Daum, and Gilbert C. Klose, *The American Petroleum Industry: The Age of Energy, 1899–1959* (Evanston, Ill.: Northwestern University Press, 1963), 389.

67. Mid-Continent Oil and Gas Association, "Investigation of Imports of Petroleum."

68. Ise, *Oil Policy,* 451–86.

69. API to the President of the United States, January 20, 1925, box 37, series 7, Sun Company Collection, Hagley Museum and Library.

70. Preface, Committee of Eleven, "Report to the FOCB," 1925, box 20, series 1D, Sun Company Collection, Hagley Museum and Library; J. Howard Pew to Mark L. Requa, November 20, 1925, box 23, series 1D, Sun Company Collection. Also see Atlantic Refining, "Memorandum on Conservation," December 5, 1925, box 20, Series 1D, Sun Company Collection; Giebelhaus, *Business and Government,* 130–31.

71. Doherty's comments in writing, Federal Oil Conservation Board, *Record of Public Hearings, May 27* (Washington D.C, 1926), 51. Doherty, who had been active in the founding of the API, continued to disagree with other industry leaders even after they accepted unitization. In 1931, proclaiming himself "out of step" with the other API directors, he resigned from the organization; *New York Times,* December, 27, 1939, 1.

72. "Copy of Federal Oil Conservation Board's Letter to Producers," Federal Oil Conservation Board, Official, Commerce Papers, HHPL.

73. J. E. Pew to J. H. Pew, June 13, 1925, FOCB, box 22, series 1D, Sun Company Collection, Hagley Museum and Library; press release, FOCB to Morning Papers, July 29, 1925, box 453, Oil 1925, FOCB, Commerce Papers, HHPL.

74. For a copy of the ten surveys and for summaries in which responses have been "digested," see box 453, Oil 1925, FOCB, Commerce Papers, HHPL.

75. FOCB, *Public Hearings, February 10 and 11,* 8–9, 10–18, 19–24.

76. Ibid., 24–33.

77. Philip Kates, "Oil Production Under State Control," address before State Bar of Oklahoma, December 19, 1930, p. 9, Oil Matters, Presidential Papers, HHPL; Williamson et al., *Age of Energy,* 321–22.

78. FOCB, *Public Hearings, February 10 and 11,* 71–72. Ironically, Marland, who headed Continental Oil, later became the governor of Oklahoma and a founder of the Interstate Oil Compact Commission; see John Joseph Matthews, *Life and Death of an Oilman: The Career of E. W. Marland* (Norman: University of Oklahoma Press, 1951).

79. Henrietta M. Larson and Kenneth W. Porter, *History of Humble Oil and Refining: A Study in Industrial Growth* (New York: Harper and Brothers, 1959), 267. Humble started securing large leases only after the company committed itself to supporting production controls.

80. FOCB, *Public Hearings, February 10 and 11,* 109–30.

81. Ibid., 35–42.

82. Ibid., 150–54.

83. Ibid., 143–54.

84. Williamson et al., *Age of Energy,* 323.

85. Carl Coke Rister, *Oil! Titan of the Southwest* (Norman: University of Oklahoma, 1949), 255–69; Williamson et al., *Age of Energy,* 321–29. For a photograph of derricks in sight of the Oklahoma Capitol, see Robert O. Anderson, *Fundamentals of the Petroleum Industry* (Norman: University of Oklahoma Press, 1984), 124.

86. Pratt, *Growth of a Refining Region,* 227.

87. Institute of Petroleum Technologists, *Petroleum: Twenty-Five Years Retrospect, 1910–1935* (London: Aldine House, 1935), 69; J. E. Brantly, "Hydraulic Rotary-Drilling System" in American Petroleum Institute, *History of Petroleum Engineering* (New York: API, 1961), 363–65; Patrick J.

Nicholson, *Mr. Jim: The Biography of James Smither Abercrombie* (Houston: Gulf Publishing, 1983), 234–35.

88. Press release marking the anniversary of Wild Mary, Mary Sudik No. 1, box 26, Cities Service Collection, Western History Collection, University of Oklahoma. Also see Rister, *Oil!* 259–60; Kenny A. Franks, *The Oklahoma Petroleum Industry* (Norman: University of Oklahoma, 1980).

89. Press release marking the anniversary of Wild Mary, Mary Sudik No. 1, box 26, Cities Service Collection, Western History Collection, University of Oklahoma.

90. Walace Hawkins and Chas. B. Wallace, "Oil and Gas Well Fire and Blowout Control," folder 1, box 42, George A. Hill, Jr., Papers, Houston Metropolitan Research Center.

91. National Oil Policy Committee of the Petroleum Industry War Council, "Chronology of U.S. Oil Conservation Movement, 1924–1935," 1944, box 80, series 4, Sun Company Collection, Hagley Museum and Library; "Report of the Committee of Nine," *API Bulletin* 9 (February 9, 1928), 1–7; Report of the Committee on Gas Conservation, *API Bulletin* 9 (January 6, 1928), 1–5; Leonard M. Fanning, *The Story of the American Petroleum Institute: A Study and Report (With Personal Reminiscences)* (New York: World Petroleum Policies, 1959), 101–25.

92. Williamson et al., *Age of Energy*, 545; Section of Mineral Law, American Bar Association, *Legal History of Conservation of Oil and Gas* (Chicago: American Bar Association, 1938), 232–33; interview, Judge F. W. Fischer by R. W. Hayes, May 21, 1955, tape no. 172, Oral History of the Oil Industry, Center for American History, University of Texas at Austin.

93. Interview, Judge F. W. Fischer by R. W. Hayes, tape no. 172; press release, Department of the Interior, Oil-General, Ray Lyman Wilbur Collection; HHPL; statement of the Oil States Advisory Committee, House Committee on the Judiciary, *Conservation of Oil and Gas and Protection of American Sources*, 72d Cong., 1st sess., May 14, 1932, 10–13.

94. Railroad Commission of Texas, *Conservation of the Oil and Gas Resources of the State: Defining "Waste" and Empowering the Railroad Commission to Make and Enforce Regulations With Reference to the Same* (Austin: 1919); Railroad Commission of Texas, Oil and Gas Division, General Order, related to the conservation and prevention of waste of crude petroleum, April 4, 1931, C. R. Austin Oil Company Papers, Drawer 1B, Panhandle Plains Historical Museum, West Texas A&M; "Digest of Act of the Legislature of Texas Relating to the Conservation of Oil and Gas," August 18, 1931, Oil Matters, Presidential Papers, HHPL.

95. Williamson et al., *Age of Energy*, 540–48; American Bar Association, *Legal History of Conservation of Oil and Gas*, 180–84, 232–35. Also see David F. Prindle, *Petroleum Politics and the Texas Railroad Commission* (Austin: University of Texas Press, 1981); William R. Childs, "The Transformation of the Railroad Commission of Texas, 1917–1940: Business-Government Relations and the Importance of Personality, Agency Culture, and Regional Differences," *Business History Review* 65 (summer 1991): 284–344.

96. Anderson, *Petroleum Industry*, 29.

97. For colorful stories about practices in the field, see James A. Clark, *The Fabulous East Texas Oil Field* (Los Angeles: Atlantic Richfield, c. 1970); Harry Harter, *East Texas Oil Parade* (San Antonio, Naylor Co., 1934); Ruel R. McDaniel, *Some Ran Hot* (Dallas: Regional Press, 1939).

98. "Report of the Texas Gasoline Tax Evasion Committee," December 1932, box 64, series 1D, Sun Company Collection, Hagley Museum and Library.

99. Minutes, Ohio Code Meeting, box 64, series 1D, Sun Company Collection, Hagley Museum and Library.

100. "Report of the Texas Gasoline Tax Evasion Committee," December 1932, box 64, series 1D, Sun Company Collection, Hagley Museum and Library; clippings on tax evasion distributed to members as *Gas Tax News,* box 64, series 1D; R. G. Guthrie to W. D. Mason, April 7, 1932, box 147, series 1. Also see "Tax Evasion is Menace to Government," *National Petroleum News* 24 (April 24, 1932): 1; Fanning, *American Petroleum Institute;* "Chronology of U.S. Oil Conservation Movement;" Federal Oil Conservation Board, *Report of the Federal Oil Conservation Board to the President of the United States,* report no. 5 (Washington, D.C.: Government Printing Office, 1932).

101. See Williamson et al., *Age of Energy*, 537–61; Gerald D. Nash, *United States Oil Policy,*

1890–1964: Business and Government in the Twentieth Century (Pittsburgh: University of Pittsburgh Press, 1968); Samuel B. Pettengill, Hot Oil: The Problem of Petroleum (New York: Economic Forum Co., 1936); M. Murphy Blakely, Conservation of Oil and Gas: A Legal History (Chicago: American Bar Association, 1949); John W. Frey, "The Interstate Oil Compact," in Energy Resources and National Policy, ed. National Resources Committee (Washington: Government Printing Office, 1939).

102. Frey, "The Interstate Oil Compact," 397–401.

103. For an example of such arguments: David M. Neuberger, "A Call to the Chemist to Purge Industry of Its Contamination of Our Coast and Inland Waters," Chemical Age 31 (October 1923), reprint, n.p.

Chapter 7. Success and Failure in the Oil Fields

1. U.S. Corps of Engineers, Pollution Affecting Navigation or Commerce on Navigable Waters, Report of the Secretary of War to the House of Representatives, 69th Cong., 1st sess., June 7, 1926, H. Doc. 417, 2.

2. API, "Survey of Oil Pollution Conditions in the United States," 1927, 4, box 35, series 1D, Sun Company Collection, Hagley Museum and Library.

3. Ibid., 5.

4. Murray Campbell and Harrison Hatton, Herbert Dow: Pioneer in Creative Chemistry (New York: Appleton-Century-Crofts, 1951), 25.

5. "Oil Wells Near Sand Springs Yield Brine for New Chemical Plant," National Petroleum News 18 (August 4, 1926): 91–92.

6. U.S. Bureau of Mines, The Disposal of Oil Field Brines, Report of Investigation no. 2945, prepared by Ludwig Schmidt and J. M. Devine (Washington, D.C.: Bureau of Mines, 1929); "Disposal of Oil-Field Waste Studied by Bureau Engineers," Oil and Gas Journal (Aug. 22, 1929): 110; L. R. Young, "The Latest in Salt Water Disposal," Oil Bulletin (April 1929): 360–62; "Disposal of Brine in Oil Fields," Petroleum Times (Oct. 19, 1929): 756; S. W. Oberg, "Methods of Waste Water Disposal," Oil and Gas Journal (Sept. 5, 1929): 76; H. C. Ferry, "Problems of Waste Water Solved by Southern California Operators," Oil and Field Engineering 56 (March 1929): 46–47; "Solving Oil-Field Waste-Disposal Problem," Engineering News Record (May 16, 1929): 795–98; Walter Humphreys, "Disposal of Oil-Field Waste," Petroleum Engineer (April 1930): 120, 122, 125.

7. Ludwig Schmidt, "Report of Inspection Trip through West Texas Fields," October 10, 1928, box 106, Records of Research Groups, Bartlesville Petroleum Research Center, Bureau of Mines, RG 70-4, National Archives, Southwest Region, Fort Worth.

8. J. G. Pew to M. B. Sweeney, July 1, 1930, East Texas Water Encroachment, box 29, series 3, Sun Company Collection, Hagley Museum and Library. For a similar example involving rice growers, see Wilbur F. Cloud, Petroleum Production (Norman: University of Oklahoma, 1937), 501.

9. J. E. Pew to J. G. Pew, August 1, 1930, East Texas Water Encroachment, box 29, series 3, Sun Company Collection, Hagley Museum and Library.

10. Pew to Sweeney, July 1, 1930.

11. A. H. Wiebe, J. G. Burr, and H. E. Faubion, "The Problem of Stream Pollution in Texas with Special Reference to Salt Water from the Oil Fields," Transactions of the American Fisheries Society 64 (1934): 81–86. The authors, all with the Texas Game, Fish, and Oyster Commission, identify how long various species of freshwater fish survive in various concentrations of salt water.

12. Ogden S. Jones, Fresh Water Protection from Pollution Arising in the Oil Fields (Lawrence: University of Kansas Publications, 1950), 7–9, 69–71.

13. M. H. Shanahan, "Texas Salt Water Problem Discussed," Oil and Gas Journal (Apr. 16, 1931): 65, 90; L. G. E. Bignell, "Salt Water Situation Already Serious in Texas Field," Oil and Gas Journal (July 9, 1931): 17, 120; D. H. Stormont, "East Texas Operators to Consider Disposal Plan," Oil and Gas Journal 40 (Dec. 11, 1931): 24.

14. H. K. Arnold (Humble Oil) to Members of the Anti-Pollution Committee, June 28, 1933,

folder 7, box 12, George A. Hill, Jr., Collection, Houston Metropolitan Research Center; letter with attachments, Arnold to George Hill (Houston Oil), August 2, 1933, folder 7, box 12, George A. Hill, Jr., Collection.

15. Con R. Gladney (Chairman, Anti-Pollution Committee) to Contributors to East Texas Anti-Pollution Work, June 28, 1935, folder 7, box 12, George A. Hill, Jr., Collection.

16. H. K. Arnold to George Hill, March 13, 1935, folder 7, box 12, George A. Hill, Jr., Collection.

17. Phillips Petroleum, "Pollution: Its Relation to the Oil Company and to the Welfare of Your Community," folder 7, box 12, George A. Hill, Jr. Collection.

18. "East Texas Salt Water Disposal Plan Considered," no date, folder 7, box 12, George A. Hill, Jr., Collection.

19. Ibid.

20. Michael W. Caplinger and Philip W. Ross, *The Historic Petroleum Industry in the Allegheny National Forest,* prepared for the USDA Forest Service (Morgantown: West Virginia University, 1993), 69–70.

21. U.S. Bureau of Mines, *History of Water Flooding of Oil Sands in Oklahoma,* Report of Investigation no. 3728, prepared by D. B. Taliaferro and David M. Logan (Washington, D.C.: Bureau of Mines, 1943), 5; Stanley C. Herold, *Oil Well Drainage* (Stanford, Calif.: Stanford University Press, 1941, 316; Robert S. Bossler, *Oil Fields Rejuvenated,* Bulletin no. 56 (Harrisburg, Pa.: Pennsylvania Bureau of Topographic and Geological Survey, 1922).

22. John R. Suman, "Evolution by Companies," in American Petroleum Institute, *History of Petroleum Engineering* (Dallas: API, 1961), 81; E. DeGolyer, "Concepts on the Occurrence of Oil and Gas," in American Petroleum Institute, *History of Petroleum Engineering,* 30; David Trax, "Repressurizing Practice in the Mid-Continent," presented to a meeting of the Mid-Continent District of the API, Oct. 7–8, 1929, box 38, series 7, Sun Company Collection, Hagley Museum and Library; Wentworth H. Osgood, *Increasing the Recovery of Petroleum,* 2 vols. (New York: McGraw-Hill, 1930).

23. Edward W. Constant II, "Cause or Consequence: Science, Technology, and the Regulatory Change in the Oil Business in Texas, 1930–1975" *Technology and Culture* 30 (April 1989): 426–55.

24. H. K. Arnold to George Hill, November 7, 1933, and progress report, East Texas Joint Salt Water Disposal Experiment, February 14, 1936, folder 7, box 12, George A. Hill, Jr., Collection, Houston Metropolitan Research Center.

25. "General History of the Bartlesville Station, 1934–1935," box 17, Records of the Petroleum Experiment Center, Bureau of Mines, RG 70-2, National Archives, Southwest Region, Fort Worth; Jones, *Fresh Water Protection,* 7; L. G. E. Bignell, "Salt Water Disposal Wells Solution of Problems for Operators in Kansas," *Oil and Gas Journal* (Sept. 5, 1935).

26. Ludwig Schmidt to F. R. Frye, January 28, 1936, box 21, Records of the Petroleum Experiment Center, Bureau of Mines, RG 70-2, National Archives, Southwest Region, Fort Worth; status reports, Lloyd Christianson to Sam Taylor, box 21, Records of the Petroleum Experiment Center, Bureau of Mines, RG 70-2, National Archives, Southwest Region, Fort Worth; U.S. Bureau of Mines, *Typical Oil-Field Brine Conditioning Systems for Preparing Brine for Subsurface Injection,* Report of Investigation no. 3434, prepared by S. Taylor, C. J. Wilhelm, and W. C. Holliman (Washington, D.C.: Bureau of Mines, 1939).

27. "Supplemental Progress Report of Brine Disposal," ca. 1935, box 22, Records of Research Groups, Bartlesville Petroleum Research Center, Bureau of Mines, RG 70-4, National Archives, Southwest Region, Fort Worth. Of the 325 wells used in water flooding operations, only 136 reinjected brine; the rest injected fresh water.

28. Ludwig Schmidt to R. A. Cattell, Nov. 15, 1938, box 21, Records of Research Groups, Bartlesville Petroleum Research Center, Bureau of Mines, RG 70-4, National Archives, Southwest Region, Fort Worth.

29. "Data on Production and Number of Wells—East Texas Field," East Texas Water Encroachment, box 29, series 3, Sun Company Collection, Hagley Museum and Library. The increase in salt water was probably higher because some companies, fearing that the Railroad Com-

mission would eventually shut down wells producing mostly water, purposely kept their figures low; see F. E. Heath to J. Edgar Pew, September 27, 1937, box 29, series 3, Sun Company Collection.

30. M. B. Sweeney to J. Edgar Pew, Aug. 3, 1937, East Texas Water Encroachment, box 29, series 3, Sun Company Collection, Hagley Museum and Library; "East Texas Water Disposal Plant Attracts Interest," *Oil and Gas Journal* (Dec. 8, 1938): 50, 52.

31. East Texas Salt Water Disposal Corporation, *Salt Water Disposal in the East Texas Field,* (Austin: University of Texas Petroleum Extension Service, 1953), 3; H. H. King, "East Texas Experiment Shows Feasibility of Returning Salt to Producing Formation," *Oil Weekly* (Nov. 28, 1938): 23–34; M. J. Leahy, "Disposal of Oil Field Brines Acute Problem in East Texas," *Oil Weekly* (Jan. 8, 1940): 19–20, 22–24.

32. East Texas Salt Water Disposal Corporation, *Salt Water Disposal,* 3–4.

33. "Plans for Creation and Operation of the East Texas Salt Water Disposal Company," ca. December 1941, box 8, series 2, Sun Company Collection, Hagley Museum and Library; R. Ingran, "Financing of East Texas Salt Water Disposal Plan Complete," *Oil and Gas Journal* 41 (Aug. 1942): 31.

34. W. S. Morris, "Results of Water Injection in Woodbine Reservoir of the East Texas Field," *API Proceedings: Division of Production* 27 (1947): 36–35; East Texas Salt Water Disposal Corporation, *Salt Water Disposal,* 7.

35. Rodney P. Carlisle and August W. Giebelhaus, *The Bartlesville Energy Center: The Federal Government in Petroleum Research, 1918–1983* (Washington, D.C.: U.S. Department of Energy, 1984), 52.

36. Leon W. Dupuy to H. C. Fowler, May 7, 1951, box 18, Records of the Petroleum Experiment Center, Bureau of Mines, RG 70-2, National Archives, Southwest Region, Fort Worth. Also see Arkansas Gas and Oil Commission, *Secondary Recovery of Petroleum in Arkansas—A Survey,* prepared by George H. Fancher and Donald K. Mackay (El Dorado, Ark.: Arkansas Gas and Oil Commission, 1946).

37. Robert O. Anderson, *Fundamentals of the Petroleum Industry* (Norman: University of Oklahoma, 1984), 25. Also see Keith L. Miller, "Plucking the Apple without Rooting Up the Tree: Environmental Concern in the Production of Prairie State Petroleum," *Illinois Historical Journal* 84 (autumn 1991): 161–72.

38. Norman Hurd Ricker, "The Autobiography of Norman Hurd Ricker," unpublished manuscript, 43–48, box 23, Norman Hurd Ricker Papers, Woodson Research Center, Rice University.

39. Henrietta M. Larson, Evelyn H. Knowlton, and Charles S. Popple, *New Horizons, 1927–1950: History of the Standard Oil Company, New Jersey* (New York: Harper and Row, 1971), 74; Ricker, "Autobiography," 47–48.

40. George Elliot Sweet, "History of Geophysical Prospecting," notes, box 21, Cities Service Oil Company Collection, Western History Collection, University of Oklahoma.

41. Louis A. Allaud and Maurice H. Martin, *Schlumberger: The History of a Technique* (New York: Wiley: 1977). Also see Geoffrey C. Bowker, *Science on the Run: Information Management and Industrial Geophysics at Schlumberger, 1920–1940* (Cambridge, Mass.: MIT Press, 1994).

42. H. A. C. Smith, "Summary of Projects at the Petroleum Experiment Station," Bartlesville, Oklahoma, box 3, Records of the Petroleum Experiment Center, Bureau of Mines, RG 70-2, National Archives, Southwest Region, Fort Worth.

43. For a clear summary of all Texas regulations, complete with sample forms, see Railroad Commission of Texas, *Texas Oil and Gas Handbook: A Guide to State Conservation Regulations* (Austin: R. W. Byram and Co., 1958). In the 1930s and 1940s, as the rules and regulations were being put into place, producers only had access to the orders issued by conservation agencies.

44. Edward W. Constant II, "Science In Society: Petroleum Engineers and the Oil Fraternity in Texas, 1925–65" *Social Studies of Science* 19 (August 1989): 461. For a good example of how contemporary works integrated the notion of conservation into the training of petroleum engineers, see M. Albertson, "Conservation," in *Elements of the Petroleum Industry,* ed. E. DeGolyer (New York: American Institute of Mining and Metallurgical Engineers, 1940), 279–88; Albertson de-

fines "conservation" as being equivalent to "maximum use," with "present needs for petroleum and gas" having a greater claim on sources of supplies than future needs.

45. Anderson, *Petroleum Industry*, 140.

46. Discussion of paper, Harvey Hardison, "Overcoming Crooked Holes," presented at meeting of the Texas-Louisiana-Arkansas section of the Division of Development and Production Engineering, April 3, 1929, box 38, series 7, Sun Company Collection, Hagley Museum and Library.

47. J. N. Pew to Elmer Sperry, July 10, 1925, Sperry, box 24, series 1D, Sun Company Collection, Hagley Museum and Library.

48. Harvey Hardison, "Overcoming Crooked Holes." For mention of when companies started measuring deviations from the vertical, see Suman, "Evolution by Companies," in American Petroleum Institute, *History of Petroleum Engineering*, 76, 82, and 86.

49. J. E. Brantly, "Hydraulic Rotary Drilling Systems," in American Petroleum Institute, *History of Petroleum Engineering*, 421; Institute of Petroleum Technologists, *Twenty-Five Years Retrospect, 1910–1935*, 70.

50. H. C. Miller and G. B. Shea, "Recent Progress in Petroleum Development and Production," U.S. Bureau of Mines, 1940, folder 2, Oil and Gas Industry (vertical file), Southwest Collection, Texas Tech University; "Directional Drilling as a Routine Method of Development," *Petroleum Engineer* 12 (October 1941): 38–41.

51. For mention of a later scandal involving directional drilling in the East Texas field, see David F. Prindle, *Petroleum Politics and the Texas Railroad Commission* (Austin: University of Texas Press, 1981), 81–94.

52. Edward W. Constant II, "The Cult of MER: or Why There is a Collective in Your Consciousness," *Business and Economic History* 22 (fall 1993).

53. J. H. Hedges to R. E. Tarbett (Senior Sanitary Engineer, U.S. Public Health Service), August 11, 1938, Box 21, Records of the Petroleum Experiment Center, Bureau of Mines, RG 70-2, National Archives, Southwest Region, Fort Worth.

54. Jones, *Fresh Water Protection*, 45–67.

55. Charles C. Williams and Charles K. Bayne, *Ground-Water Conditions in Elm Creek Valley, Barber County, Kansas*, Bulletin 64 (Topeka, Kansas: State Geological Survey of Kansas, 1946).

56. Olin Culberson, "Speech to Oil Compact Commission," September 1952, Speeches, Oil and Gas Resources of Texas, Box 3G313, Olin Culberson Papers, Center for American History, University of Texas at Austin.

57. *Brown et al. v. Lundell*, Texas Court of Civil Appeals, Amarillo, no. 6925, March 21, 1960, *Oil and Gas Reporter*, v. 15, 1960. Also see William M. Cotton, "Some Legal Problems Encountered in Lease Operations After Discovery of Oil," *West Texas Oil Lifting Short Course* (Lubbock: Texas Tech University, 1959), 228–34.

58. Henry Lewelling and Monte Kaplan, "What to do about Salt Water," in *Frontiers in Petroleum Engineering* (Dallas: Petroleum Engineering Publishing Co., 1960), 57.

59. Research Committee, Interstate Oil Compact Commission, *Production and Disposal of Oil Field Brines in the United States of America* (Oklahoma City: Interstate Oil Compact Commission, 1960), 64.

60. Kentucky Geological Survey, *Effects of Greensburg Oil Field Brines on the Streams, Wells, and Springs of the Upper Green River Basin, Kentucky*, Report of Investigations 2, prepared by R. A. Krieger and G. E. Hendrickson (Lexington: University of Kentucky College of Arts and Sciences, 1960).

61. Interstate Oil Compact, *Disposal of Oil Field Brines*, 64.

62. Jerry T. Thornhill, "Pollution Control Activities of the Water Pollution Control Board and Texas Water Commission," *West Texas Oil Lifting Short Course* (Lubbock: Texas Tech University, 1962), 172.

63. John Opie, *Ogallala: Water for a Dry Land* (Lincoln: University of Nebraska, 1993), 32.

64. John R. Sheaffer and Leonard A. Stevens, *Future Water* (New York: William Morrow and Co., 1983), 174. Also see Donald E. Green, *Land of the Underground Rain: Irrigation in the Texas High Plains, 1910–1970* (Austin: University of Texas Press, 1973).

65. L. G. McMillion, "Hydrological Aspects of Disposal of Oil-Field Brines in Texas," *Ground Water* 3 (October 1965): 36–42.

66. Roy D. Page, "Pollution Control for Oil Field Brines," *Drill Bit* 15 (December 1967): 32–36.

67. "Crack Down on Oil Country Pollution," *Petroleum Engineer* 39 (July 1967): 33–38.

68. See A. Raschke et al., "Let Engineering Solve Salt-Pollution Problems," *Oil and Gas Journal* 63 (August 9, 1965): 75–79; "Salt Water Disposal and Oil field Water Conservation," *Petroleum Engineering and Services* 30 (No. 4, 1967): 22–28; Roy D. Payne, "Salt Water Pollution Problems in Texas," *Journal of Petroleum Technology* (November 1966): 1401–10; "Salt Water Disposal Pits," *Drill Bit* 40 (November 1967): 18–19.

69. Carlisle and Giebelhaus, *Bartlesville*, 52.

70. Constant, "Science In Society," 461.

71. Carlisle and Giebelhaus, *Bartlesville*, 48–49.

72. H. C. Fowler, trip report to the Midyear Meeting of the Interstate Oil Compact Commission, May 30 to June 2, 1956, Box 92, Records of Research Groups, Bartlesville Petroleum Research Center, Bureau of Mines, RG 70-4, National Archives, Southwest Region, Fort Worth.

73. Carlisle and Giebelhaus, *Bartlesville*, 58; J. Wade Watkins to R. A. Cattell, March 19, 1949, monthly letters, Box 89, Records of Research Groups, Bartlesville Petroleum Research Center, Bureau of Mines, RG 70-4, National Archives, Southwest Region, Fort Worth.

74. Trip report to 1957 Nuclear Conference, Robert W. Geehan, Box 92, Records of Research Groups, Bartlesville Petroleum Research Center, Bureau of Mines, RG 70-4, National Archives, Southwest Region,

75. Research Committee of the Oil Compact Commission, *Underground Storage of Liquid Petroleum Hydrocarbons in the United States* (Oklahoma City, Okla.: Interstate Oil Compact Commission, 1956).

76. J. W. Watkins to J. E. Crawford, August 5, 1959, Waste Disposal, Box 24, Records of Research Groups, Bartlesville Petroleum Research Center, Bureau of Mines, RG 70-4, National Archives, Southwest Region, Fort Worth; H. C. Fowler to J. W. Watkins, August 7, 1959, Box 24, Records of Research Groups, Bartlesville Petroleum Research Center, Bureau of Mines, RG 70-4, National Archives, Southwest Region, Fort Worth.

77. James T. Ramey, "Recent Nuclear Developments of Interest to the Petroleum Industry," remarks to the National Petroleum Council, March 25, 1965, folder API 1964–65, box 46, Bernard Majewski Collection, American Heritage Center, University of Wyoming; Carlisle and Giebelhaus, *Bartlesville*, 58.

78. T. M. Garland and Frank Parrish, Jr., *Sources of Water for Water Flooding*, prepared for the North Texas Oil and Gas Association by the Wichita Falls Petroleum Field Office (Wichita Falls, Tex.: U.S. Bureau of Mines, 1958).

Chapter 8. Eliminating Corrosion and Monitoring Flows

1. See Ulick R. Evans, *The Corrosion of Metals* (London: Edward Arnold and Co., 1924), 1–11. Evans traces various theories back to 1788.

2. Ibid., 6.

3. Erick Larson, *Pipe Corrosion and Coatings* (New York: American Gas Journal, 1938), 31.

4. Some oil companies did hire chemists and chemical engineers to study corrosion problems. For example, in 1917 Gulf hired Dr. F. M. Siebert to establish a research group that would study "emulsions, corrosions, and muds"; John R. Suman, "Evolution by Companies" in API, *History of Petroleum Engineering* (Dallas: API, 1961), 85.

5. F. N. Speller, *Abatement of Corrosion in Central Heating Systems*, Bureau of Mines Technical Paper no. 236 (Washington, D.C.: Government Printing Office, 1919).

6. B. C. McPherson (Traffic Manager) to J. H. Pew, 1925, Tank Cars, box 85, series 1, Sun Company Collection, Hagley Library and Museum.

7. Frank N. Speller, *Corrosion—Causes and Prevention* (New York: McGraw-Hill, 1926), 1.

8. Ibid.

9. Interview of A. J. Thaman and J. H. Anderson by Maude Ross, January 29, 1960, tape no. 216, Oral History of the Oil Industry, Center for American History, University of Texas at Austin.

10. Speller, *Corrosion,* 3rd ed. (1951), 573.

11. E. C. Kincaide, "Corrosion of Pipelines," *API Bulletin* 8 (January 31, 1927): 350–52. This issue is a record of all papers presented at the Seventh Annual Meeting of the API.

12. "Discussion," *API Bulletin* 8 (January 31, 1927): 366.

13. C. R. Weidner, "Protecting and Reconditioning Underground Pipelines," *API Bulletin* 8 (January 31, 1927), 352–56; R. Van Mills, "Corrosion in Oil and Gas Wells," *API Bulletin* 8 (January 31, 1927), 365–66.

14. "Practical Methods of Combating Corrosion in Petroleum Refining," *Chemical and Metallurgical Engineering* 33 (October 1926): 628–30.

15. Interview of A. J. Thaman and J. H. Anderson by Maude Ross.

16. One producer reported pulling rods once or twice or month due to sulfur corrosion; see interview of E. I. Thompson by M. C. Boatright, September 3, 1953, tape no. 66, Oral History of the Oil Industry, Center for American History, University of Texas at Austin.

17. "Oil Industry's Greatest Foe, Corrosion, Intimately Studied by Expert," *National Petroleum News* 18 (September 29, 1926): 69–70.

18. Speller, *Corrosion,* 3rd ed., 563.

19. James H. Wescott, *Oil: Its Conservation and Waste* (New York: Beacon, 1930), 194.

20. Suman, "Evolution by Companies," 80. Also see Benson M. Kingston, *Acidizing Handbook* (Houston, Tex.: Gulf Publishing, 1936).

21. C. E. Beecher and H. C. Fowler, "Production Techniques and Control," in API, *History of Petroleum Engineering,* 757–58. Although acid treatments replaced nitroglycerin as the more effective way to loosen limestone formations, nitroglycerin continued to be important in loosening formations that acid could not dissolve, and both methods remained competitive until hydraulic sand fracturing emerged in the late 1940s as the more effective method.

22. "Extensive Study of Underground Corrosion and Its Prevention Conducted by Government," *Oil and Gas Journal* 31 (September 1, 1932): 16.

23. Chairman, API Committee on Corrosion to Sun Oil, April 24, 1928, API, series 1, box 35, Sun Company Collection, Hagley Museum and Library.

24. "Wasteful Economy," *Oil and Gas Journal* 31 (September 1, 1932): 16.

25. Progress reports, box 3, Records of the Petroleum Experiment Center, Bureau of Mines, RG 70-2, National Archives, Southwest Region, Fort Worth.

26. "Extensive Study of Underground Corrosion," 16.

27. Speller, *Corrosion,* 3rd ed., 628.

28. "Pipeline Corrosion—And Its Control," *Oil and Gas Journal* 57 (January 28, 1959), E32–33.

29. Henrietta M. Larson and Kenneth W. Porter, *History of Humble Oil and Refining: A Study in Industrial Growth* (New York: Harper and Brothers, 1959), 645.

30. Marshall Parker, *Pipe Line Corrosion and Cathodic Protection* (Houston: Gulf Publishing, 1954), 63.

31. API, *Proceedings of the Division of Transportation, Twenty-Sixth Annual Meeting,* November 11, 1946, 90; Frank N. Speller, "Corrosion Research and Abatement—Yesterday and Today," *Corrosion* 1 (March 1945): 1–13; "Committee Activities," *Corrosion* 1 (March 1945): 38–45.

32. "Distributing Gasoline Through Sun Oil's 730-Mile Line," *Pipeline News* 3 (October 1931): 30–35.

33. F. L. Hadley to J. N. Pew, Jr., January 26, 1933, box 78, series 1D, Sun Company Collection, Hagley Museum and Library.

34. W. C. Kinsolving to R. W. Pack, August 22, 1938, box 45, series 3, Sun Company Collection, Hagley Museum and Library.

35. "Repairing Leaks in River Crossings," *Oil and Gas Journal* 30 (March 17, 1932), 48H.

36. C. C. Bledsoe, "Recent Advances in Pipe-Line Construction," *API Proceedings* v. 33 (1953), part V, Division of Transportation, 58.

37. John L. Loos, *Oil on Stream: A History of the Interstate Oil Pipe Line Company, 1909–1959* (Baton Rouge: Louisiana State University Press, 1959), 332–33.

38. Interview of Ralph B. McLaughlin by W. A. Owen, Tape 189, May 28, 1956, Oral History of the Oil Industry, Center for American History, University of Texas at Austin.

39. Hubbert L. O'Brien, *Petroleum Tankage and Transmission* (East Chicago, Indiana: Graver Tank & Mfg. Co., 1951), 16.

40. Loos, *Oil On Stream,* 343–46.

41. "Hundred Years of Petroleum Storage," *Oil and Gas Journal* 57 (January 28, 1959): E36.

42. Hubbert L. O'Brien, *Petroleum Tankage and Transportation* (East Chicago, Ill.: Graver Tank and Manufacturing Co.), 21.

43. Loos, *Oil on Stream,* 343–44.

44. Brochure, *Fauna Pump Station,* no date., Misc. Brochures, box Number 5, Cities Service Oil Company Collection, Western History Collection, University of Oklahoma.

45. "National Gas Pipeline History," no date., Amarillo Pipeline, box 13, Cities Service Oil Company Collection; "Natural Gas Pipelining," *Oil and Gas Journal* 57 (January 28, 1959): E8–9.

46. George S. Wolpert, Jr., *U.S. Oil Pipelines: An Examination of How Oil Pipe Lines Operate and the Current Public Policy Issues Concerning Their Ownership* (Washington, D.C.: API, 1979), 10, 19–23.

47. "Communication Performs Vital Part," *Oil and Gas Journal* 28 (August 29, 1929): T81.

48. "Communication and Control," *Oil and Gas Journal* 57 (January 28, 1959): E22.

49. "Automatic Control Stops Flow When Pipe Line Breaks," *Oil Weekly* 70 (June 26, 1933): 24–25; "Automatic Measuring and Recording at Main Pipe Line Stations," *Oil Field Engineering* 5 (January 1, 1929): 14–16.

50. Letter to R. A. Cattell, January 21, 1938, box 18, Records of the Petroleum Experiment Center, Bureau of Mines, RG 70-2, National Archives, Southwest Region, Fort Worth; H. H. Anderson, "Recent Developments in Pipe-line Technology," paper presented to Group Session on Transportation, Twenty-Sixth Annual Meeting of the API, November 12, 1946, folder 3, box 41, George A. Hill, Jr., Collection, Houston Metropolitan Research Center. Customs authorities in some countries did not accept the use of meters until the late 1950s: P. De Hall, "Transport and Distribution," in *Modern Petroleum Technology* (London: Institute of Petroleum, 1954), 669.

51. "Controls and Communication," *Oil and Gas Journal* 57 (January 28, 1959), E23; *Humble Way* (January-February, 1952), 23.

52. J. R. Ellis, "A Microwave-Controlled Pump Station," *API Proceedings* v. 32 (1952), part V, Division of Transportation, 29; Max T. Nigh, "Pipeline Machinery With Self-Control," *Oil and Gas Journal* 55 (July 8, 1957): 90–105.

53. J. Edgar Pew to T. C. Stauffer, April 9, 1940, Geophysical, box 58, series 3, Sun Company Collection, Hagley Museum and Library.

54. Loos, *Oil On Stream,* 344–45.

55. D. R. Patterson, "Petroleum Automation," in *Frontiers in Petroleum Engineering* (Dallas, Texas: Petroleum Engineering Publishing Company, 1960), 18. This volume contains a series of lectures delivered at the University of Texas at Austin between September 1958 and January 1959.

56. Ohio River Valley Water Sanitation Commission, *Preventing Stream Pollution from Oil Pipeline Breaks* (Cincinnati, Ohio: Ohio River Valley Water Sanitation Commission, 1950), 18.

57. Ibid.

58. API, Bulletin of Recommended Practices for Bulk Liquid Loss Control in Service Stations, third draft, 1950, box 166, folder 1950 API, Bernard Majewski Collection, American Heritage Center, University of Wyoming.

Chapter Nine. Creating a Pollution Control Manual

1. U.S. Corps of Engineers, *Pollution Affecting Navigation or Commerce on Navigable Waters,* H. Doc. 417, Report of the Secretary of War to the House of Representatives, 69th Cong., 1st sess., 1926.

2. H.R. 10903, 69th cong., 1st sess., 1926. For an assessment of the committee as an effort to

deflect regulation, see Lacey Walker, "Chapter II: Executive and Administrative Activities," in *American Petroleum Institute and Its Activities*, p. 7, folder "1942 API," box 45, Bernard Majewski Collection, American Heritage Center, University of Wyoming.

3. American Petroleum Institute, Survey of Oil Pollution Conditions in the U.S. with Remedial Measures Recommended (New York: API, 1927), attachment to letter, J. H. Pew to W. R. Boyd, June 21, 1928, box 35, series 10, Sun Company Collection, Hagley Museum and Library.

4. D. V. Stroop, "Chapter VIII: Department of Engineering" in *American Petroleum Institute and Its Activities*, pp. 5–6, folder "1942 API," box 45, Bernard Majewski Collection, American Heritage Center.

5. House Committee on Rivers and Harbors, *Hearings on a Bill to Amend the Oil Pollution Act of 1924*, 71st Cong., 2d sess, May 2, 3, and 26, 1930, 33.

6. Ibid., 22–24.

7. API, *Waste Water Containing Oil*, vol. 1 of *Manual on the Disposal of Refinery Wastes*, 1st ed. (New York: API, 1930); API, *Waste Gases or Vapors*, vol. 2 of *Manual*, 1st ed. (New York: API, 1931); API, *Chemical Wastes*, vol. 3 of *Manual*, 1st ed. (New York: API, 1935), API, *Solid Wastes*, vol. 6 of *Manual*, 1st ed. (Washington, D.C.: API, 1963). For the early tentative draft of "Solid Wastes," vol. 4, see box 54, series 1D, Sun Company Collection, Hagley Museum and Library. Volumes 4 and 5 cover sampling procedures and were released in the 1950s.

8. Norman Hurd Ricker, unpublished autobiography, 45–46, box 23, Norman Hurd Ricker Papers, Woodson Research Center, Rice University.

9. "Revivifying Fuller's Earth in Modern Filter Plant," *Chemical and Metallurgical Engineer* 33 (August 1926): 472–73.

10. Corps of Engineers, *Pollution*, 3.

11. Vladimir A. Kalichevsky and Bert A. Stagner, *Chemical Refining of Petroleum: The Action of Various Refining Agents and Chemicals on Petroleum and Its Products* (New York: Chemical Catalog Co., 1933), 96–111; *Composite Catalog of Oil Refinery Equipment and Process Handbook*, various editions (Houston, Texas: Gulf Publishing, 1931–39).

12. Kalichevsky and Stagner, *Chemical Refining of Petroleum*, 95.

13. Jeanne Marie Logsdon, "Organizational Responses to Environmental Issues: Oil Refining Companies and Air Pollution" (Ph.D. diss.: University of California, Berkeley, 1983), 366.

14. API, *Chemical Wastes*, appendix 1.

15. D. V. Stroop (Chair, API Committee on Disposal of Refinery Wastes) to J. Howard Pew (Sun Oil), H. R. Gallagher (Consolidated Oil), and H. M. Dawes (Pure Oil), December 17, 1932, box 65, series 1D, Sun Company Collection, Hagley Museum and Library; A. E. Pew, Jr., to Stroop, December 23, 1932, box 65, series 1D, Sun Company Collection. Also see, Joel A. Tarr, "Industrial Wastes and Public Health: Some Historic Notes, Part I, 1876–1932," *American Journal of Public Health* 75 (1985), 1059–67.

16. API, "Solid Wastes," pp. 4–6, box 54, series 1D, Sun Company Collection, Hagley Museum and Library.

17. API, *Waste Gases or Vapors*, 7.

18. Ibid., 9.

19. Interview of A. J. Thaman and J. H. Anderson by Maude Ross, January 29, 1960, tape no. 216, Oral History of the Petroleum Industry, Center for American History, University of Texas at Austin.

20. API, "Solid Wastes," 4–6, appendix 2, box 54, series 1D, Sun Company Collection, Hagley Museum and Library.

21. API, *Waste Water Containing Oil*, 4–9.

22. E. F. Eldridge, *Industrial Waste Treatment Practice* (New York: McGraw Hill, 1942), 340–41.

23. The various volumes of the manual were printed on loose sheets of paper and often stapled together; it was not a carefully designed publication.

24. "Lake Charles Operations," c. 1966, Lake Charles, box 23, Cities Service Oil Company Collection, Western History Collection, University of Oklahoma.

25. "Refinery Locations and Crude-Oil Capacities," supplement to *Oil and Gas Journal* 48 (Sept. 21, 1950); Standard Oil (N. J.), Marine Dept., "Oil Refineries and Principle Oil Fields of the United States," January 1, 1927.

26. John L. Enos, *Petroleum Progress and Profits: A History of Process Innovations* (Cambridge, Mass.: MIT Press, 1962), 221. For an overview of refinery advances in this period, see Harold F. Williamson, Ralph L. Andreano, Arnold R. Daum, and Gilbert C. Klose, *The American Petroleum Industry: The Age of Energy, 1899–1959* (Evanston, Ill.: Northwestern University Press, 1963), 603–47.

27. Williamson et al., *Age of Energy*, 635–42.

28. "Modified Regeneration of Catalyst," report no. 6, September 19, 1935, box 22, series 5, Sun Company Collection, Hagley Museum and Library. For the growing sophistication of auxiliary equipment, see *Composite Catalog of Oil Refinery Equipment*.

29. M. G. Van Voorhis, "The Houdry Process," *National Petroleum News* 31 (August 23, 1939), unpaged reprint; E. B. Badger and Sons, "Manufacture of Aviation Gasoline by the Houdry Process," box 2, series 5, Sun Company Collection, Hagley Museum and Library. Also see Enos, *Petroleum Progress*, 163–86; August W. Giebelhaus, *Business and Government in Industry: A Case Study of Sun Oil, 1876–1945* (Greenwich, Conn.: JAI Press, 1980), 169–92.

30. Henrietta Larson, Evelyn H. Knowlton, and Charles S. Popple, *New Horizons, 1927–1950: History of the Standard Oil Company (New Jersey)* (New York: Harper and Row, 1971), 166–68.

31. Enos, *Petroleum Progress*, 222; Williamson et al., *Age of Energy*, 624.

32. See list of session titles from API, Division of Refining, *API Proceedings*, vol. 34, 1954.

33. Ray Benson, "Safety in the Operation of HF Alkylation Plants," in *Refinery Operation and Maintenance* (Cleveland, Ohio: National Petroleum News, 1945), 12–15.

34. Roy F. Weston, Robert G. Merman, and Joseph G. DeMann, "Waste Disposal Problems in the Petroleum Industry," in *Industrial Wastes: Their Disposal and Treatment*, ed. Willem Rudolfs (New York: Reinhold, 1953), 431–41; Roy F. Weston and R. G. Merman, "The Chemical Flocculation of a Refinery Waste," Division of Refining, *API Proceedings*, vol. 34, 1954, pp. 207–25.

35. Humble Oil, *The Humble Way* (November-December, 1952), 17.

36. Elton B. Tucker, "Removal of Sulfur Compounds From Petroleum Naphthas By Catalytic Decomposition with Bauxite" (Ph.D. diss.: Columbia University, 1928).

37. Progress report of manufacturing department, 1945, box 81, series 10, Sun Company Collection, Hagley Museum and Library.

38. "Estimated Cost to the Public Health Service of a Study of Oil Refinery Wastes and Their Polluting Effects," c. 1942, box 94, Records of Research Groups, Bartlesville Petroleum Research Center, Bureau of Mines, RG 70-4, National Archives, Southwest Region, Fort Worth.

39. *The Humble Way* (January-February 1952): 20.

40. Weston et al., "Waste Disposal Problems in the Petroleum Industry," 424.

41. W. B. Hart, *Industrial Waste Disposal for Petroleum Refineries and Allied Plants* (Cleveland: Petroleum Processing, 1947), 8–11.

42. See Edward J. Cleary, *The ORSANCO Story: Water Quality Management in the Ohio Valley under an Interstate Compact* (Baltimore: Resources For the Future, John Hopkins University Press, 1967).

43. "Refiners' Confab," *Oil and Gas Journal* 48 (May 4, 1950): 55–56.

44. F. W. Mohlman, "Twenty-Five Years of Activated Sludge" in Langdon Pearse, ed., *Modern Sewage Disposal* (Lancaster, Pa.: Lancaster Press, 1938), 68–84.

45. R. J. Austin, W. F. Meehan, and J. D. Stockham, "Biological Oxidation of Oil-Containing Wastes," *Sewage and Industrial Waste* 28 (1954): 316–18; Weston, Merman, and DeMann, "Waste Disposal Problems of the Petroleum Industry," 419–49.

46. Alex D. McRae, William K. Ross, and Allan A. Sheppard, "Biological Oxidation of Phenolic Waste Water," in Division of Refining, *API Proceedings* 36 (1956), 320–31.

47. J. B. Davis, *Petroleum Microbiology* (New York: Elsevier, 1967), 354.

48. "Pollution Bill Now Law," *Engineering News Record* 157 (July 5, 1956): 28.

49. See David Stradling, *Smokestacks and Progressives: Environmentalists, Engineers, and Air*

Quality in America, 1881–1951 (Baltimore: Johns Hopkins University Press, 1999); Joel A. Tarr and Kenneth E. Koons, "Railroad Smoke Control," in *Energy Transport: Historical Perspectives on Policy Issues*, ed. George H. Daniels and Mark H. Rose (Beverly Hills: Sage Publications, 1982), 71–91; Joel A. Tarr and Bill C. Lamperes, "Changing Fuel Behavior and Energy Transitions: The Pittsburgh Smoke Control Movement, 1940–1950," *Journal of Social History* 14 (summer 1981): 571–72; R. Dale Grinder, "The Battle for Clean Air: The Smoke Problem in Post-Civil War America," in *Pollution and Reform in American Cities, 1870–1930*, ed. Martin Melosi (Austin: University of Texas Press, 1980), 83–104.

50. Lynne P. Snyder, "The Death-Dealing Smog over Donora, Pennsylvania: Demanding a Legal Response: Industrial Air Pollution, Public Health Policy, and the Politics of Expertise, 1948–1949," *Environmental History Review* 18 (spring 1994): 117–41.

51. Stanford Research Institute, *The Smog Problem in Los Angeles County* (Los Angeles: Western Oil and Gas Association, 1954), 7. Also see James E. Krier and Edmund Ursin, *Pollution and Policy: A Case Essay on California and Federal Experience with Motor Vehicle Air Pollution, 1940–1975* (Berkeley: University of California Press, 1977); Logsdon, "Organizational Responses;" Marvin Brienes, "The Fight Against Smog In Los Angeles, 1943–57" (Ph.D. diss., University of California, Davis, 1975).

52. Logsdon, "Organizational Responses," 421.

53. "Oil's Sulfur Boomlet," *Oil and Gas Journal* 50 (January 14, 1953): 61–64.

54. John W. Newton, "Opening Remarks of the Chairman to the Session on Smoke and Fumes," Division of Refining, *API Proceedings* 35 (1955), 143; Logsdon, "Organizational Responses," 356, 377, 387, 394, 406, 415.

55. "Sulfur From H_2S," *Oil and Gas Journal* 50 (March 31, 1952): 84–85, 95.

56. "Oil's Ugly Duckling," *Oil and Gas Journal* 51 (January 14, 1952): 59.

57. Logsdon, "Organizational Responses," 366. The Humble Oil refinery in Baytown, Texas, which still burned some of its acid sludge up to the mid-1950s, was taking the lead in developing hydro-refining.

58. Krier and Ursin, *Pollution and Policy*, 78–79.

59. Ibid., 6, 81.

60. Stanford Research Institute, *The Smog Problem*, 7–20.

61. R. N. Creek, "Evaporation Losses Held Near Zero," *Oil and Gas Journal* 50 (May 10, 1951): 129–37.

62. Logsdon, "Organizational Responses," 99–102.

63. Vance N. Jenkins, "Status Report on the Research Program of the Smoke and Fumes Committee," Division of Refining, *API Proceedings* 34 (1954), 233–38; W. H. Claussen, "The Smoke and Fumes Committee," Division of Refining, *API Proceedings* 35 (1955), 145–54.

64. L. C. Kemp, Jr., "Review of the Smoke and Fumes Committee Research Program," Division of Refining, *API Proceedings* 36 (1956), 287.

65. Robert W. Van Dolah, "Trip Report to Third National Air Pollution Symposium," April 18–20, 1955, May 11, 1955, Box 18, Records of Petroleum Experiment Center, Bureau of Mines, RG 70-2, National Archives, Southwest Region, Fort Worth; Krier and Ursin, *Pollution and Policy*, 85–87.

66. R. B. Ripley, "Congress and Clean Air: The Issue of Enforcement, 1963," in *Congress and Urban Problems*, ed. F. Cleveland (Washington, D.C.: Brookings Institute, 1963), 229–31.

67. Barton H. Eccleston to C. C. Ward, July 5, 1956, Box 24, Records of Research Groups, Bartlesville Petroleum Research Center, Bureau of Mines, RG 70-4, National Archives, Southwest Region, Fort Worth.

68. Dolah, "Trip Report."

69. *Encyclopedia Americana*, 1991 international ed., s. v. "mass spectrometer."

70. Klaus H. Altgelt and T. H. Gouw, eds., *Chromatography in Petroleum Analysis* (New York: Marcel Dekker: 1979).

71. Logsdon, "Organizational Responses," 99–102.

72. A. Berk (Chief, Industry Water Lab, Eastern Experimental Station) to R. Corey (Chief, Di-

vision of Solid Fuels), May 23, 1957, Box 92, Records of Research Groups, Bartlesville Petroleum Research Center, Bureau of Mines, RG 70-4, National Archives, Southwest Region, Fort Worth.

73. API, *Waste Gases and Particulate Matter,* vol. 2 of *Manual on the Disposal of Refinery Wastes,* 4th ed. (New York: API, 1952), 9.

74. API, *Sampling and Analysis of Waste Water,* vol. 4 of *Manual on the Disposal of Refinery Wastes,* 1st ed. (New York: API, 1953); *Sampling and Analysis of Waste Gases and Particulate Matter,* vol. 5 of *Manual on the Disposal of Refinery Wastes,* 1st ed. (New York: API, 1954).

75. John D. Frame, "Can Refinery Pollution Control Have a Payout?" Division of Refining, *API Proceedings* 40 (1960), 295–301.

76. Port of Houston Industrial Map, 1945, Industrial Maps, box 16, J. Russell Wait Collection, Woodson Research Center, Rice University.

77. *Directory of Oil Refineries and Field Processing Plants* (Tulsa, Okla.: *Oil and Gas Journal,* 1952), 7–8.

78. W. A. Quebedeaux to Jean Paul Bradshaw, December 7, 1950, folder 7, box 7, Dr. W. A. Quebedeaux Environmental Collection, Houston Metropolitan Research Center; W. A. Quebedeaux," Air and Stream Pollution Control in Harris County, Texas," *Public Health Reports* 69 (September 1954): 836–40.

79. Comments of John Latchford (Texas Water Quality Board), "The Houston Ship Channel," folder 8, box 5, Citizens Environmental Coalition Collection, Houston Metropolitan Research Center.

80. Photograph album, "My Trip Down the Houston Ship Channel," box 3S28, Olin Culberson Papers, Center for American History, University of Texas at Austin.

81. "Before It's Too Late—Make the Ship Channel Safe," *Houston Press,* February 18, 1960, 1.

82. W. A. Quebedeaux, "The Relationship of Water Pollution to Conservation," January 20, 1961, folder 11, box 3, Dr. W. A. Quebedeaux Environmental Collection. Also see W. A. Quebedeaux, "Active Prosecution as the Key to Air Pollution Control," paper presented at the 1957 Meeting of the Air Pollution Control Association, folder 10, box 3.

83. John C. Esposito and Larry J. Silverman, *Vanishing Air: The Ralph Nader Study Group Report on Air Pollution* (New York: Grossman Publishers, 1970), 200–203.

84. "How Humble Combats Water and Air Pollution," *Oil and Gas Journal* 64 (March 28, 1966): 132–36.

85. Senate Subcommittee on Air and Water Pollution, *Hearings on Water Pollution,* 89th Cong., 1st sess., June 23 and June 24, 1965, 912.

86. "Pollution: Congressional Target for Oil to Watch," *Oil and Gas Journal* 63 (May 24, 1965): 49–52.

87. W. L. Stewart, Jr., "Management's Responsibility in Air Pollution," presented to the Sixth Industrial Health Conference, Houston, Texas, Oct. 2, 1953, folder "1950 API," box 166, Bernard Majewski Collection, American Heritage Center, University of Wyoming.

88. For a discussion that frames T.N.E.C. and its charges as one more set of charges in a series of similar accusations that dates back to the days of Standard Oil, see Roger M. Olien and Diana Davids Olien, *Oil and Ideology: The Cultural Creation of the American Petroleum Industry* (Chapel Hill: University of North Carolina Press, 2000).

89. John Scoville and Noel Sargent, eds., *Fact and Fancy in the T.N.E.C. Monographs* (New York: National Association of Manufacturers, 1942), 675–760.

90. Williamson et al., *Age of Energy,* 674, 697–702.

91. API brochure, *The Key to Enduring Progress,* 1947.

92. Eugene V. Rostow, *A National Policy for the Oil Industry* (New Haven: Yale University Press, 1948); boxes 41 and 42, George A. Hill, Jr., Collection, Houston Metropolitan Research Center, which contain Hill's files related to his tenure on the API Board of Directors from 1945 to 1949. For an analysis of the controversy, see Leslie Cookenboo, Jr., *Crude Oil Pipe Lines and Competition in the Oil Industry* (Cambridge, Mass.: Harvard University Press, 1955), 76–94.

93. For more on the federal regulation of oil and gas markets, see Gerald D. Nash, *United States Oil Policy, 1890–1964: Business and Government in the Twentieth Century* (Pittsburgh: Uni-

versity of Pittsburgh Press, 1968); Anthony Copp, *Regulating Competition in Oil: Government Intervention in the U.S. Refining Industry* (College Station: Texas A&M University Press, 1976); Richard H. K. Vietor, *Energy Policy in America Since 1945: A Study of Business-Government Relations* (Cambridge: Cambridge University Press, 1984); and James Castaneda, *Regulated Enterprises: Natural Gas Pipelines and Northeastern Markets, 1938–1954* (Columbus: Ohio State University Press, 1993).

94. Ripley, "Congress and Clean Air," 226.

95. Logsdon, "Organizational Responses," 57.

96. "API Sets Up Division to Study Air," *Oil and Gas Journal* 63 (November 15, 1965): 110–111.

97. Logsdon, "Organizational Responses," 49.

Chapter 10. The Ocean Ignored As Tankers Grow

1. *Preliminary Conference on Oil Pollution of Navigable Waters*, June 8–16 (Washington, D.C.:, Government Printing Office, 1926).

2. G. A. B. King, *Tanker Practice: The Construction, Operation, and Maintenance of Tankers* (London: Maritime Press, 1969), 113.

3. Ibid., 92–93.

4. Ibid., 100.

5. King, *Tanker Practice*, 118.

6. James E. Moss, *Character and Control of Sea Pollution by Oil* (Washington: API, 1963), 31–36.

7. J. H. Pew to R. W. Pack, August 3, 1936, Pollution by Sun Boats, box 28, series 3, Sun Company Collection, Hagley Museum and Library. Pew advocates building an earthen tank capable of receiving ballast water at one of the company's loading terminals so that Sun ships would not have any excuse to discharge their ballast near the shore.

8. U.S. Bureau of Mines, *Pollution By Oil of the Coast Waters of the United States*, preliminary report, Sept. 1923, 90.

9. House Committee on Rivers and Harbors, *Hearings on a Bill to Amend the Oil Pollution Act of 1924*, 71st Cong., 2d sess, May 2, 3, and 26, 1930, 24.

10. Department of Commerce to Charles Crotty, December 17, 1936, Oil Drilling in the Bay, box 19, J. Russell Wait Collection, Woodson Research Center, Rice University.

11. American Petroleum Institute, Survey of Oil Pollution Conditions in the U.S. with Remedial Measures Recommended (New York: API, 1927), p. 5, attachment to letter, J. H. Pew to W. R. Boyd, June 21, 1928, box 35, series 10, Sun Company Collection, Hagley Museum and Library.

12. C. L. Boyle to J. Edgar Pew, August 25, 1936, Pollution by Sun Boats, box 28, series 3, Sun Company Collection, Hagley Museum and Library.

13. Donelson Caffrey to R. W. Pack, June 17, 1937, Pollution by Sun Boats, box 28, series 3, Sun Company Collection, Hagley Museum and Library; C. R. Innis to J. E. Pew, June 17, 1938, Pollution by Sun Boats, box 28, series 3, Sun Company Collection; J. E. Pew to T. L. Foster, December 5, 1938, Pollution by Sun Boats, box 28, series 3, Sun Company Collection.

14. King, *Tanker Practice*, 113–14.

15. Ronald B. Mitchell, *Intentional Oil Pollution at Sea: Environmental Policy and Treaty Compliance* (Cambridge, Mass.: MIT Press, 1994), 83–85.

16. Ibid.

17. United Nations, *Pollution of the Sea by Oil: Results of an Inquiry by the United Nations Secretariat* (New York: United Nations, 1956).

18. Ibid., 108–9; R. L. Meley to R. W. Pack, August 1 1936, Pollution by Sun Boats, box 28, series 3, Sun Company Collection, Hagley Museum and Library.

19. API, Division of Transportation, *The History of the Tanker Corrosion Research Project* (Washington, D.C.: API, ca. 1965).

20. Rene De La Pedraja, *The Rise and Decline of U.S. Merchant Shipping in the Twentieth Century* (New York: Twayne, 1992), 120.

21. Harold F. Williamson, Ralph L. Andreano, Arnold R. Daum, and Gilbert C. Klose, *The*

American Petroleum Industry: The Age of Energy, 1899–1959 (Evanston, Ill.: Northwestern University Press, 1963), 802.

22. Nicholas Fraser, Philip Jacobson, Mark Ottaway, and Lewis Chester, *Aristotle Onassis* (J. B. Lippincott Co., 1977), 101–5. Also see Senate Committee on Government Operations, *Hearings on the Sale of Government-Owned Surplus Tanker Vessels,* 82nd Cong., 2d sess., 1952.

23. Pedraja, *U.S. Merchant Shipping,* 134.

24. Fraser et al., *Aristotle Onassis,* 102–5.

25. Ibid., 157–59.

26. Moss, *Sea Pollution by Oil,* 30.

27. In 1959, the United States had to import approximately 10 percent of its oil. Williamson et al., *Age of Energy,* 810–821.

28. Mitchell, *Intentional Oil Pollution,* 86. Also see Sonia Zaide Pritchard, *Oil Pollution Control* (London: Croom Helm, 1987).

29. R. R. Churchill and A. V. Lowe, *The Law of the Sea* (Manchester, England: Manchester University Press, 1983), 18. The original name of the organization was the Intergovernmental Maritime Consultative Organization (IMCO).

30. U.S. Senate, Committee on Commerce, *Hearings on Wetland Acquisition and Oil Pollution of the Sea,* 87th Cong., 1st sess., July 31, 1961, 7.

31. American Petroleum Institute, *Oil Pollution Survey of the United States Atlantic Coast,* prepared by John V. Dennis (New York, D.C.: API, 1959), C1. Also see American Petroleum Institute, *Oil Pollution Survey of the Great Lakes,* prepared by John V. Dennis (New York, D.C.: API, 1960).

32. Dennis, *Oil Pollution Survey of the United States Atlantic Coast,* C1-D1.

33. Ibid., F27.

34. Ibid., C2.

35. Moss, *Sea Pollution by Oil,* 46; R. Maybourn, "Operational Pollution from Tankers and Other Vessels," in *The Prevention of Oil Pollution,* ed. J. Wardley-Smith (New York: Wiley, 1979), 120–121.

36. Fraser et al., *Aristotle Onassis,* 117–23.

37. Mitchell, *Intentional Oil Pollution,* 226–27.

38. Senate Committee on Commerce, *Hearings on the Navigable Waters Safety and Environmental Quality Act,* 92nd Cong., 1st sess., September 22–24, 1971, 303.

39. R. M. M'Gonigle and Mark W. Zacher, *Pollution, Politics, and International Law: Tankers at Sea* (Berkeley: University of California Press, 1979), 99.

40. Mitchell, *Intentional Oil Pollution,* 87.

41. "Statistics of the American Tanker Fleet," April 15, 1937, box 47, series 4, Sun Company Collection, Hagley Museum and Library. All tonnage listed in this chapter is deadweight tonnage given in long tons of 2,240 pounds. For the amount of crude a ship can carry, assume seven barrels to the long ton, though the exact amount depends on the specific gravity of the crude and the temperature.

42. J. Bes, *Tanker Chartering and Management* (Amsterdam: C. de Boer, 1956), 12–29.

43. Fraser et al., *Aristotle Onassis,* 157–59; U.S. Maritime Administration, *A Statistical Analysis of the World's Merchant Fleets as of December 31, 1956* (Washington, D.C.: Government Printing Office), 1.

44. National Research Council, *Tanker Spills: Prevention By Design* (Washington, D.C., National Academy Press, 1991), 32.

45. Moss, *Sea Pollution by Oil,* 23.

46. Theodore E. Swigart, "Handling Oil and Gas in the Field," in American Petroleum Institute, *History of Petroleum Engineering* (New York: API, 1961), 964; John Ise, *The United States Oil Policy* (New Haven: Yale University Press, 1926), 85–90.

47. M. S. Russell, "Life Story of James Cunningham Russell," manuscript, 1964, petroleum exhibit files, drawer 2, folder "J. C. Russell life story," National Museum of American History, Smithsonian.

48. Henrietta M. Larson, Evelyn H. Knowlton, and Charles S. Popple, *New Horizons, 1927–50:*

History of Standard Oil Company (New Jersey) (New York: Harper and Row, 1971), 138. In a photo spread between pages 452 and 453, a photograph of derricks in Lake Maracaibo reflects the experience with offshore drilling that Standard Oil had gained by World War II.

49. Swigart, "Handling Oil and Gas in the Field," 964. Also see Kenny A. Franks and Paul F. Lambert, *Early Louisiana and Arkansas Oil* (College Station, Tex.: Texas A&M University Press, 1982), 206–8; Robert O. Anderson, *Fundamentals of the Petroleum Industry* (Norman: University of Oklahoma Press, 1984), 144–48.

50. Franks and Lambert, *Early Louisiana and Arkansas Oil,* 208; Theron Wasson, "Creole Field, Gulf of Mexico, Coast of Louisiana," in *Structure of Typical American Oil Fields,* vol. 3, ed. J.V. Howell (Tulsa, Okla.: American Association of Petroleum Geologists, 1948), 281.

51. "Directional Drilling as a Routine Method of Development,"*Petroleum Engineer* 12 (October 1941): 38–41; Anderson, *Petroleum Industry,* 144–48.

52. F. J. Hortig, "Jurisdictional, Administrative, and Technical Problems Related to the Establishment of California Coastal and Offshore Boundaries," in *The Law of the Sea: Offshore Boundaries and Zones,* ed. Lewis M. Alexander (Columbus: Ohio State University Press, 1967): 230–241.

53. Port of Houston Industrial Map, 1945, Industrial Maps, box 16, J. Russell Wait Collection, Woodson Research Center, Rice University.

54. Charles Crotty (Ass't Director, Port of Houston) to J. H. Walker (State Land Commissioner), June 17, 1936, Oil Drilling in the Bay, box 19, J. Russell Wait Collection.

55. T. H. Jackson (Corps of Engineers) to Charles Crotty, July 28, 1936, Oil Drilling in the Bay, box 19, J. Russell Wait Collection.

56. U.S. Bureau of Mines, *Pollution by Oil of the Coast Waters of the United States,* preliminary report (Washington, D.C., 1923), 144.

57. Charles Crotty to Captain Manuel J. Asensio, March 4, 1938, Oil Drilling in the Bay, box 19, J. Russell Wait Collection.

58. *Look* (November 22, 1938): 47.

59. Clippings, *Galveston Daily News,* March 1, 1941, and *Houston Chronicle,* June 18, 1941, Oil Drilling in the Bay, box 19, J. Russell Wait Collection.

60. Marcer H. Parks and James C. Posgate, "Offshore Drilling and Development," preprint of paper, folder "1948 API," box 178, Bernard Majewski Collection, American Heritage Center, University of Wyoming. Also see John Samuel Ezell, *Innovations in Energy: The Story of Kerr-McGee* (Norman: University of Oklahoma, 1971).

61. Louisiana Department of Conservation, *Development of Louisiana's Offshore Oil and Gas Reserves* (Baton Rouge: Louisiana Department of Conservation, 1956), 1.

62. "History," Continental Oil, folder "Gulf of Mexico," box 32, Continental Oil Company Collection, American Heritage Center, University of Wyoming; Robert E. Hardwicke, "The Tidelands and Oil," *Atlantic Monthly* 183 (June 1949): 21–26.

63. U.S. Congress, Senate Committee on Interior and Insular Affairs, *Hearings on Submerged Lands,* 82d Cong., 1s sess., Feb. 19–22, 1951; U.S. Congress, Committee on the Judiciary, *Claimed Authority for Seizure and Lease of Certain Submerged Lands,* 82d Cong., 2d sess., June 12, 1952; Ernest E. Bartley, *The Tidelands Oil Controversy* (Austin: University of Texas Press, 1953); API, "Seventh Progress Report of the API Tidelands Study Committee," folder 1953–55 API, box 165, Bernard Majewski Collection; Arthur H. Dean, "The Law of the Sea Conference, 1958–60, and Its Aftermath" in *Law of the Sea,* ed. Alexander, 244–64.

64. Henry Lewelling, "Offshore's Challenge to the Engineer," *Frontiers in Petroleum Engineering* (Dallas, Tex.: Petroleum Engineering Publishing Company, 1960), 57–62.

65. "California Loses Offshore Oil Suit," *New York Times,* May 18, 1965, 1. The ruling was based on international law that designated a bay as inland waters only if the mouth of a bay was less than twenty-four miles across.

66. "Rules and Regulations Governing Drilling and Producing Operations in Coastal Waters," 1965, box 3G69, Frances Tarlton Farenthold Papers, Center for American History, University of Texas at Austin.

67. Moss, *Sea Pollution by Oil,* 45–46.

Chapter 11. Crude Awakening

1. Samuel P. Hays, *Beauty, Health, and Permanence: Environmental Politics in the United States, 1955–1985* (New York: Cambridge University Press, 1987), 13–39.

2. Frank N. Ikard, Report of the President, Minutes of the Meeting of API Board of Directors, April 23, 1966, p. 76, folder "1963–1965 API," box 46, Bernard Majewski Collection, American Heritage Center, University of Wyoming; Report on Air and Water Conservation Expenditures of the Petroleum Industry of the United States, August 1968, folder "API Board of Directors Correspondence, 1968," box 29, series 6, Sun Company Collection, Hagley Museum and Library.

3. E. L. Steiniger, Report of the Committee on Air and Water Conservation, Minutes of the API Board of Directors Meeting, Sept. 27, 1967, pp. 8–11, folder 2, 3X315, Morgan Jones Davis Papers, Center for American History, University of Texas at Austin.

4. Press release, Coordinating Research Council, "Water Conservation in General," Nov. 13, 1967, API press interviews, box 32, series 6, Sun Company Collection, Hagley Museum and Library.

5. John J. Scott, Report of the Committee on Air and Water Conservation, "Minutes of the API Board of Directors Meeting," Apr. 28, 1967, pp. 18–29, folder 2, 3X315, Morgan Jones Davis Papers.

6. Harold W. Kennedy and Martin E. Weekes, "Control of Automobile Emission—California Experience and the Federal Legislation," in *Air Pollution Control*, ed. Clark C. Havighurst (Dobbs Ferry: Oceana Publications, 1969), 101–18; James E. Krier and Edmund Ursin, *Pollution and Policy: A Case Essay on California and Federal Experience with Motor Vehicle Air Pollution, 1940–1975* (Berkeley: University of California Press, 1977).

7. Robert Martin and Lloyd Symington, "A Guide to the Air Quality Act of 1967," in *Air Pollution Control*, ed. Clark C. Havighurst (Dobbs Ferry: Oceana Publications, 1969), 43–78; Randall B. Ripley, "Congress and Clean Air: The Issue of Enforcement, 1963," in *Congress and Urban Problems*, ed. F. N. Cleaveland (Washington, D.C.: Brookings Institute, 1969).

8. P. N. Gammelgard, Report of the Committee on Air and Water Conservation, Minutes of the API Executive Committee, White Sulfur Springs, West Virginia, API, September 27, 1967, p. 33, box 3X315, Morgan Jones Davis Papers, Center for American History, University of Texas at Austin.

9. Ibid., p. 39.

10. Press release, Coordinating Research Council, Nov. 13, 1967, "Sulfur in Petroleum Fuels," 1–4, API Press Interviews, box 32, series 6, Sun Company Collection, Hagley Museum and Library. Also see Eric J. Cassel, "The Health Effects of Air Pollution and Their Implications for Control," in *Air Pollution Control*, ed. Clark C. Havighurst (Dobbs Ferry: Oceana Publications, 1969), 21–30.

11. "East Coast Resid Imports Flunk U.S. Sulfur Test," *Oil and Gas Journal* 65 (May 22, 1967): 75; "Cutting Down the Reek of Pollution," *Business Week* (January 28, 1967): 79–84.

12. "Where Oil Stands in War on SO_2," *Oil and Gas Journal* 65 (July 3, 1967): 27–28.

13. Ibid. For more on the New York City pollution problem, see John Esposito and Larry J. Silverman, *Vanishing Air: The Ralph Nader Study Group Report on Air Pollution* (New York: Grossman Publishers, 1970), 203–23.

14. Esposito and Silverman, *Vanishing Air*, 97–99.

15. Not all companies adopted the technology because refineries that processed low-sulfur crudes often found it more economical to use other methods for removing sulfur. See James G. Speight, *The Chemistry and Technology of Petroleum* (New York: Marcel Dekker, 1991), 509, 622.

16. "Where Oil Stands in War on SO2," 28.

17. "Cutting Down the Reek of Pollution," 79–80.

18. Ibid; Harold F. Elkin, "Petroleum Refining," in *Engineering Control of Air Pollution*, ed. Arthur Stern, vol. 4, *Air Pollution* (New York: Academic Press, 1977), 822–25.

19. Marshall Sittig, *Petroleum Refining Industry: Energy Saving and Environmental Control* (Park Ridge, N. J.: Noyes Data Corporation, 1978), 70–73.

20. "Where Oil Stands in War on SO_2," 28.

21. P. N. Gammelgard, Report of the Committee for Air and Water Conservation, Minutes of the API Executive Committee, White Sulfur Springs, West Virginia, API, September 27, 1967, pp. 39–40, box 3X315, Morgan Jones Davis Papers, Center for American History, University of Texas at Austin.

22. E. E. Weaver, "Effects of Tetraethyl Lead on Catalyst Life and Efficiency in Customer Type Vehicle Operation," paper, International Automotive Engineering Congress, Detroit, Mich., January 13–17, 1969.

23. Press release, Coordinating Research Council, "Lead Content of Gasoline," Nov. 13, 1967, API press interviews, box 32, series 6, Sun Company Collection, Hagley Museum and Library; H. F. Elkin, "Notes on Muskie Subcommittee Transcripts," March 4, 1967, box 632, series 02, Sun Company Collection; W. A. Burhouse to API Committee for Air and Water Conservation, March 7, 1967, box 632, series 02, Sun Company Collection.

24. Robert W. Van Dolah, Trip Report to Third National Air Pollution Symposium, April 18–20, 1955, May 11, 1955, Box 18, Bartlesville Records, Bureau of Mines, RG 70-2, NARS, Fort Worth. Also see Krier and Ursin, *Pollution and Policy*, 97–100.

25. Examples of research from the Society of Automotive Engineers (SAE) in the late 1950s and early 1960s include the following: E. F. Hill, W. A. Cannon, and C. E. Welling, "Single Cylinder Engine Testing of Hydrocarbon Oxidation Catalysts," SAE Meeting, paper no.174, Seattle, August 1957; E. F. Hill, W. A. Cannon, and C. E. Welling, "Oxidation Catalysts Reduce Hydrocarbons in Automobile Exhaust Gas," *SAE Journal* (January 1958), p. 36; W. A. Cannon and C. E. Welling, "The Application of Vanadia-Alumina Catalysts for the Oxidation of Exhaust Hydrocarbons," SAE Meeting, paper no. 29T, Detroit, Michigan, January 1959; D. L. Davis and G. F. Onishi, "Catalytic Converter Development Problems," Presented at the SAE National Meeting, Paper No. 486F, Detroit, Michigan, March, 1962. SAE meeting papers can be found at Kettering University (formerly General Motors Institute).

26. W. A. Bachman, "The Verdict on '68 Autos: Not So Favorable For Oil," *Oil and Gas Journal* 65 (September 25, 1967): 69–71; "The Unleaded-Gasoline Tab: $4.2 Billion," *Oil and Gas Journal* 65 (May 22, 1967): 74–75.

27. Press release, Coordinating Research Council, "Lead Content in Gasoline," Nov. 13, 1967, API press interviews, box 32, series 6, Sun Company Collection, Hagley Museum and Library.

28. William D. Hurley, *Environmental Legislation* (Springfield, Ill.: Charles C. Thomas, 1971), 37.

29. Jeanne Marie Logsdon, "Organizational Responses to Environmental Issues: Oil Refining Companies and Air Pollution" (Ph.D. diss., University of California, Berkeley, 1983), 58–61. Also see John E. O'Fallon, "Deficiencies in the Air Quality Act of 1967" in *Air Pollution Control*, ed. Clark C. Havighurst (Dobbs Ferry: Oceana Publications, 1969), 79–100.

30. "Oil Pollution Act is Found Crippled," *New York Times*, April 16, 1967; H. F. Elkin to D. W. Ferguson, June 9, 1967, box 656, series 2, Sun Company Collection, Hagley Museum and Library.

31. Richard Petrow, *In the Wake of the Torrey Canyon* (New York: David McKay Co., 1968), 1–18.

32. Ibid., 33–49.

33. Ibid., 83–100.

34. "Troops in Britain Battle Oil Slick," *New York Times*, March 26, 1967, 17; J. E. Smith, ed., *"Torrey Canyon" Pollution and Marine Life: A Report by the Plymouth Laboratory of the Marine Biological Association of the United Kingdom* (Cambridge: Cambridge University Press: 1968).

35. "Jets Bomb Grounded Tanker Off Cornwall," *New York Times*, March 29, 1967, 1; Petrow, *Torrey Canyon*, 101–12; Norman Polind, "Effects of Torrey Canyon Pollution on Marine Life," in *Oil on the Sea: Proceedings of a Symposium Sponsored by MIT and Woods Hole Oceanographic Institution*, ed. David P. Hoult (New York: Plenum Press, 1969), 1–4; Alison Jolly, "Wreck of the Torrey Canyon—Epilogue," *Animal Kingdom* (February 1968): 6–14. The French, in a slightly different strategy, responded by spraying a chalky substance on the slick, trying to sink the crude before it reached the continental shelf.

36. Memorandum, Lyndon B. Johnson to Secretary of the Interior and Secretary of Transportation, May 26, 1967, HE 8-4, box 24, Subject Files, White House Central Files, Lyndon B. Johnson Presidential Library (LBJ Library).

37. Briefing on President's Memorandum, May 26, 1967, Pricing Files, Oil Pollution, box 15, Aides Files—Robson-Ross, White House Central Files, LBJ Library.

38. U.S. Secretary of the Interior and U.S. Secretary of Transportation, Oil Pollution: A Report to the President, February 1968, 2, Oil Pollution, Aides Files—Gaither, White House Central Files, LBJ Library.

39. Ibid., 12–17, 45–51.

40. K. E. Biglane, "A History of Major Oil Spill Incidents," in *Proceedings of a Joint Conference on Prevention and Control of Oil Spills* (New York: American Petroleum Institute and Federal Water Pollution Control Administration, 1969), 141.

41. Press Release, "Statement by the President Announcing Approval of National Contingency Plan for Combating Oil Spills," no date; "National Multi-Agency Oil and Hazardous Materials Pollution Contingency Plan," September 1968, HE 8-4, box 24, Subject Files, White House Central Files, LBJ Library.

42. P. N. Gammelgard, Report of the Committee for Air and Water Conservation, Minutes of the API Executive Committee, 1967, p. 38; "Remarks by W. T. Askew," Associated Petroleum Industries of Pennsylvania, May 22, 1969, box 178, series 1, Sun Company Collection, Hagley Museum and Library.

43. Ernest Cotton, "Oil Spill Cooperative Program," in *Proceedings of a Joint Conference,* 141.

44. J. V. Langstron, "Training Program to Improve Well Control Operations," in *Proceedings of a Joint Conference,* 137.

45. Walter D. Tower, *The Story of Oil* (New York: Appleton, 1920), 165.

46. Paul R. Waddell and Robert F. Niven, *Sign of the 76: The Fabulous Life and Times of the Union Oil Company of California* (Los Angeles: Union Oil Co., 1976), 381–82.

47. Harry L. Franklin, "Investigation of Recent Oil Pollution of Pacific Ocean Waters Opposite the City of Santa Barbara, California," box 3G69, Frances Tarlton Farenthold Papers, Center for American History, University of Texas at Austin.

48. Waddell and Niven, *Sign of the 76,* 251–52.

49. Franklin, "Investigation of Recent Oil Pollution."

50. For a detailed description of the blowout, see Carol Steinhart and John Steinhart, *Blowout: A Case Study of the Santa Barbara Oil Spill* (Belmont, Calif.: Duxbury Press, 1972).

51. David P. Hoult, "The Santa Barbara Oil Spill," in *Oil on the Sea,* 21. Steinhart and Steinhart put the figure at 3,000 to 30,000 barrels per day: Steinhart and Steinhart, *Blowout,* 74.

52. Franklin, "Investigation of Recent Oil Pollution."

53. Ibid.; Steinhart and Steinhart, *Blowout,* 7–15.

54. "Oil Leak Plugged and Cleanup Begins," *New York Times,* February 9, 1969, 1; Franklin, "Investigation of Recent Oil Pollution."

55. Steinhart and Steinhart, *Blowout,* 69, 77–88; Waddell and Niven, *Sign of the 76,* 384; "Sun Platform Set Up, GOO Outmaneuvered," undated clipping, box 52, series 8, Sun Company Collection, Hagley Museum and Library.

56. Kerryn King, "Opening Remarks," in *Proceedings of a Joint Conference,* 3.

57. L. P. Haxby, "Industry Research and Response Plans," in *Proceedings of a Joint Conference,* 11–13.

58. Alleb Cywin, "Federal Research and Development Program for Oil Spills," in *Proceedings of a Joint Conference,* 16–22.

59. Federal Water Pollution Control Administration, *Manpower and Training Needs in Water Pollution Control* (Washington, D.C.: U.S. Department of the Interior, 1967), 5.

60. House Committee on Merchant Marine and Fisheries, *Hearings on Oil Pollution: Bills to Amend the Oil Pollution Act, 1924, for the Purpose of Controlling Oil Pollution from Vessels,* 91st Cong., 1st sess., Feb. 25–27, March 11–13, 18, 26–28, and Apr. 1, 1969; U.S. Coast Guard, *Legal, Economic, and Technical Aspects of Liability and Financial Responsibility As Related to Oil Pollution,*

prepared by the Program of Policy Studies in Science and Technology, George Washington University (Washington, D.C.: George Washington University, 1970).

61. Hurley, *Environmental Legislation*, 16–20.

62. House, *Hearings on Oil Pollution*, 1969, 156–58.

63. Hurley, *Environmental Legislation*, 17.

64. Russell V. Randle, "The Oil Pollution Act of 1990: Its Provisions, Intents, and Effects," in *Oil Pollution Deskbook* (Washington, D.C.: Environmental Law Reporter, 1991), 3–18. The Water Quality Improvement Act of 1970 was eventually incorporated into the Federal Water Pollution Control Amendments of 1972.

65. National Environmental Policy Act of 1969, 42 U.S.C. §4321, Public Law 91–190.

66. William H. Rogers, "Where Environmental Law and Biology Meet: Of Panda's Thumbs, Statutory Sleepers, and Effective Law," *University of Colorado Law Review* 65 (1993): 58.

67. "Alaska Soberer on Oil Discovery," *New York Times*, January 6, 1958, 132.

68. "Alaska Oil Lands Reserved for Navy," *New York Times*, February 28, 1923, 7; Charles E. Bowles, *The Petroleum Industry* (Kansas City, Mo.: Schooley Printing, 1921), 29. Also see P. H. Ray, *Expedition to Point Barrow, Alaska* (Washington, D.C.: United States Army, Signal Corps, 1884).

69. Kenneth Harris, *The Wildcatter: A Portrait of Robert O. Anderson* (New York: Weidenfeld and Nicolson, 1987), 78.

70. Ibid., 79–82.

71. Charles J. Cicchetti, *Alaskan Oil: Alternative Routes and Markets* (Baltimore: Resources for the Future, Johns Hopkins University Press, 1972), 2.

72. "Two Oil Projects Move Ahead," *New York Times*, February 11, 1969, 41.

73. Peter A. Coates, *The Trans-Alaska Pipeline Controversy: Technology, Conservation, and the Frontier* (Bethlehem, Pa.: Lehigh University Press, 1991), 176–89.

74. Ibid., 180; "Alaska: Closer to Cashing Oil's Riches," *Business Week* (April 25, 1972): 78–79; "Tundra," *The Encyclopedia of the Environment*, ed. Ruth A. Eblen and William R. Eblen (New York: Houghton Mifflin Co., 1994), 743–46.

75. Coates, *Trans-Alaska Pipeline Controversy*, 70; "Hickel Highway Made of Ice, Critics Say It's Not Very Nice," *Houston Chronicle*, May 17, 1970; Walter Hickel, "Alaska the Magnificent," October 1974, unidentified magazine article, Alaska Pipeline Clippings Binder, box 4, Mr. and Mrs. Emmott Environmental Collection, Houston Metropolitan Research Center.

76. U.S. Bureau of Mines, Recommendations on D-2 Lands, Alaska Native Claims Settlement Act, July 1972, 1, folder "Bureau of Mines," box 997, Allen F. Agnew Collection, American Heritage Center, University of Wyoming.

77. Coates, *Trans-Alaska Pipeline Controversy*, 176–89; Mary Clay Berry, *The Alaska Pipeline: The Politics of Oil and Native Land Claims* (Bloomington: Indiana University Press, 1975), 108; "Hickel Lists Snags Facing Trans-Alaskan Oil Pipeline," *Christian Science Monitor*, Nov. 28, 1969, B10.

78. Coates, *Trans-Alaska Pipeline Controversy*, 178; Walter Hickel, *Who Owns America?* (Englewood Cliffs, N.J.: Prentice-Hall, 1971).

79. Secretary of the Interior to Thomas McClary, in reply to letter of Aug. 4, 1969, folder "States—Alaska—Oil pipeline," box 28, David D. Dominick Collection, American Heritage Center, University of Wyoming. Dominick was in charge of the Federal Water Pollution Control Agency.

80. Earl N. Kari to Robert S. Burd, Sept. 3, 1969, folder "States—Alaska—Oil pipeline," box 28, David D. Dominick Collection; presentation to the Department of Interior Hearings, James A. Anderagg, Chief, Branch of Environmental Health, August 29, 1969, folder "States—Alaska—Oil pipeline," box 28, David D. Dominick Collection.

81. Earl N. Kari to James L. Agee, Sept. 25, 1969, folder "States—Alaska—Oil pipeline," box 28, David D. Dominick Collection; Raymond Morris to Deputy Regional Director, FWPCA, Oct. 31, 1969, folder "States—Alaska—Oil pipeline," box 28, David D. Dominick Collection.

82. James L. Agee to Frank M. Covington, Nov. 21, 1969, folder "States—Alaska—Oil pipeline," box 28, David D. Dominick Collection

83. "Oil Companies Press for Pipeline Release," *Christian Science Monitor,* January 10, 1970, 5; news release, "Probe Permafrost Effects on Alaska Pipeline Construction," January 12, 1970, Geological Survey, folder "States—Alaska—Oil pipeline;" news release, Office of the Secretary, Dept. of the Interior, January 14, 1970, folder "States—Alaska—Oil pipeline," box 28, David D. Dominick Collection.

84. Berry, *Alaska Pipeline,* 117–23; Robert S. Burd to Eugene Jensen, Dec. 9, 1970, folder "States—Alaska—Oil pipeline," box 28, David D. Dominick Collection

85. Anthony Sampson, *The Seven Sisters: The Great Oil Companies and the World They Made* (New York: Viking Press, 1975), 178–79; "Tanker Leaves to Conquer Fabled Northwest Passage," *New York Times,* August 25, 1970, 1; Harris, *The Wildcatter,* 91–92.

86. Chief, EPA Federal Activities Branch to David D. Dominick, Acting Commission, EPA Water Quality Office, March 2, 1971, folder "States—Alaska—Oil pipeline," box 28, David D. Dominick Collection; Dominick to William Ruckelshaus, Administrator, EPA, March 2, 1971, folder "States—Alaska—Oil pipeline," box 28, David D. Dominick Collection; William Ruckelshaus to Rogers C. B. Morton, Secretary of the Interior, March 12, 1971, folder "States—Alaska—Oil pipeline," box 28, David D. Dominick Collection; news release, "Ruckelshaus Asks Further Study of Alaska Pipeline," EPA, March 15, 1971, folder "States—Alaska—Oil pipeline," box 28, David D. Dominick Collection; Kenneth Biglane to Chairman, Interior Task Force on Alaskan Oil Development, May 11, 1971, folder "States—Alaska—Oil pipeline," box 28, David D. Dominick Collection; "Trans-Alaskan Pipeline: Impact Study Receives Bad Reviews," *Science* 171 (March 19, 1971): 1130; Coates, *Trans-Alaska Pipeline Controversy,* 199.

87. Berry, *Alaska Pipeline,* 121–41; Coates, *Trans-Alaska Pipeline Controversy,* 227.

88. "Alaska: Closer to Cashing Oil's Riches," *Business Week* (April 25, 1972): 78; "Alaska Pipeline Report Blasted," May 5, 1972, newspaper clipping, Alaska Pipeline Clippings Binder, box 4, Mr. and Mrs. Emmott Environmental Collection, Houston Metropolitan Research Center; Coates, *Trans-Alaska Pipeline Controversy,* 227.

89. "Senators Urge Piping Alaska Oil to Midwest," *Houston Chronicle,* April 16, 1973, n.p., Alaska Pipeline Clippings Binder, box 4, Mr. and Mrs. Emmott Environmental Collection, Houston Metropolitan Research Center; Cicchetti, *Alaskan Oil,* 58–85. The law that eventually authorized construction of the pipeline prohibited the export of oil without the permission of the president and approval of Congress.

90. Coates, *Trans-Alaska Pipeline Controversy,* 227; Berry, *Alaska Pipeline,* 228; "Interior Issues Alyeska Right-of-Way Permit," *Oil and Gas Journal* 72 (January 28, 1974): 100.

91. Harris, *The Wildcatter,* 95; "Alaska Line Develops New Technology," *Oil and Gas Journal* 75 (November 21, 1977): 95–111; Alyeska Pipeline Service Company, *Trans Alaska Pipeline System Facts* (Alyeska, 1994).

92. As it turned out, the elevated pipe did not represent a barrier to caribou movements.

93. Ibid.; "Pipeline Contractors Gain Arctic Construction Experience," *Oil and Gas Journal* 74 (February 23, 1976): 108–16; "Stabilizing Pipeline Piling in Arctic Permafrost Regions," *Houston Chronicle,* October 29, 1973, clipping, Alaska Pipeline Clippings Binder, box 4, Mr. and Mrs. Emmott Environmental Collection, Houston Metropolitan Research Center.

94. "Alyeska Speeds Rebuilding of Damaged Pump Station," *Oil and Gas Journal* 75 (November 21, 1977): 83–85; "Prudhoe Bay Operators Lay Plans for Mammoth Seawater Injection Project," *Oil and Gas Journal* 78 (February 25, 1980): 80–88; Alyeska, *Facts,* 9–13.

95. "Tankering Oil From Valdez Seen Safe," *Oil and Gas Journal,* 75 (June 13, 1977): 38–40; U.S. Maritime Administration, *A Statistical Analysis of the World's Merchant Fleets* (Washington, D.C.: Government Printing Office, 1972).

96. "Tankers of 500,000 DWT Seen Feasible," *Oil and Gas Journal* 65 (May 22, 1967): 79; "The New Breed of Tankers," *Petroleum Today* (Fall 1971): 16; "Texas Sizes up Its Super Stakes Offshore," *Houston* (November 1972): 26, clipping, superport binder, box 4, Mr. and Mrs. Emmott Environmental Collection, Houston Metropolitan Research Center; "Gulf Superports

Await the Green Light," *Business Week* (May 26, 1973), clipping, superport binder, box 4, Mr. and Mrs. Emmott Environmental Collection, Houston Metropolitan Research Center; "Million-DWT Tanker Foreseen for 1980," *Oil and Gas Journal* 72 (January 28, 1974): 92–93.

97. "Superports Will Be Built But Where?" *Houston Chronicle,* July 23, 1972, clipping, superport binder, box 4, Mr. and Mrs. Emmott Environmental Collection, Houston Metropolitan Research Center.

98. Peter A. Bradford, *Fragile Structures: A Story of Oil Refineries, National Security, and the Coast of Maine* (New York: Harper's Magazine Press, 1975), 244–83.

99. Max Blumer et al., "A Small Oil Spill," *Environment* 13 (March 1971): 2–21.

100. "Why Delaware Acted," *Open Soundings* (April 1972): 3, clipping, superport binder, box 4, Mr. and Mrs. Emmott Environmental Collection, Houston Metropolitan Research Center.

101. "E. Canada to Get Another Supertanker Terminal," *Oil and Gas Journal* 69 (November 29, 1971): 29.

102. National Research Council, *Tanker Spills: Prevention By Design* (Washington, D.C.: National Academy Press, 1991), 35–36.

103. "State Lacks Pipeline Maps to Check Coast Pollution," *Corpus Christi Caller,* December 23, 1969, clipping, box 3G69, Frances Tarlton Farenthold Papers, Center for American History, University of Texas at Austin; "Texas Must Build Offshore or Do a Lot of Dredging," *Houston Chronicle,* July 23, 1972, clipping, superport binder, box 4, Mr. and Mrs. Emmott Environmental Collection, Houston Metropolitan Research Center.

104. "A Superport for Texas," *The Texas Observer* (August 13, 1971): 1–5, clipping, superport binder, box 4, Mr. and Mrs. Emmott Environmental Collection, Houston Metropolitan Research Center.

105. "Ecologists Tries to Prove Oil, Superports Mix," *Houston Post,* March 3, 1973, clipping, superport binder, box 4, Mr. and Mrs. Emmott Environmental Collection, Houston Metropolitan Research Center.

106. "Superports are One Step Closer," *Houston Chronicle,* December 26, 1974, clipping, superport binder, box 4, Mr. and Mrs. Emmott Environmental Collection, Houston Metropolitan Research Center.

107. "World Tanker Doldrums May Last to Mid-1980s," *Oil and Gas Journal* 75 (Aug. 29, 1977): 21; "DOT Chief Signs Licenses for LOOP, Seadock," *Oil and Gas Journal* 75 (Jan. 24, 1977): 28; "Seadock Seeking Replacement for Mobil," *Oil and Gas Journal* 75 (June 20, 1977): 53; "Texas Port Authority Taking its Last Gasps," *Oil and Gas Journal* 78 (March 31, 1980): 55; "LOOPs Environmental Impact Closely Studied," *Oil and Gas Journal* 78 (Sept. 22, 1980): 55; "Oil Superport Not Doing a Super Business," *Dallas Times Herald,* March 18, 1984, n.p., clipping, superport binder, box 4, Mr. and Mrs. Emmott Environmental Collection, Houston Metropolitan Research Center; National Research Council, *Tanker Spills,* 36.

108. Richard J. Bigda, "Here's How Oil Planners Can Live With the New Environmental Rules," *Oil and Gas Journal* 69 (July 19, 1971): 66–68; Harris, *The Wildcatter,* 90–91. Also see Jack Raymond, *Robert O. Anderson: Oil Man/Environmentalist* (Aspen: Aspen Institute for Humanistic Studies, 1988).

109. Paul Ehrlich, *The Population Bomb* (New York: Ballantine, 1968); Barry Commoner, *The Closing Circle: Nature, Man, and Technology* (New York: Knopf, 1971); Donella Meadows, Dennis L. Meadows, Jorgen Randers, and William W. Behrens III, *The Limits to Growth: A Report for the Club of Rome's Project on the Predicament of Mankind* (New York: Universe Books, 1972); E. F. Schumacher, *Small is Beautiful: A Study of Economics As If People Mattered* (London, Blond and Briggs, 1973). For an analysis of how these and other works helped shape an environmental "style of thought," see Patricia D'Andrade, "Saving the Environment: Science and Social Action" (Ph.D. diss., City University of New York, 1993).

110. Ronald Coase, "The Problem of Social Cost," *Journal of Law and Economics* 3 (October 1960): 1–44; Kenneth J. Arrow, "Criteria for Social Investment," *Water Resources Research* 1 (1965): 1–8. For the general origins of this line of thought, see Arthur C. Pigou, *The Economics of Welfare* (New York: St. Martin's Press, 1932).

Chapter 12. Redefining Efficiency

1. David Howard Davis, *American Environmental Politics* (Chicago: Nelson Hall, 1998), 16–17.

2. U.S. EPA, *The Challenge of the Environment: A Primer on EPA's Statutory Authority,* (Washington, D.C.: Government Printing Office, 1972).

3. Frank F. Skillern, *Environmental Protection: The Legal Framework* (New York: McGraw-Hill, 1981), 138, 172–74.

4. Significant pieces of legislation include the Federal Insecticide, Fungicides, and Rodenticides Act of 1972; the Safe Drinking Water Act of 1974; the Toxic Substances Control Act of 1976; the Resource Conservation and Recovery Act (RCRA) of 1976; and the Comprehensive Environmental Response, Compensation, and Liability Act (CERCLA) of 1980.

5. Jeanne Marie Logsdon, "Organizational Responses to Environmental Issues: Oil Refining Companies and Air Pollution" (Ph.D. diss., University of California, Berkeley, 1983), 327–51.

6. See Gregg Kerlin and Daniel Rabovsky, *Cracking Down: Oil Refining and Pollution Control* (New York: Council on Economic Priorities, 1975). Kerlin and Rabovsky argue that strong local enforcement was the major factor in encouraging refineries to invest in pollution control technology.

7. Summary of winning projects, Petroleum Engineering's meritorious awards program, Environment, box 11, Cities Service Collection, Western History Collection, University of Oklahoma.

8. Kerlin and Rabovsky, *Cracking Down,* 7–8.

9. Ibid., 54–56.

10. Leland R. Johnson, *The Headwaters District: A History of the Pittsburgh District, U.S. Army Corps of Engineers* (Washington, D.C.: U.S. Army Corps of Engineers, 1978), 298.

11. U.S. EPA, *Toward Cleaner Water* (Washington D.C.: Government Printing Office, 1974); Skillern, *Environmental Protection,* 144–87.

12. U.S. EPA, *Development Document for Proposed Effluent Limitations Guidelines and New Source Performance Standards for the Petroleum Refining Point Source Category* (Washington, D.C.: Government Printing Office, 1974), 11–12.

13. Robert W. Dellinger, "Incorporation of the Priority Pollutants into Effluent Guideline Division Documents," in *Proceedings of the Second Open Forum on Management of Petroleum Refinery Wastewater,* ed. Francis S. Manning, prepared for the EPA (Ada, Okla.: Robert S. Kerr Environmental Research Laboratory, EPA, 1978), 112.

14. EPA, *Effluent Limitations Guidelines,* 1974, 12–13; API, *Analysis of the 1972 API-EPA Raw Waste Load Survey Data,* prepared by Brown and Root, Publication no. 4200 (Washington, D.C.: API, 1974).

15. EPA, *Effluent Limitations Guidelines,* 1974, 179.

16. Ibid., 3.

17. James Spata to Robert E. Denham, April 4, 1972, "Cases filed under the Refuse Act of 1899," box 7, Office of General Enforcement, RG 412, Records of the EPA, National Archives, College Park.

18. EPA, *Effluent Limitations Guidelines,* 1974, 4–8.

19. National Commission on Water Quality, *Petroleum Refining Industry: Technology and Costs of Wastewater Control,* prepared by Engineering-Science, Inc. (Washington, D.C.: National Commission on Water Quality, 1975), I6-I9. Chromium was included in the list because industrial plants often used it to prevent cooling water from being corrosive. Also see Joan Norris Boothe, *Cleaning Up: The Cost of Refinery Pollution Control* (New York: Council on Economic Priorities, 1975).

20. EPA, *Effluent Limitations Guidelines,* 1974, 4–8.

21. EPA, *Toward Cleaner Water,* 1–2.

22. C. Howard Hardesty, Jr., "Report of the Committee for Air and Water," Minutes of the API Board of Directors, Nov. 15, 1971, folder "API Annual Meeting, 1971," series 6, box 30, Sun Company Collection, Hagley Museum and Library.

23. National Commission on Water Quality, *Technology and Costs of Wastewater Control,* I4-

l16. In 1977, Congress added a new class of standards, those based on the best conventional technology (BCT), that replaced BAT for conventional pollutants. In 1978, the EPA designated oil as a conventional pollutant; U.S. EPA, *Development Document for Proposed Effluent Limitations Guidelines, New Source Performance Standards, and Pretreatment Standards for the Petroleum Refining Point Source Category* (Washington, D.C.: Government Printing Office, 1979), 253.

24. Joe G. Moore, Jr., "What Does July 1, 1977 Mean Now?" in *Proceedings of the Second Open Forum,* ed. Manning, 3–11.

25. Dellinger, "Priority Pollutants," 112.

26. Dwight G. Ballinger, "EPA's Analytical Development Program for Priority Pollutants," in *Proceedings of the Second Open Forum,* ed. Manning, 157–61.

27. National Commission on Water Quality, *Technology and Costs of Wastewater Control,* I4 to I5.

28. Irv Kornfeld and Jay G. Kremer, "Petroleum Refinery Discharges to a Large Sanitation District," in Manning, *Proceedings of the Second Open Forum,* 304–12.

29. Kerlin and Rabovsky, *Cracking Down,* 168; Kornfeld and Kremer, "Sanitation District," 307–8.

30. EPA, *Refinery Effluent Water Treatment Plant Using Activated Carbon,* prepared by Gary C. Loop (Washington, D.C.: Government Printing Office, 1975). Also see Kerlin and Rabovsky, *Cracking Down,* 169.

31. EPA, *Challenge of the Environment;* Skillern, "*Environmental Protection,* 83–143.

32. Skillern, *Environmental Protection,* 107.

33. Nicholas Iammartino, "Detroit's Catalyst Choices," *Chemical Engineering* 80 (November 26, 1973): 24–36; Donel R. Olson, "The Control of Motor Vehicle Emissions," in *Engineering Control of Air Pollution,* 620–623.

34. E. E. Weaver, "Effects of Tetraethyl Lead on Catalyst Life and Efficiency in Customer Type Vehicle Operation" (paper presented at International Automotive Engineering Congress, Detroit, Mich., January 13–17, 1969); "Safety, Clean-Air, Highway Bills Passed," *Oil and Gas Journal* 68 (December 28, 1970): 66–67.

35. Joseph C. Robert, *Ethyl: A History of the Corporation and the People Who Made It* (Charlottesville: University Press of Virginia, 1984), 211–52; E. F. Hill, W. A. Cannon and C. E. Welling, "Single Cylinder Engine Testing of Hydrocarbon Oxidation Catalysts" (paper no. 174 presented at the SAE Meeting, Seattle, August 1957); E. F. Hill, W. A. Cannon and C. E. Welling, "Oxidation Catalysts Reduce Hydrocarbons in Automobile Exhaust Gas," *SAE Journal* (January 1958), 36; W. A. Cannon and C. E. Welling, "The Application of Vanadia-Alumina Catalysts for the Oxidation of Exhaust Hydrocarbons" (paper no. 29T presented at the SAE Meeting, Detroit, Michigan, January 1959); D. L. Davis and G. F. Onishi, "Catalytic Converter Development Problems," (paper no. 486F presented at the SAE National Meeting, Detroit, Michigan, March 1962).

36. "Industry Proposal for Joint Research on Air Pollution," Edwin M. Zimmerman (Assistant Attorney General, Antitrust Department) to Lawrence E. Levinson (Deputy Special Counsel to the President), May 21, 1967, HE 8-1, Air Pollution, Subject Files, White House Central Files, LBJ Library; "U.S. Charges Auto Makers Plot to Delay Fume Curbs," *New York Times,* Jan. 11, 1969, 1; Robert, *Ethyl,* 211–21; William D. Hurley, *Environmental Legislation* (Springfield, Ill.: Charles C. Thomas, 1971), 53.

37. W. D. Preston and H. A. Toulmin, "The Automobile Powerplant and its Fuel Requirements, 1972–1982," folder "Research and Development," box 13, series 6, Sun Company Collection, Hagley Museum and Library.

38. Advertisement, Fluor Corporation, *Oil and Gas Journal* 69 (February 15, 1971): 21–23. The Foster Wheeler Corporation placed a similar type of advertisement on pp. 58–59.

39. "EPA Sticks to Present Lead-Out Timetable," *Oil and Gas Journal* 72 (May 20, 1974): 36.

40. "Low-lead, No-lead Gasoline Units Set by Cities, Ashland," *Oil and Gas Journal* 69 (February 15, 1971): 45; Paul R. Waddell and Robert F. Niven, *Sign of the 76: The Fabulous Life and Times of the Union Oil Company of California* (Los Angeles: Union Oil Co., 1976), 314–16; table,

"Who's Moving to Reduce or Eliminate Lead in Gasoline," *Oil and Gas Journal* 69 (Oct. 11, 1971): 35; "No-Lead Will Magnify Gasoline Woes," *Oil and Gas Journal* 72 (January 21, 1974): 46–47.

41. "Smooth-Sailing Refiners Eye Storm Clouds in US," *Oil and Gas Journal* 73 (Sept. 15, 1975): 79–82; "EPA Sticks to Present Lead-Out Timetable," *Oil and Gas Journal* 72 (May 20, 1974): 36; Douglas Costle to Oswall Newell, Jr., Nov. 16, 1977, General Correspondence, box 260, Office of the Administrator, RG 412, Records of the EPA, National Archives, College Park.

42. "U.S. Lead Entitlements Program Urged," *Oil and Gas Journal* 80 (May 31, 1982): 177–78.

43. "Refiners Split on Lead Phasedown Plan," *Oil and Gas Journal* 80 (Sept. 13, 1982): 35–36.

44. "Lead Entitlement Programs Urged," *Oil and Gas Journal* (May 31, 1982): 177; "Refiners Split on Lead Phasedown Plan," *Oil and Gas Journal* (Sept. 13, 1982): 45; "U.S. EPA Proposes 'Banking' of Lead Usage Rights," *Oil and Gas Journal* (Jan 14, 1985): 45; Lily Whiteman, "Trades to Remember: The Lead Phasedown," *EPA Journal* 18 (May/June 1992): 38–39; Barry D. Nussbaum, "Phasing Down Lead in Gasoline in the U.S.: Mandates, Incentives, Trading and Banking," in OECD, *Climate Change: Designing a Tradeable Permit System* (Paris: Organization for Economic Co-Operation and Development, 1992), 20.

45. "EPA Pushing Air Emissions Trade-Off Policy," *Oil and Gas Journal* 74 (Nov. 22, 1976): 72.

46. Richard A. Liroff, *Air Pollution Offsets: Trading, Selling, and Banking* (Washington, D.C.: Conservation Foundation, 1986). Also see Richard A. Liroff, *Reforming Air Pollution Regulation: The Toil and Trouble of EPA's Bubble* (Washington, D.C. Conservation Foundation, 1986).

47. "EPA's New Emissions Policy Flayed," *Oil and Gas Journal* 75 (April 4, 1977): 52–53.

48. "Exxon, Unocal Detail Air Pollution Initiatives," *Oil and Gas Journal* 88 (June 25, 1990): 16–17.

49. "Court's Air Quality Ruling Hits New Industry," *Oil and Gas Journal* 70 (Nov. 13, 1972): 116; Skillern, *Environmental Protection*, 115–20. The EPA eventually addressed the problem of air quality deterioration by defining three classes of attainment areas, each with their own rules and amounts of deterioration allowed.

50. U.S. Environmental Protection Agency, "Emission Offset Interpretative Ruling," 44 *Fed. Reg.* 3274-82 (1979); Brent M. Haddad, "Marketable Permits and Pollution Charges: Two Case Studies," in *The Environment Goes to Market: The Implementation of Economic Incentives for Pollution Control* (Washington, D.C.: National Academy of Public Administration, 1994), 31; U.S. General Accounting Office, *A Market Approach to Air Pollution Control Could Reduce Compliance Costs Without Jeopardizing Clean Air Goals* (Washington, D.C.: U.S. Government Printing Office, 1982); Paul R. Portney, "Air Pollution Policy" in *Public Policies for Environmental Protection*, ed. Paul Portney (Washington, D.C.: Resources for the Future, 1990), 27–96.

51. David V. Bubenick, ed., *Acid Rain Information Book* (Park Ridge, N. J.: Noyes Publications, 1984); Ellis B. Cowling, "Acid Precipitation in Historic Perspective," *Environmental Science and Technology* 16 (February 1982): 110a-123a; Daniel J. Fiorino, *Making Environmental Policy* (Berkeley: University of California Press, 1995), 166, 189; Gary C. Bryner, *Blue Skies, Green Politics: The Clean Air Act of 1990* (Washington, D.C.: CQ Press, 1993), 121–59.

52. "EPA Invites States to Demand Scarce Fuels," *Oil and Gas Journal* 69 (April 19, 1971): 80; "Sulfur Still Declining in Most U.S. Fuels," *Oil and Gas Journal* 69 (September 20, 1971): 84; "Federal Court Tangles Clean Air Rules," *Oil and Gas Journal* 70 (June 5, 1972): 44.

53. Harold F. Elkin, "Petroleum Refining," in *Engineering Control of Air Pollution*, ed. Arthur Stern, vol. 4 of *Air Pollution* (New York: Academic Press, 1977), 822–25; "API Blasts Planned Air-Quality Standards, Urges Changes," *Oil and Gas Journal* 69 (March 12, 1971): 180; "EPA's Clean-Air Stance Challenged," *Oil and Gas Journal* 69 (May 24, 1971): 43; James P. Tomany, *Air Pollution: The Emissions, the Regulations, and the Controls* (New York: American Elsevier, 1975), 47; Also see Richard H. K. Vietor, *Environmental Politics and the Coal Coalition* (College Station: Texas A&M University Press, 1980); Vietor notes that in the early 1970s many oil companies invested heavily in coal companies, making those oil companies concerned with how sulfur limits affected their investment in coal as well.

54. "The Environmental Craze: Will It Strangle Energy," *Oil and Gas Journal* 69 (March 15, 1971): 23; "EPA Invites States to Demand Scarce Fuels," *Oil and Gas Journal* 69 (April 19, 1971): 80;

"Clean-Air Drive Creating Impossible Oil-Supply Job," *Oil and Gas Journal* 70 (November 13, 1972): 93.

55. "Rivals to Testify on Sulfur in Air," *New York Times,* August 2, 1981, section 23, 1.

56. "Environmentalist Dispute Priorities," *New York Times,* May 12, 1978, section 2, 1.

57. "Here's What's Being Done to Combat Sulfur-Oxide Air Pollution," *Oil and Gas Journal* 68 (May 11, 1970): 63–67.

58. Vietor, *Environmental Politics,* 251–55. Vietor notes that environmentalists pushed companies to develop scrubber technology so that the need to strip-mine low-sulfur western coal would be reduced.

59. U.S. EPA, *Acid Rain Program: Overview* (EPA 430-F-92-019, April 1996); Eileen Claussen, "Acid Rain: The Strategy," *EPA Journal* 17 (January/February 1991): 21–23; Paul L. Joskow and Richard Schmalensee, "The Political Economy of Market-Based Environmental Policy: The U.S. Acid Rain Program," *Journal of Law and Economics* 41 (April 1998): 37–83; Barry D. Solomon, "Five Years of Interstate SO_2 Allowance Trading: Geographic Patterns and Potential Cost Savings," *The Electricity Journal* 11 (May 1998): 58–70.

60. J. M. Lents and P. Leyden, "RECLAIM: Los Angeles' New Market-Based Smog Cleanup Program," *Journal of the Air and Waste Management Association* 46 (1996), 196–205.

61. Barry D. Solomon and Hugh S. Gorman, "State-Level Air Emissions Trading: The Michigan and Illinois Models," *Journal of the Air and Waste Management Association* 48 (1998), 1156–65.

62. Dutch Committee for Long-Term Environmental Policy, eds., *The Environment: Towards a Sustainable Future* (Boston: Kluwer Academic Publishers, 1993), 3–16. For an early expression of this idea, see Allen V. Kneese, Robert U. Ayres, and Ralph C. D'Arge, "Economics and the Environment: A Material Balance Approach," in *The Economics of Pollution,* ed. Harold Wolzin (Morristown, N.J.: General Learning Press, 1974), 44; the authors pointed to the need for environmental standards that would become "fixed constraints" and "part of the overall framework within which voluntary market exchanges take place."

63. Many historians of technology have examined the interplay between technological development and the influence of various social groups on that development. For works that place these issues in perspective, see David F. Noble, "Social Choice in Machine Design: The Case of Automatically Controlled Machine Tools," in *Case Studies on the Labor Process,* ed. Andrew Zimbalist (New York: Monthly Review Press, 1979), 18–50; Weibe Bijker, Thomas P. Hughes, and Trevor J. Pinch, eds., *The Social Construction of Technological Systems: New Directions in the Sociology and History of Technology* (Cambridge, Mass.: MIT Press, 1987); George Basalla, *The Evolution of Technology* (New York: Cambridge University Press, 1988); Merritt Roe Smith and Leo Marx, eds., *Does Technology Drive History: The Dilemma of Technological Determinism* (Cambridge, Mass.: MIT Press, 1994). For a discussion of the effect of regulations, though not environmental regulations, on the development of technological systems, see Thomas P. Hughes, *Networks of Power: Electrification in Western Society, 1880–1930* (Baltimore: Johns Hopkins University Press, 1983).

64. Marshall Sittig, *Petroleum Refining Industry: Energy Saving and Environmental Control* (Park Ridge, N.J.: Noyes Data Corporation, 1978), 77–101; "A Look at Sulfur-Recovery Costs," *Oil and Gas Journal* 72 (March 18, 1974): 120.

65. Joyce A. Ober, "Sulfur" in U.S. Geological Survey, *Mineral Yearbook* (Online version, 1998). In the case of sulfur, the price per ton has hovered around $30 since 1992, mainly because producers of mined sulfur have cut back their production to maintain prices.

66. Clifford S. Russell, *Residual Management in Industry: A Case Study of Petroleum Refining* (Baltimore: Resources for the Future, Johns Hopkins University Press, 1973), 130–131; Logsdon, "Organizational Responses," 94–96.

67. James H. Gary and Glenn E. Handwerk, *Petroleum Refining: Technology and Economics* (New York: Marcel Dekker, 1994), 293–96.

68. "California Standard Will Install Hydrotreating Units at Two Refineries," *Oil and Gas Journal* 72 (Jan. 21, 1974): 72–73. This article describes a sulfur recovery plant capable of produc-

ing 300 tons of sulfur per day, which implies that tail gases amount to approximately .05 x 300 tons of sulfur/day x 2 lbs SO_2/S = 120 tons/SO_2 per day.

69. R. L. Duprey, "The Status of SOx Emission Limitations," *Chemical and Engineering Progress* 68 (Feb. 1972): 70–76. The recommendation came in the form of Reasonable Available Control Technology (RACT) standards, which eventually become the expected standards for existing sources in all nonattainment areas

70. "Cost, Air Regulations Affect Process Choice," *Oil and Gas Journal* 73 (August 18, 1975): 109–12; "There are Ways to Smoother Operation of Sulfur Plants," *Oil and Gas Journal* 74 (November 15, 1976): 55–60; Elkin, "Petroleum Refining," 832–33.

71. "New Process Removes Stack-Gas Sulfur," *Oil and Gas Journal* 68 (July 13, 1970): 49–50; "Sulfur Recovery Process 99.9% Effective," *Oil and Gas Journal* 69 (September 20, 1971): 84. Gary and Handwerk, *Petroleum Refining*, 296.

72. "Dry-Bed Processes Can Cover a Wide Application Range," *Oil and Gas Journal* 73 (August 25, 1975): 96–103; "Tail Gas Process Makes Fertilizer," *Oil and Gas Journal* 78 (October 20, 1980): 132.

73. Boothe, *Cleaning Up*, 43, 62. Boothe estimates that meeting federal standards for both water and air quality would cost about seventeen cents per barrel, which would be less than one cent per gallon of gasoline.

74. Gary and Handwerk, *Petroleum Refining*, 297.

Chapter 13. Environmental Objectives and the Evolution of Tankers

1. "Horizon to Horizon," *Environment* 13 (March 1971): 13–21.

2. For a table summarizing the movement of oil over various tanker routes: Economics Department, Sun Oil, *Analysis of World Tank Ship Fleet*, August 1968, Summary, Table Two, box 52, series 8, Sun Oil Records, Hagley Museum and Library.

3. National Petroleum Council, "An Interim Report on Current Key Issues Relating to Environmental Conservation: The Oil and Gas Industries, 15–19, attachment, H. F. Elkin to W. T. Askew et al., June 24, 1970, folder "Air and water pollution," box 28, series 6, Sun Oil Records.

4. For example, in 1976, twenty-nine incidents resulted in approximately 1.5 million barrels of oil spilling into the sea. If all incidents resulted in the greatest clean-up cost possible, $14 million per liable vessel, the total cost of all spills to the industry would have been about $400 million. Through insurance, that amount could be spread over the 10 billion barrels of oil transported that year, representing a fraction of a cent per gallon transported. See "World Tanker Spills Top 1.3 million bbl in First 9 months," *Oil and Gas Journal* 75 (Dec. 12, 1977): 33; U.S. Coast Guard, prepared by the Program of Policy Studies in Science and Technology, George Washington University, *Legal, Economic, and Technical Aspects of Liability and Financial Responsibility As Related to Oil Pollution* (Washington, D.C.: George Washington University, 1970); Martin T. Katzman, *Chemical Catastrophes, Regulating Environmental Risk Through Pollution Liability Insurance* (Homewood, Ill,: Richard D. Irwin, 1985).

5. The requirement was first passed as part of the Water Quality Improvement Act of 1970, which was then superseded by and incorporated in the Federal Water Pollution Control Amendments of 1972.

6. EPA, *Oil Spills and Pollution Reports: July 1974 to October 1974*, prepared by Floyd A. DeWitt, Jr., and Penelope Melvin (Washington, D.C.: Government Printing Office, 1975). The series has been variously titled *Oil Spill and Oil Pollution Reports* (1974–1978), *Oil Pollution Reports* (1978), and *Oil Pollution Abstracts* (after 1979). Also see Inter-Governmental Maritime Consultative Organization, *Tanker Casualties Report* (London: IMCO, 1978); "An Analysis of Outflows Due to Tanker Accidents," A Note by the United States of America, Jan. 1972; American Institute of Biological Sciences, *Sources, Effects, and Sinks of Hydrocabons in the Aquatic Environment* proceedings of a symposium (Washington, D.C.: American University, 1976).

7. For various estimates, see Working Group on the Study of Critical Environmental Problems, *Man's Impact on the Global Environment: Report of Critical Environmental Problems* (Cambridge, Mass.: MIT, 1970); National Academy of Sciences, *Petroleum in the Marine Environment*

(Washington, D.C.: Government Printing Office, 1975), 6; J. D. Porricelli, V. F. Keith, and R. I. Storch, "Tankers and the Ecology," *S.N.A.M.E. Transactions*, 1971, 172.

8. For more on the problem of waste crankcase oil, see EPA, *Waste Oil Study*, prepared as a report to Congress (Washington, D.C.: EPA, 1974); Norman J. Weinstein, *Waste Oil Recycling and Disposal*," prepared for EPA by Recon Systems, Inc. (Washington D.C.: Government Printing Office, 1974).

9. Senate Committee on Commerce, *Hearings on the Navigable Waters Safety and Environmental Quality Act*, 92d Cong., 1st sess., September 22–24, 1971.

10. Dennis Livingston, "Oil on the Seas: Two Cheers for a New Treaty," *Environment* 16 (September 1974): 38–43.

11. U.S. Coast Guard, *Legal, Economic, and Technical Aspects of Liability and Financial Responsibility as Related to Oil Pollution*, 18–4.

12. Ronald B. Mitchell, *Intentional Oil Pollution at Sea: Environmental Policy and Treaty Compliance* (Cambridge, Mass.: MIT Press, 1994), 221–56. One study (Working Group, *Man's Impact on the Global Environment*) concluded that 90 percent of all intentional discharges came from 20 percent of the tankers; even if true, the policy can still be considered a failure.

13. Clipping, "Officials Ponder Residue on Beach at Mansfield Cut," May 27, 1970, box 3G69, Francis Tarlton Farenthold Papers, Center for American History, University of Texas at Austin.

14. Senate Committee, *Hearings on the Navigable Waters*, 301–302. J. D. Porricelli, V. F. Keith, and R. I. Storch, "Tankers and the Ecology" (*S.N.A.M.E. Transactions*, 1971) is entered into the record of the hearings as pp. 295–328.

15. Ibid., 315–16.

16. Ibid., 312–15.

17. Ibid., 315–19.

18. Ibid., 307–11. Also see R. C. Page and A. Ward Gardner, *Petroleum Tankship Safety* (London: Maritime Press, 1971).

19. Edward Wenk, Jr., *Making Waves: Engineering, Politics, and the Social Management of Technology* (Urbana: University of Illinois Press, 1995).

20. Mitchell, *Intentional Oil Pollution*, 93.

21. R. M. M'Gonigle and Mark W. Zacher, *Pollution, Politics, and International Law: Tankers at Sea* (Berkeley: University of California Press, 1979), 107–14.

22. R. Maybourn, "Operational Pollution from Tankers and Other Vessels," in *The Prevention of Oil Pollution*, ed. J. Wardley-Smith (New York: Wiley, 1979), 123–24; W. G. Waters II, T. D. Heaver, and T. Verrier, *Oil Pollution From Tanker Operations: Causes, Costs, and Controls* (Vancouver: Centre For Transportation Studies, 1980), 22–24.

23. Ibid., 124.

24. "Sun Ship's First Ecology Class Tanker Delivered to SOHIO," *Marine Engineering Log* 82 (December 1977): 87.

25. Mitchell, *Intentional Oil Pollution*, 272.

26. Fredrik Zachariasen, *Oil Pollution in the Sea: Problems for Future Work* (Arlington, Va.: Institute for Defense Analysis, 1968), 19–11; Federal Water Pollution Control Administration, *Oil Tagging System Study* (Washington, D.C.: Government Printing Office, 1970).

27. Nick J. Malueg and Daniel F. Krawczyk, "Tracing Oil as a Pollutant in Water," *Journal of Petroleum Technology* 25 (March 1973): 243–48; "Equipment Fingerprints Crude Oil," *Oil and Gas Journal* 76 (February 20, 1978): 103–7. For examples of incidents that pushed the development of this technology, see correspondence, folder "Pollution—petroleum-hazardous material," box 21, David D. Dominick Collection, American Heritage Center, University of Wyoming; correspondence, folder 331-17-13, box 17, 3m33b, Frances Tarlton Farenthold Papers, Center for American History, University of Texas at Austin.

28. "Houston Firm Uses Space Skills to Check on Pollution" and "NASA Agrees to Aid State Pollution Fight," newspaper clippings, box 3G69, Frances Tarlton Farenthold Papers.

29. U.S. Congress, Office of Technology Assessment, *Coping With an Oiled Sea: An Analysis of Oil Spill Response Technologies* (Washington, D.C.: Government Printing Office, 1990), 4. Only

three oil spills involving tankers have been larger than the *Amoco Cadiz* spill: in 1979, a collision in the Caribbean between the *Atlantic Empress* and *Aegean Captain* resulted in a 300,000-ton spill (2 million barrels); in 1983, the *Castillo de Bellvera* caught fire off of Cape Town, South Africa, resulting in a 250,000-ton spill (1.7 million barrels); and in 1991, the *AMT Summer* also spilled about 250,000 tons in the open ocean off the coast of Angola. As another point of comparison, the grounding of the *Exxon Valdez* in 1989 released less than one-seventh of the crude spilled by the *Amoco Cadiz* but did so in a more confined area.

30. Noël Mostert, *Supership* (New York: Warner Books, 1974), 135–36.

31. General Accounting Office, *Tanker and Oil Transfer Operations on the Delaware River and Bay,* prepared by the U.S. Coast Guard (Washington, D.C.: Government Printing Office, 1977). Also see W. O. Gray, "Accidental Spills from Tankers and Other Vessels," in *The Prevention of Oil Pollution,* ed. Wardley-Smith; Inter-Governmental Maritime Consultative Organization, *Tanker Casualties Report* (London: IMCO, 1978); *Elements of Oil-Tanker Transportation* (Tulsa, Okla.: PennWell Books, 1982), 203.

32. David Fairhall and Phillip Jordan, *The Wreck of the Amoco Cadiz* (New York: Stein and Day, 1980), 75–80.

33. Ibid., 80–97.

34. House Committee on Merchant Marine and Fisheries, Subcommittee on Water Resources, *Hearings on the Blowout of the Mexican Oil Well IXTOC 1,* 96th Cong. 2d sess., 1980, statement of Steven C. Mahood.

35. Ibid.

36. Ibid., "Weather Delays Final Try For Ixtoc Kill," *Oil and Gas Journal* 78 (March 3, 1980): 40.

37. Charles Perrow, *Normal Accidents: Living With High Risk Technologies* (New York: Basic Books, 1984). Also see William H. Rogers, "Where Environmental Law and Biology Meet: Of Panda's Thumbs, Statutory Sleepers, and Effective Law," *University of Colorado Law Review* 65 (1993): 25–75; in a critique of the Oil Pollution Act of 1990, Rogers suggests that portions of the act attempt to regulate actions that are impossible to regulate.

38. Office of Technology Assessment, *Coping With an Oiled Sea,* 1–9.

39. Senate Committee on Public Works, *Analysis of the Decision of the Florida District Court in the Case of* American Waterways Operators, Inc., et al. v. Askew, 92nd Cong., 2d sess., 1972.

40. U.S. Senate, Committee on Environment and Public Works, *Hearings on the Oil Pollution Liability and Compensation Act,* 99th Cong., 2d sess., Sept. 10–12. Also see Environmental Law Institute, prepared for the API, *Assessing Natural Resource Damages From Hazardous Substances: An Examination of Existing Assessment Approaches* (Washington, D.C.: API, 1987).

41. Thomas A. Birkland, *After Disaster: Agenda Setting, Public Policy, and Focusing Events* (Washington, D.C.: Georgetown University Press, 1997).

42. Edward Cowen, *Oil and Water: The Torrey Canyon Disaster* (Philadelphia: J. B. Lippincott, 1968); Fairhall and Jordan, *The Wreck of the Amoco Cadiz.*

43. U.S. House of Representatives, Committee on Merchant Marine and Fisheries, *Hearing on Oil Pollution and Compensation,* 101st Cong., 1st sess, May 11, 1989; Public Law 101 380 (OPA 1990), title 1, section 1001–1020.

44. Public Law 101 380 (OPA 1990), title 4, section 4115 prohibits single-hull tankers from trading in U.S. waters by 2010; exceptions extend that date to 2015 for double-sided tankers and for single-hull tankers unloading in deepwater lightering areas or in the deepwater port off the shore of Louisiana.

45. MARPOL regulations 13G and 13F, adopted in 1992, cover the phaseout of single-hull tankers over 5,000 deadweight tons.

46. "World Tanker Doldrums May Last Until the Mid-1980s," *Oil and Gas Journal* 75 (Aug. 29, 1977): 28.

47. Organisation for Economic Co-operation and Development, *Pipeline and Tankers: A Report On the Use of Pipelines and the Transport of Oil By Tankers* (Paris: OCED, 1961).

48. National Research Council, *Tanker Spills: Prevention by Design* (Washington, D.C.: National Academy Press, 1991), p. xviii; 9,000 tons out of 450 million tons transported.

49. National Research Council, *Double Hull Tanker Legislation: An Assessment of the Oil Pollution Act of 1990* (Washington, D.C.: National Academy Press, 1998), 139.

50. National Research Council, *Double Hull Tanker Legislation,* 119.

51. National Research Council, *Tanker Spills,* 56.

52. United States Congress, House Committee on Merchant Marine and Fisheries, Subcommittee on Coast Guard and Navigation, *Hearings on Topics Concerning the Exxon Valdez Spill into Prince William Sound,* 101st Cong., 1st sess., April 6, 1989, 83.

53. National Research Council, *Double Hull Tanker Legislation,* 139.

54. National Research Council, *Tanker Spills,* 170

55. Wenk, *Making Waves,* 154–59.

56. "Supreme Court to Consider Who Can Set Rules for Oil Tankers," *New York Times,* Sept. 10 1999 (online edition).

Chapter 14. Closing the Loop

1. Marvin L. Wood, "EPA's Research in the Refining Category," in *Proceedings of the Second Open Forum on Management* of Petroleum Refinery Wastewater, ed. Francis S. Manning, prepared for the EPA (Ada, Okla.: Robert S. Kerr Environmental Research Laboratory, EPA, 1978), 23.

2. Joel A. Tarr, *The Search for the Ultimate Sink: Urban Pollution in Historical Perspective* (Akron, Ohio: University of Akron Press, 1996).

3. Hugh S. Gorman, "Manufacturing Brownfields: The Case of Neville Township, Pennsylvania, 1899–1989," *Technology and Culture* 38 (July 1997): 539–74.

4. Hugh S. Gorman, "Efficiency, Environmental Quality, and Oil Field Brines: The Success and Failure of Pollution Control by Self-Regulation," *Business History Review* 73 (winter 1999): 601–40.

5. "EPA Outlines Hazardous-Waste Disposal Rules," *Oil and Gas Journal* 78 (March 3, 1980): 40.

6. Joe G. Moore, Jr., "What Does July 1, 1977 Mean Now?" in *Proceedings of the Second Open Forum on Management of Petroleum Refinery Wastewater,* ed. Francis S. Manning (Ada, Okla.: Robert S. Kerr Environmental Research Laboratory, EPA, 1978), 10. For a contemporary introduction to RCRA, see EPA, *RCRA Orientation Manual* (Washington, D.C.: Government Printing Office, 1986). Also see Richard C. Fortuna and David J. Lennett, *Hazardous Waste Regulations, the New Era* (New York: McGraw-Hill, 1987).

7. "Landfarming Shows Promise for Refinery Waste Disposal, "*Oil and Gas Journal* 77 (May 14, 1979): 108–16; Carl E. Adams, "The Economics of Handling Refinery Sludges," in *Proceedings of the Second Open Forum,* ed. Manning, 506.

8. Dudley J. Burton and K. Ravishankar, *Treatment of Hazardous Petrochemical and Petroleum Wastes: Current, New, and Emerging Technologies* (Park Ridge, N.J.: Noyes Publications, 1989): 3.

9. "EPA Sees Problems with Muds, Brines Environment Study," *Oil and Gas Journal* (Nov. 3, 1980): 42. For a pre-RCRA examination of toxic chemicals used in oil fields, see U.S. EPA, Office of Toxic Substances, *Environmental Aspects of Chemical Use in Well-Drilling Operations* (Washington, D.C.: EPA, 1975).

10. Robert L. Carlson, "RCRA Overview," in *Hazardous Materials Management,* ed. Doy Cox (Rockville, Md.: Institute of Hazardous Materials Management, 1995): 111–31.

11. Examples I have seen include a small refinery (now closed) in Alma, Michigan, and a medium-sized refinery in Lemont, Illinois; the records of the Michigan DEQ and the Illinois EPA describe the groundwater concerns and the monitoring and control systems installed.

12. Examples of sediment-related problems I have examined include those for a small refinery (now closed) in Alma, Michigan, and for a small refinery in Superior, Wisconsin; the records of the Michigan DEQ and the Wisconsin DNR document the concerns.

13. U.S. EPA, *Musts for USTs: A Summary of the New Regulations for Underground Storage*

Tank Systems, September 1998; Elizabeth Philips, "Managing Underground Storage Tanks," in *Hazardous Materials Management,* ed. Cox, 149–66.

14. For example, see Santa Clara Valley Water District, "Investigation of MtBE Occurrence Associated with Operating UST Systems," July 22, 1999 (posted on http://www.scvwd.dst.ca.us/wtrqual/pdf).

15. CBS aired a segment about MtBE on its television program *60 Minutes* on January 16, 2000, which emphasized the lack of communication between people in air- and water-related programs.

16. National Transportation Safety Board, pipeline accident report no. NTSB-PAR-771-1, "United Gas Pipeline Co. 20" Pipeline Rupture and Fire, Aug. 9, 1976, Cartwright, Louisiana, 9–10.

17. Ibid.

18. U.S. Department of Transportation, Office of Pipeline Safety *Common Ground: Study of One Call Systems and Best Practice Damage Prevention Systems* (August 1999).

19. Cheryl Trench, prepared for the Association of American Pipelines, *The U.S. Oil Pipeline Safety Performance* (May 1999), 2.

20. Trench, *U.S. Oil Pipeline Safety Performance,* 3–14.

21. "Texas Was Marred By 42 Spills Over 10,000 Gallons Last Year," *Houston Post,* May 23, 1993, A20.

22. Ibid.

23. Virgil Johnston, "Pipeline Casing Value Studied," *Oil and Gas Journal* (Oct. 27, 1980): 59–63.

24. Ryuki Takatsu and Shigeru Kokan, "Automatic Pipeline Control Advances in Japan," *Oil and Gas Journal* 75 (Jan. 17, 1977): 54–59; Earl Seaton, "Pipelines Laid After 6-Year Struggle," *Oil and Gas Journal* (Feb. 11, 1980): 119–20; Stuart B. Eynon, "Line Leak-Detection Methods Updated," *Oil and Gas Journal* (Sept. 15, 1980): 205–6. For the issue seen from a slightly different perspective, see F. A. Inkley, *Oil Loss Control In the Petroleum Industry* (New York: John Wiley and Sons, 1985).

25. U.S. House of Representatives, Public Works and Transportation Committee, *Colonial Pipeline Rupture,* 103rd Cong., 1st sess., May 18, 1993.

26. National Research Council, Transportation Research Board, *Pipelines and Public Safety,* special report 219 (National Research Council: 1988), 2; Trench, *U.S. Oil Pipeline Safety Performance,* 1.

27. Office of Pipeline Safety, American Petroleum Institute, and Interstate National Gas Association of America, progress report, *Beyond Compliance: Creating a Responsible Regulatory Environment that Promote Excellence, Innovation, and Efficiency* (May 1999).

28. "EPA Keeps Eye on Waste Wells, Official Says," December 8, 1971, clipping, injection well binder, box 2, Mr. and Mrs. Emmot Environmental Collection, Houston Metropolitan Research Center.

29. "Deep Well Injection is Effective for Waste Disposal," *Environmental Science and Technology* 2 (June 1968): 405–10; "Permit Data," Exhibit V, Table of Injection Wells, folder 331-28-19, box 28, 3m37a, Frances Tarlton Farenthold Papers, Center for American History, University of Texas at Austin.

30. "EPA Keeps Eye on Waste Wells, Official Says," December 8, 1971, clipping, injection well binder, box 2, Mr. and Mrs. Emmot Environmental Collection, Houston Metropolitan Research Center.

31. U.S. EPA, *Brine Disposal Treatment Practices Relating to the Oil Production Industry,* prepared by George W. Reid et al. (Norman: School of Civil Engineering and Environmental Engineering, University of Oklahoma, 1974). Also see David Keith Todd and Daniel E. Orren McNulty, *Polluted Groundwater: A Review of Significant Literature* (Port Washington, N.Y.: Water Information Center, Inc., 1976), 34–39.

32. Safe Drinking Water Act, 1974, P.L. 93–523, sections 1421–1425.

33. See "Pollution, Railroad Commission Correspondence," box 3G69, Frances Tarlton Far-

enthold Papers, Center for American History, University of Texas at Austin. Also see clippings, folder 16, box 26, J. M. Heiser Environmental Collection, Houston Metropolitan Research Center.

34. Reid et al., *Brine Disposal,* 14–15.

35. David W. Miller, ed., *Waste Disposal Effects on Groundwater* (Berkeley, Calif.: Premier Press, 1980), 301.

36. Statement of C. A. Clark, Railroad Commission Hearing, June 30, 1967, "Pollution, Railroad Commission Correspondence," box 3G69, Frances Tarlton Farenthold Papers, Center for American History, University of Texas at Austin; Special Study by the Bay Drilling Committee On Disposal of Oil and Gas Well Produced Salt Water from Saxet Area, box 3G69, Frances Tarlton Farenthold Papers; Mrs. James Scott to Lenford Williams, October 28, 1968, box 3G69, Frances Tarlton Farenthold Papers.

37. "Rail Panel OK'd Discharges of Brine, Parks Man Testifies," *Houston Chronicle,* February 16, 1972, n. p., clipping, folder 16, box 26, J. M. Heiser Environmental Collection, Houston Metropolitan Research Center.

38. "EPA: 18 States Need Revised UIC Regs," *Oil and Gas Journal* 75 (Dec. 19, 1977), 26–27; U.S. Environmental Protection Agency, prepared by the Office of Water Supply and the Office of Solid Waste Management Programs, *The Report to Congress: Waste Disposal Practices and their Effect on Ground Water* (1977).

39. "IOCC Renews Attack on Injection Regs," *Oil and Gas Journal* 75 (Jan. 24, 1977); "EPA: 18 States Need Revised UIC Regs," *Oil and Gas Journal* 75 (Dec. 19, 1977): 26–27.

40. "EPA, API Differ Over the Cost of Injection Rules," *Oil and Gas Journal* 78 (Aug. 25, 1980): 54.

41. "EPA Announces Plans," *Ground Water Newsletter* 6 (March 31, 1977): 1; "API Study Sees Big Costs in UIC Implementation," *Oil and Gas Journal* 78 (Aug. 4, 1978): 34–35.

42. "IOCC Renews Attack on Injection Regs," *Oil and Gas Journal* 75 (Jan. 24, 1977).

43. "GAO Finds Brine Still Contaminates Aquifers," *Oil and Gas Journal* 87 (Oct. 16, 1989), 38.

44. Michael Barcelona et al., *Handbook of Groundwater Protection* (New York: Science Information Resource Center, 1988), 5.

45. U.S. Geological Survey, *Geohydrology and Water Quality in Northern Portage County, Ohio in Relation to Deep-Well Brine Injection,* prepared by Sandra M. Eberts, Water Resources Investigation Report 90-4158 (Washington, D.C.: U.S. Geological Survey, 1990). In this study, investigators used the proportion of bromide and chloride in samples to distinguish between potential sources of elevated chloride, but such studies represent a major research effort involving numerous test wells and considerable expense.

46. Texas Department of Agriculture, *Agricultural Land and Water Contamination From Injection Wells, Disposal Pits, and Abandoned Wells Used in Oil and Gas Production* (Austin: Texas Dept. of Agriculture, 1985).

47. Texas Department of Agriculture, *Agricultural Land and Water Contamination,* 10.

48. Problems with poorly plugged abandoned wells have emerged throughout the United States: "Oil Slick: Profits Abroad and Poison at Home: Big Petroleum Ships Out, Leaving Behind a Big Mess," *New York Times,* July 31, 1994, C3; "Big Trouble for Tiny New York Oil Industry," *New York Times,* January 10, 2000, 1.

49. "Pennsylvania Oil and Gas Association Seeks Brine Treatment Grant," *Oil Daily,* May 21, 1987, 3; Tom Stewart-Gordon, "Industry Eyes New Standards for Injection Wells," *Oil Daily,* June 26, 1992.

50. Editorial, "EPA Drafts Tougher Injection Well Rules," *Oil and Gas Journal* 91 (June 14 1993): 11.

51. Southwestern Legal Foundation, *Economics of the Petroleum Industry: New Ideas, New Methods, New Developments,* vol. 1 (Houston, Tex.: Gulf Publishing, 1963), 174–196; vol. 2 (1964), 153–76; vol. 3 (1965), 137–50.

52. Robert O. Anderson, *Fundamentals of the Petroleum Industry* (Norman: University of Ok-

lahoma Press, 1984), 174–77; Lloyd E. Elkins, "Overview: 25 Years of Professional Development and Production," *Journal of Petroleum Technology* 25 (December 1973): 1337–41.

53. John Opie, *Ogallala: Water for a Dry Land* (Lincoln: University of Nebraska Press, 1993), 180.

54. For example, see R. F. Brown, D. C. Signor, and W. W. Wood, *Artificial Ground-Water Recharge as a Water-Management Technique on the Southern High Plains of Texas and New Mexico*, report no. 220 (Austin: Texas Department of Water Resources, 1978); Ann E. Bell and Shelly Morrison, *Analytical Study of the Ogallala Aquifer in Yoakum County*, Texas Department of Water Resources, report no. 221 (Austin: Texas Department of Water Resources, 1978).

55. "Unit Waterflood Will Yield Production and Environmental Benefits," *Oil and Gas Journal* 75 (Feb. 28, 1977): 119–21.

56. *Oklahoma Water Resources Board et al. v. Texas County Irrigation and Water Resources Association, Inc.*, Oklahoma Supreme Court, Dec. 26, 1984, no. 56355, *Oil and Gas Reporter* 88, report no. 2 (10–86): 2–117 to 2–181.

57. Opie, *Ogallala*, 175–86.

58. Thomas C. Schelling, "The Cost of Combating Global Warming: Facing the Tradeoffs," *Foreign Affairs* 76 (November/December 1997): 8–13; Richard N. Cooper, "Toward a Real Global Warming Treaty" *Foreign Affairs* 77 (March/April 1998): 66–79; Stuart Eizenstat, "Stick with Kyoto: A Sound Start on Global Warming," *Foreign Affairs* 77 (May/June 1998): 119–21. Congress's reaction to the 1977 Kyoto protocol, stating that the United States will not agree to binding targets unless developing nations such as India and China do as well, suggests that reaching consensus on this general issue will difficult.

Bibliography

Archives and Manuscript Collections

American Heritage Center, University of Wyoming (Laramie, Wyoming)
    Allen F. Agnew Collection
    Continental Oil Company Collection
    David D. Dominick Collection
    Bernard Majewski Collection
    George Otis Smith Collection
Center for American History, University of Texas at Austin (Austin, Texas)
    Walter Benona Sharp Papers
    Morgan Jones Davis Papers
    Frances Tarlton Farenthold Papers
    Oral History of the Oil Industry
    Olin Culberson Papers
    E. O. Thompson Papers
Hagley Museum and Library (Wilmington, Delaware)
    Sun Company Collection
    Pamphlet Collection
Herbert Hoover Presidential Library (West Branch, Iowa)
    Presidential Papers
    Commerce Papers
    Ray Lyman Wilbur Collection
Houston Metropolitan Research Center (Houston, Texas)
    George A. Hill, Jr., Collection
    Dr. W. A. Quebedeaux Environmental Collection
    Mr. and Mrs. Emmott Environmental Collection
    Citizens' Environmental Coalition Collection
    J. M. Heiser Environmental Collection
    J. S. Cullinan Collection
Lyndon B. Johnson Presidential Library (Austin,Texas)
    Aides Files, White House Central Files
    Subject Files, White House Central Files
National Archives, College Park (College Park, Maryland)
    Records of the Environmental Protection Agency, RG 412

National Archives, Southwest Region (NARS) (Fort Worth, Texas)
   Records of the Petroleum Experiment Center, Bureau of Mines, RG 70-2
   Records of Research Groups, Bartlesville Petroleum Research Center, Bureau of Mines, RG 70-4
Panhandle Plains Historical Museum, West Texas A&M (Amarillo, Texas)
   C. R. Austin Oil Company Papers
   Transcripts of Interviews
Permian Basin Petroleum Museum (Midland, Texas)
   Joseph Graybeal Collection
   Transcripts of Interviews
Southwest Collection, Texas Tech University (Lubbock, Texas)
   West Texas Oil Lifting Short Course
   Oil and Gas Industry File
Western History Collection, University of Oklahoma (Norman, Oklahoma)
   Cities Service Oil Company Collection
   Engineering Bulletins
Woodson Research Center, Rice University (Houston, Texas)
   J. Russell Wait Collection
   Norman Hurd Ricker Papers

Periodicals Used As Primary Sources

API Bulletin
API Proceedings
Chemical Age
*Chemical and Metallurgical
   Engineering*
Corrosion
Drill Bit
Engineering News Record
Humble Way
Industrial Standardization

*Journal of Petroleum
   Technology*
National Petroleum News
New York Times
Oil and Gas Journal
Oil and Gas Reporter
Oil Bulletin
Oil Weekly
Petroleum Age
Petroleum Engineer

*Petroleum Engineering and
   Services*
Petroleum Times
Pipe Line News
Sewage and Industrial Waste
*Transactions of the American
   Institute of Metallurgical
   and Mining Engineers
   (AIMME Transactions)*

Government Documents

Arkansas Gas and Oil Commission. *Secondary Recovery of Petroleum in Arkansas—A Survey.* Prepared by George H. Fancher and Donald K. Mackay. El Dorado: Arkansas Gas and Oil Commission, 1946.

Kentucky Geological Survey. *Effects of Greensburg Oil Field Brines on the Streams, Wells, and Springs of the Upper Green River Basin, Kentucky.* Prepared by R. A. Krieger and G. E. Hendrickson, Report of Investigation 2. Frankfort: Kentucky Geological Survey, 1960.

Louisiana Department of Conservation. *Development of Louisiana's Offshore Oil and Gas Reserves.* Baton Rouge: Louisiana Department of Conservation, 1956.

National Commission on Water Quality. *Petroleum Refining Industry: Technology and Costs of Wastewater Control.* Prepared by Engineering-Science, Inc. Washington D.C.: National Commission on Water Quality, 1975.

Preliminary Conference on Oil Pollution of Navigable Waters. June 8–16. Washington, D.C.: Government Printing Office, 1926.

President's Panel on Oil Spills. *The Oil Spill Problem: First Report of the President's Panel on Oil Spills.* Washington, D.C.: Government Printing Office, 1969.

Railroad Commission of Texas. *Texas Oil and Gas Handbook: A Guide to State Conservation Regulations.* Austin: R. W. Byram and Co., 1958.

Texas Department of Agriculture. *Agricultural Land and Water Contamination From Injection Wells, Disposal Pits, and Abandoned Wells Used in Oil and Gas Production.* Austin: Texas Dept. of Agriculture, 1985.

U.S. Bureau of Fisheries. *Danger to Fisheries From Oil and Tar Pollution of Waters.* Prepared by J. S. Gutsell and printed as appendix 7 to *Report of the U.S. Commissioner of Fisheries.* Washington, D.C.: Government Printing Office, 1921.

U.S. Bureau of Mines. *Abatement of Corrosion in Central Heating Systems.* Prepared by F. N. Speller. Technical Paper no. 236. Washington, D.C.: Government Printing Office, 1919.

U.S. Bureau of Mines. *Evaporation Losses of Petroleum in the Mid-Continent Field.* Prepared by J. H. Wiggins. Bulletin no. 200. Washington, D.C.: Government Printing Office, 1921.

U.S. Bureau of Mines. *Experimental Studies on the Effects of Ethyl Gasoline and Its Combustion Products: Report of the Bureau of Mines to the General Motors Research Corporation and the Ethyl Gasoline Corporation.* Prepared by R. R. Sayers et al. Washington, D.C.: Government Printing Office, 1927.

U.S. Bureau of Mines. *Extinguishing and Preventing Oil and Gas Fires.* Prepared by C. P. Bowie. Bulletin no. 170. Washington, D.C.: Government Printing Office, 1920.

U.S. Bureau of Mines. *History of Water Flooding of Oil Sands in Oklahoma.* Prepared by D. B. Taliaferro and David M. Logan. Report of Investigation no. 3728. Washington, D.C.: U.S. Bureau of Mines, 1943.

U.S. Bureau of Mines. *Investigation of Toxic Gasses from Mexican Crude and Other High-Sulfur Petroleum Products.* Prepared jointly with the API. Bulletin no. 231. Washington, D.C.: Government Printing Office, 1925.

U.S. Bureau of Mines. *Methods of Decreasing Evaporation Losses of Petroleum.* Prepared by J. H. Wiggins. Technical Paper no. 318. Washington, D.C.: Bureau of Mines, 1923.

U.S. Bureau of Mines. *Methods for Increasing the Recovery of Oil From Sands.* Prepared by J. O. Lewis. Bulletin no. 148. Washington, D.C.: Government Printing Office, 1917.

U.S. Bureau of Mines. *Methods of Shutting off Water in Oil and Gas Fields.* Prepared by F. B. Tough. Bulletin no. 163. Washington, D.C.: Government Printing Office, 1918.

U.S. Bureau of Mines. *Pollution By Oil of the Coast Waters of the United States.* Preliminary report. Washington D.C.: Bureau of Mines, 1923.

U.S. Bureau of Mines. *Prospecting and Testing for Oil and Gas.* Prepared by R.E. Collom. Bulletin no. 201. Washington, D.C.: Government Printing Office, 1916.

U.S. Bureau of Mines. *The Disposal of Oil Field Brines.* Prepared by Ludwig Schmidt and J. M. Devine. Report of Investigation no. 2945. Washington, D.C.: Bureau of Mines, 1929.

U.S. Bureau of Mines. *Typical Oil-Field Brine Conditioning Systems for Preparing Brine for Subsurface Injection.* Prepared by S. Taylor, C. J. Wilhelm, and W. C. Holliman. Report of Investigation no. 3434. Washington, D.C.: Bureau of Mines, 1939.

U.S. Coast Guard. *Legal, Economic, and Technical Aspects of Liability and Financial Responsibility as Related to Oil Pollution.* Prepared by the Program of Policy Studies in Science and Technology, George Washington University. Washington, D.C.: George Washington University, 1970.

U.S. Congress. House Committee on Foreign Affairs. *Hearing on Oil Pollution of Navigable Waters.* 67th Cong., 2d sess., February 15–18, 1922.

U.S. Congress. House Committee on Interior and Insular Affairs, Subcommittee on the Environment. *Hearings on Deepwater Port Facilities.* Washington, D.C.: Government Printing Office, 1974.

U.S. Congress. House Committee on Merchant Marine and Fisheries. *Hearing on Oil Pollution and Compensation,* 101st Cong., 1st sess., May 11, 1989.

U.S. Congress. House Committee on Merchant Marine and Fisheries. *Hearings on Oil Pollution: Bills to Amend the Oil Pollution Act, 1924, for the Purpose of Controlling Oil Pollution from Vessels.* 91st Cong., 1st sess., 1969.

U.S. Congress. House Committee on Merchant Marine and Fisheries. *Hearings on the Blowout of the Mexican Oil Well IXTOC 1.* 96th Cong., 2d sess., 1980.

U.S. Congress. House Committee on Public Works and Transportation. *Colonial Pipeline Rupture,* 103rd Cong., 1st sess., May 18, 1993.

U.S. Congress. House Committee on Rivers and Harbors. *Hearings on a Bill to Amend the Oil Pollution Act of 1924.* 71st Cong., 2d sess., May 2, 3, and 26, 1930.

U.S. Congress. House Committee on Rivers and Harbors. *Hearings on the Pollution of Navigable Waters.* 67th Cong., 2d sess., Oct. 25, 1921.

U.S. Congress. House Committee on Rivers and Harbors. *Hearings on the Pollution of Navigable Waters.* 67th Cong., 2d sess., December 7 and 8, 1921.

U.S. Congress. House Committee on Rivers and Harbors. *Hearings on the Pollution of Navigable Waters.* 68th Cong., 1st sess., January 23–30, 1924.

U.S. Congress. House Committee on the Judiciary. *Conservation of Oil and Gas and Protection of American Sources.* 72d Cong., 1st sess., May 14, 1932.

U.S. Congress. Office of Technology Assessment. *Coping With an Oiled Sea: An Analysis of Oil Spill Response Technologies.* Washington, D.C.: Government Printing Office, 1990.

U.S. Congress. Office of Technology Assessment. *Oil Transportation by Tankers: An Analysis of Marine Pollution and Safety Measures.* Washington, D.C.: Government Printing Office, 1975.

U.S. Congress. Senate Committee on Commerce. *Hearing on the Pollution of Navigable Waters,* 68th Cong., 1st sess., January 9, 1924.

U.S. Congress. Senate Committee on Commerce. *Hearings on the Navigable Waters Safety and Environmental Quality Act.* 92nd Cong., 1st sess., 1971.

U.S. Congress. Senate Committee on Commerce. *Hearings on Wetland Acquisition and Oil Pollution of the Sea.* 87th Cong., 1st sess., July 31, 1961.

U.S. Congress. Senate Committee on Environment and Public Works. *Hearings on the Oil Pollution Liability and Compensation Act.* 99th Cong., 2d sess., Sept. 10–12.

U.S. Congress. Senate Committee on Environment and Public Works. *Three Recent Oil Spills: Hearing before the Subcommittee on Environmental Protection on Oil Spills in the Coastal Waters of Rhode Island, the Delaware River, and the Houston Ship Channel.* 103d Cong., 2d sess., 1989.

U.S. Congress. Senate Committee on Government Operations. *Hearings on the Sale of Government-Owned Surplus Tanker Vessels.* 82nd Cong., 2d sess., 1952.

U.S. Congress. Senate Committee on Interior and Insular Affairs. *Hearings on Submerged Lands.* 82d Cong., 1s sess., Feb. 19–22, 1951.

U.S. Congress. Senate Subcommittee on Air and Water Pollution. *Hearings on Water Pollution,* 89th Cong., 1st sess., June 23 and June 24, 1965.

U.S. Corps of Engineers. *Pollution Affecting Navigation or Commerce on Navigable Waters.* H. Doc. 417. Report of the Secretary of War to the House of Representatives. 69th Cong., 1st sess., 1926.

U.S. Department of Commerce. *A Statistical Analysis of the World's Merchant Fleets.* Washington, D.C.: Government Printing Office, 1972.

U.S. Department of Transportation. Office of Pipeline Safety. *Common Ground: Study of One Call Systems and Best Practice Damage Prevention Systems.* August 1999.

U.S. Environmental Protection Agency. *Brine Disposal Treatment Practices Relating to the Oil Production Industry.* Prepared by George W. Reid et al. Norman: School of Civil Engineering and Environmental Engineering, University of Oklahoma, 1974.

U.S. Environmental Protection Agency. *Development Document for Proposed Effluent Limitations Guidelines and New Source Performance Standards for the Petroleum Refining Point Source Category.* Washington, D.C.: Government Printing Office, 1974.

U.S. Environmental Protection Agency. *Musts for USTs: A Summary of the New Regulations for Underground Storage Tank Systems.* September 1998.

U.S. Environmental Protection Agency. Office of Toxic Substances. *Environmental Aspects of Chemical Use in Well-Drilling Operations.* Washington, D.C.: EPA, 1975.

U.S. Environmental Protection Agency. *Oil Spills and Pollution Reports: July 1974 to October 1974.* Prepared by Floyd A. DeWitt, Jr., and Penelope Melvin. Washington, D.C.: Government Printing Office, 1975.

U.S. Environmental Protection Agency. *Proceedings of the Second Open Forum on Management of Petroleum Refinery Wastewater.* Edited by Francis S. Manning. Ada, Okla.: Robert S. Kerr Environmental Research Laboratory, EPA, 1978.

U.S. Environmental Protection Agency. *Refinery Effluent Water Treatment Plant Using Activated*

Carbon. Prepared by Gary C. Loop. Washington, D.C.: Government Printing Office, 1975.

U.S. Environmental Protection Agency. *The Challenge of the Environment: A Primer on EPA's Statutory Authority.* Washington, D.C.: Government Printing Office, 1972.

U.S. Environmental Protection Agency. *Toward Cleaner Water.* Washington, D.C.: Government Printing Office, 1974.

U.S. Federal Oil Conservation Board. *Complete Record of Public Hearings, February 10 and 11.* Washington D.C.: Government Printing Office, 1926.

U.S. Federal Oil Conservation Board. *Record of Public Hearings, May 27.* Washington D.C.: Government Printing Office, 1926.

U.S. Federal Water Pollution Control Administration. *Manpower and Training Needs in Water Pollution Control.* Washington, D.C.: Department of the Interior, 1967.

U.S. Federal Water Pollution Control Administration. *Oil Tagging System Study.* Washington, D.C.: Government Printing Office, 1970.

U.S. Federal Water Pollution Control Administration. *Proceedings of a Joint Conference on Prevention and Control of Oil Spills.* Jointly sponsored with the API. Washington, D.C.: Federal Water Pollution Control Administration, 1969.

U.S. General Accounting Office. *Drinking Water: Safeguards Are Not Preventing Contamination from Injected Oil and Gas Waste.* Washington, D.C.: General Accounting Office, 1989.

U.S. General Accounting Office. *Tanker and Oil Transfer Operations on the Delaware River and Bay.* Prepared by the U.S. Coast Guard. Washington, D.C.: Government Printing Office, 1977.

U.S. Geological Survey. *Chemical Relations of the Oil-Field Waters in San Joaquin Valley, California.* Prepared by G. Sherburne Rogers. Washington, D.C.: Government Printing Office, 1917.

U.S. Geological Survey. *Geohydrology and Water Quality in Northern Portage County, Ohio in Relation to Deep-Well Brine Injection.* Prepared by Sandra M. Eberts. Water Resources Investigation Report 90-4158. Washington, D.C.: U.S. Geological Survey, 1990.

U.S. Geological Survey. *Preliminary Evaluation of the Effects of an Abandoned Oil Refinery on the Chemical Quality of Water in the Arkansas River Valley, 1985–86.* Prepared by T. B. Spruill. Lawrence, Kans.: U.S. Geological Survey, 1990.

U.S. Maritime Administration. *A Statistical Analysis of the World's Merchant Fleet.* Washington, D.C.: Government Printing Office, 1956.

U.S. Maritime Administration. *A Statistical Analysis of the World's Merchant Fleets.* Washington, D.C.: Government Printing Office, 1972.

U.S. Public Health Service. *Oil Pollution at Bathing Beaches,* public health report. Washington, D.C.: Government Printing Office, 1924.

United Nations. *Pollution of the Sea by Oil: Results of an Inquiry by the United Nations Secretariat.* New York: United Nations, 1956.

Books, Articles, and Dissertations

Albertson, M. "Conservation." In *Elements of the Petroleum Industry,* edited by E. DeGolyer. New York: American Institute of Mining and Metallurgical Engineers, 1940.

Alchon, Guy. *The Visible Hand of Planning: Capitalism, Social Science, and the State in the 1920s.* Princeton, N. J.: Princeton University Press, 1985.

Aldrich, Mark. "Preventing 'the Needless Peril of the Coal Mine': The Bureau of Mines and the Campaign against Coal Mine Explosions, 1910–1940." *Technology and Culture* 36 (July 1995): 483–518.

Allaud, Louis A., and Maurice H. Martin. *Schlumberger: The History of a Technique.* New York: Wiley, 1977.

Allin, Colonel Benjamin Casey, III. *Reaching for the Sea.* Boston: Meador Publishing, 1956.

American Bar Association, Section of Mineral Law. *Legal History of Conservation of Oil and Gas.* Chicago: American Bar Association, 1938.

American Institute of Biological Sciences. *Sources, Effects, and Sinks of Hydrocarbons in the Aquatic Environment.* Proceedings of a symposium. Washington, D.C.: American University, 1976.

American Institute of Chemical Engineers. *The Petroleum/Petrochemical Industry and the Ecologi-*

cal Challenge. Symposium papers. New York: American Institute of Chemical Engineers, 1973.

American Petroleum Institute. *Analysis of the 1972 API-EPA Raw Waste Load Survey Data*. Prepared by Brown and Root. Publication no. 4200. Washington, D.C.: API, 1974.

American Petroleum Institute. *Assessing Natural Resource Damages From Hazardous Substances: An Examination of Existing Assessment Approaches*. Washington, D.C.: API, 1987.

American Petroleum Institute. *Fire Protection in Refineries*. New York: API, 1933.

American Petroleum Institute. *History of Petroleum Engineering*. New York: API, 1961.

American Petroleum Institute. *Manual on the Disposal of Refinery Wastes*. All editions and volumes. New York and Washington, D.C.: API, 1930–1963.

American Petroleum Institute. *Oil Pollution Survey of the Great Lakes*. Prepared by John V. Dennis. New York: API, 1960.

American Petroleum Institute. *Oil Pollution Survey of the United States Atlantic Coast*. Prepared by John V. Dennis. New York: API, 1959.

American Petroleum Institute. *Report Covering Survey of Oil Conditions in the United States*. New York: API, 1927.

American Petroleum Institute. *The History of the Tanker Corrosion Research Project*. Washington, D.C.: API, ca. 1965.

American Society for Testing and Materials. *Oil Field Subsurface Injection: Symposium Papers*. Philadelphia: American Society for Testing and Materials, 1977.

Anderson, Robert O. *Fundamentals of the Petroleum Industry*. Norman: University of Oklahoma Press, 1984.

Arrow, Kenneth J. "Criteria for Social Investment." *Water Resources Research* 1 (1965): 1–8.

Aub, Joseph C., Lawrence T. Fairhall, A. S. Minot, Paul Reznikoff, and Alice Hamilton. *Lead Poisoning*. Baltimore, Md.: Williams and Wilkins Co., 1926.

Bacon, Raymond Foss, and William Allen Hamor. *The American Petroleum Industry*. New York: McGraw-Hill, 1916.

Bahr, Betsy W. "New England Mill Engineering: Rationalizing and Reform in Textile Mill Design, 1790–1920." Ph.D. diss., University of Delaware, 1987.

Bain, Joe S. *The Economics of the Pacific Coast Petroleum Industry*. 3 vols. Berkeley: University of California Press, 1945–47.

Baker, Elijah, III. *Introduction to Steel Shipbuilding*. New York: McGraw-Hill, 1943.

Ball, Max W. *This Fascinating Oil Business*. New York: Bobbs-Merrill Company, 1940.

Bamberg, J. H. *The History of the British Petroleum Company: the Anglo-Iranian Years, 1928–1954*. Cambridge: Cambridge University Press, 1994.

Banta, Michael B. "The Regulation and Conservation of Petroleum Resources in Louisiana, 1901–1940." Ph.D. diss., Louisiana State University Press, 1982.

Barcelona, Michael, et. al. *Handbook of Groundwater Protection*. New York: Science Information Resource Center, 1988.

Bartley, Ernest R. *The Tidelands Oil Controversy*. Austin: University of Texas Press, 1953.

Basalla, George. *The Evolution of Technology*. New York: Cambridge University Press, 1988.

Batchelor, Bronson. *Stream Pollution: A Study of Proposed Federal Legislation and its Effect on the Oil Industry*. New York: API, 1937.

Beaton, Kendall. *Enterprise in Oil: A History of Shell in the United States*. New York: Appleton-Century-Crofts, 1957.

Beecher, C. E. and I. P. Parkhurst. "Effect of Dissolved Gas upon the Viscosity and Surface Tension of Crude Oil." In *Petroleum Technology and Development in 1926*. New York: American Institute of Mining and Metallurgical Engineers, 1927.

Beerstecher, Ernest, Jr. *Petroleum Microbiology*. Houston: Elsevier Press, 1954.

Bell, Harold S. *American Petroleum Refining*. New York: D. Van Nostrand, 1923.

Bell, H. W., and J. B. Kerr. *The El Dorado, Arkansas Oil and Gas Field*. Little Rock, Ark.: U.S. Bureau of Mines, 1922.

Beniger, James R. *The Control Revolution*. Cambridge, Ma.: Harvard University Press, 1986.

Bennett, Stuart. "The Industrial Instrument—Master of Industry, Servant of Management": Automatic Control of the Process Industries, 1900–1940." *Technology and Culture* 21 (1991): 69–81.

Benson, Ray. "Safety in the Operation of HF Alkylation Plants." In *Refinery Operation and Maintenance.* Cleveland, Ohio: National Petroleum News, 1945.

Benson, Robert D., and William S. Benson. *History of the Tide Water Companies.* New York: published by the authors, 1913.

Bentley, Jerome T. "The Effect of Standard Oil's Vertical Integration into Transportation on the Structure and Performance of the American Petroleum Industry, 1872–1884." Ph.D. diss., University of Pittsburgh, 1976.

Berry, Mary Clay. *The Alaska Pipeline: The Politics of Oil and Native Land Claims.* Bloomington: Indiana University Press, 1975.

Bes, J. *Tanker Chartering and Management.* Amsterdam: C. de Boer., 1956.

Bijker, Weibe, Thomas P. Hughes, and Trevor J. Pinch, eds. *The Social Construction of Technological Systems: New Directions in the Sociology and History of Technology.* Cambridge, Mass.: MIT Press, 1987.

Birkland, Thomas A. *After Disaster: Agenda Setting, Public Policy, and Focusing Events.* Washington, D.C.: Georgetown University Press, 1997.

Blair, John M. *The Control of Oil.* New York: Pantheon Books, 1976.

Blakely, Murphy M. *Conservation of Oil and Gas: A Legal History.* Chicago: American Bar Association, 1949.

Blakey, Ellen Sue. *Oil On Their Shoes: Petroleum Geology to 1918.* Tulsa, Okla.: American Association of Petroleum Geologists, 1985.

Boatright, Mody C. *Folklore of the Oil Industry.* Dallas: Southern Methodist University Press, 1963.

Boatright, Mody C., and William A. Owens. *Tales from the Derrick Floor: A People's History of the Oil Industry.* Lincoln: University of Nebraska, 1970.

Bonn, C. R. H. *The Oil Tanker.* Glasgow: Association of Engineering and Shipbuilding Draughtsmen, 1922.

Boone, Lalia Phipps. *The Petroleum Dictionary.* Norman: University of Oklahoma Press, 1952.

Boothe, Joan Norris. *Cleaning Up: The Cost of Refinery Pollution Control.* New York: Council on Economic Priorities, 1975.

Bossler, Robert S. *Oil Fields Rejuvenated.* Bulletin no. 56. Harrisburg, Pa: Pennsylvania Bureau of Topographic and Geological Survey, 1922.

Bowker, Geoffrey C. *Science on the Run: Information Management and Industrial Geophysics at Schlumberger, 1920–1940.* Cambridge, Mass.: MIT Press, 1994.

Bowles, Charles E. *The Petroleum Industry.* Kansas City, Mo.: Schooley Printing, 1921.

Bradford, Peter A. *Fragile Structures: A Story of Oil Refineries, National Security, and the Coast of Maine.* New York: Harper's Magazine Press, 1975.

Brantly, J. E. *History of Oil Well Drilling.* Houston: Gulf Publishing Company, 1971.

Bricnes, Marvin. "The Fight Against Smog in Los Angeles, 1943–57." Ph.D. diss., University of California, Davis, 1975.

Brown, Stanley H. *H. L. Hunt.* Chicago: Playboy Press, 1976.

Bryner, Gary C. *Blue Skies, Green Politics: The Clean Air Act of 1990.* Washington, D.C.: CQ Press, 1993.

Bubenick, David V., ed. *Acid Rain Information Book.* Park Ridge, N. J.: Noyes Publications, 1984.

Burton, Dudley J., and K. Ravishankar. *Treatment of Hazardous Petrochemical and Petroleum Wastes: Current, New, and Emerging Technologies.* Park Ridge, N.J.: Noyes Publications, 1989.

Cahan, David. *An Institute for an Empire: The Physikalische-Technische Reichsanstalt, 1871–1918.* Cambridge: Cambridge University Press, 1989.

Calvert, Monte. *The Mechanical Engineer in America, 1830–1910: Professional Cultures in Conflict.* Baltimore: Johns Hopkins University Press, 1967.

Campbell, Murray, and Harrison Hatton. *Herbert Dow: Pioneer in Creative Chemistry.* New York: Appleton-Century-Crofts, 1951.

Cannon, W. A., and C. E. Welling. "The Application of Vanadia-Alumina Catalysts for the Oxidation of Exhaust Hydrocarbons." Paper no. 29T presented at the SAE meeting, Detroit, Michigan, January 1959.

Caplinger, Michael W., and Philip W. Ross. *The Historic Petroleum Industry in the Allegheny National Forest.* Prepared for the USDA Forest Service. Morgantown: West Virginia University, 1993.

Carlisle, Rodney P., and August W. Giebelhaus. *Bartlesville Energy Center: The Federal Government in Petroleum Research, 1918–1983.* Washington, D.C.: U.S. Dept. of Energy, 1984.

Carlson, Robert L. "RCRA Overview." In *Hazardous Materials Management,* edited by Doy Cox. Rockville, Md.: Institute of Hazardous Materials Management, 1995.

Case, Leslie C. *Water Problems in Oil Production: An Operator's Manual.* Tulsa, Oklahoma: Petroleum Publishing Company, 1970.

Cassel, Eric J. "The Health Effects of Air Pollution and Their Implications for Control." In *Air Pollution Control,* edited by Clark C. Havighurst. Dobbs Ferry, N.Y.: Oceana Publications, 1969.

Castaneda, James. *Regulated Enterprises: Natural Gas Pipelines and Northeastern Markets, 1938–1954.* Columbus: Ohio State University Press, 1993.

Chandler, Alfred D., Jr. *The Visible Hand: The Managerial Revolution in American Business.* Cambridge, Mass.: Belknap Press, 1977.

Chase, Stuart. *The Tragedy of Waste.* New York: Macmillan Co., 1927.

Childs, William R. "The Transformation of the Railroad Commission of Texas, 1917–1940." *Business History Review* 65 (summer 1991): 284–344.

Churchill, R. R., and A. V. Lowe. *The Law of the Sea.* Manchester, England: Manchester University Press, 1983.

Cicchetti, Charles J. *Alaskan Oil: Alternate Routes and Markets.* Baltimore: Resources for the Future, John Hopkins University Press, 1972.

Clark, James A. *The Fabulous East Texas Oil Field.* Los Angeles: Atlantic Richfield, ca. 1970.

Clark, James A. and Michel T. Halbouty. *Spindletop.* New York: Random House, 1952.

Cleary, Edward J. *The ORSANCO Story: Water Quality Management in the Ohio Valley under an Interstate Compact.* Baltimore: Resources for the Future, John Hopkins University Press, 1967.

Cloud, Wilbur F. *Petroleum Production.* Norman: University of Oklahoma Press, 1937.

Coase, Ronald. "The Problem of Social Cost." *Journal of Law and Economics* 3 (October 1960): 1–44.

Coates, Peter A. *The Trans-Alaska Pipeline Controversy: Technology, Conservation, and the Frontier.* Bethlehem, Pa.: Lehigh University Press, 1991.

Cochrane, Rexmond. *Measures for Progress.* Washington, D.C.: National Bureau of Standards, 1966.

Colten, Craig E., and Peter N. Skinner. *The Road to Love Canal: Managing Industrial Waste before EPA.* Austin : University of Texas Press, 1996.

Commoner, Barry. *The Closing Circle: Nature, Man, and Technology.* New York: Knopf, 1971.

Composite Catalog of Oil Refinery Equipment and Process Handbook. Houston, Tex.: Gulf Publishing, various editions from 1931 to 1939.

Comverse, Thomas P. *Oil and Where to Find It.* Amarillo, Tex.: Russell and Cockrell, 1920.

Cone, Andrew, and Walter R. Johns. *Petrolia: A Brief History of the Pennsylvania Petroleum Region From 1859 to 1869.* New York: D. Appleton and Co., 1870.

Constant, Edward W., II. "Cause or Consequence: Science, Technology, and the Regulatory Change in the Oil Business in Texas, 1930–1975." *Technology and Culture* 30 (April 1989): 426–55.

———. "Science in Society: Petroleum Engineers and the Oil Fraternity in Texas, 1925–65." *Social Studies of Science* 19 (August 1989): 439–72.

———. "State Management of Petroleum Resources, Texas 1910–1940," in *Energy and Transport: Historical Perspectives on Policy Issues,* edited by George Daniels and Mark H. Rose. Beverly Hills: Sage Publications, 1982.

———. "The Cult of MER: or Why There is a Collective in Your Consciousness." *Business and Economic History* 22 (fall 1993).

Continental Oil Company. *CONOCO: The First One Hundred Years.* New York: Dell, 1975.

Cook, Sir Basil Kemball. "Ocean Transport." In *Petroleum: Twenty-five Years Retrospect, 1910–1935,* edited by Institute of Petroleum Technologists. London: Aldine House, 1935.

Cookenboo, Leslie, Jr. *Crude Oil Pipe Lines and Competition in the Oil Industry.* Cambridge, Mass.: Harvard University Press, 1955.

Cooper, Richard N. "Toward a Real Global Warming Treaty." *Foreign Affairs* 77 (March/April 1998): 66–79.

Copp, Anthony E. *Regulating Competition in Oil: Government Intervention in the U.S. Refining Industry, 1948–1975.* College Station, Tex.: Texas A&M University Press, 1976.

Cotton, William M. "Some Legal Problems Encountered in Lease Operations After Discovery of Oil." In *West Texas Oil Lifting Short Course.* Lubbock: Texas Tech University, 1959.

Cowdrey, Albert. "Pioneering Environmental Law: The Army Corps of Engineers and the Refuse Act." *Pacific Historical Review* 44 (August 1975): 331–39.

Cowen, Edward. *Oil and Water: The Torrey Canyon Disaster.* Philadelphia: J. B. Lippincott, 1968.

Cowling, Ellis B. "Acid Precipitation in Historic Perspective." *Environmental Science and Technology* 16 (February 1982): 110a–123a.

Cox, G. H., C. L. Dake, and G. A. Muilenburg. *Field Methods in Petroleum Geology.* New York: McGraw-Hill, 1921.

Crook, Leo. *Oil Terms: A Dictionary of Terms Used in Oil Exploration and Development.* London: Wilton House, 1975.

Cumbler, John T. "Conflict, Accommodation, and Compromise: Connecticut's Attempt to Control Industrial Wastes in the Progressive Era." *Environmental History* 5 (July 2000): 314–34.

D'Andrade, Patricia. "Saving the Environment: Science and Social Action." Ph.D. diss., City University of New York, 1993.

Davidson, Art. *In the Wake of the Exxon Valdez: The Devastating Impact of the Alaska Oil Spill.* San Francisco: Sierra Club Books, 1990.

Davis, David Howard. *American Environmental Politics.* Chicago: Nelson Hall, 1998.

Davis, J. B. *Petroleum Microbiology.* New York: Elsevier, 1967.

Day, David T., ed. *A Handbook of the Petroleum Industry.* New York: John Wiley and Sons, 1922.

Dean, Arthur H. "The Law of the Sea Conference, 1958–60, and Its Aftermath." In *The Law of the Sea: Offshore Boundaries and Zones,* edited by Lewis M. Alexander. Columbus: Ohio State University Press, 1967.

De Chazeau, Melvin G. and Alfred E. Kahn. *Integration and Competition in the Petroleum Industry.* New Haven: Yale University Press, 1959.

Dedmon, Emmett. *Challenge and Response: A Modern History of Standard Oil Company (Indiana).* Chicago: Mobium Press, 1984.

De Hall, P. "Transport and Distribution." In *Modern Petroleum Technology.* London: Institute of Petroleum, 1954.

De La Pedraja, Rene. *The Rise and Decline of U.S. Merchant Shipping in the Twentieth Century.* New York: Twayne, 1992.

Denny, Ludwell. *We Fight For Oil.* New York: Alfred A. Knopf, 1928.

Directory of Oil Refineries and Field Processing Plants. Tulsa, Okla.: Oil and Gas Journal, 1952.

Drake, Douglas. "Herbert Hoover, Ecologist: The Politics of Oil Pollution Control, 1921–1926." *Mid-America* 55 (1973): 207–28.

Dunlap, Thomas. *DDT: Scientists, Citizens, and Public Policy.* Princeton: Princeton University Press, 1981.

Dunn, Laurence. *The World's Tankers.* New York: John De Graff, 1956.

Dupree, A. Hunter. *Science in the Federal Government: A History of Policies and Activities to 1940.* Cambridge, Mass.: Belknap Press, 1987.

Duprey, R. L. "The Status of SOx Emission Limitations." *Chemical and Engineering Progress* 68 (Feb. 1972): 70–76.

Dutch Committee for Long-Term Environmental Policy, eds. *The Environment: Towards a Sustainable Future.* Boston: Kluwer Academic Publishers, 1993.

East Texas Salt Water Disposal Company. *Salt Water Disposal in the East Texas Oil Field*. Austin: University of Texas, Petroleum Extension Service, 1958.

Eaton, S. *Petroleum: A History of the Oil Region of Venango County, Pennsylvania*. Philadelphia: J. P. Skelly and Co., 1866.

Ehrlich, Paul R. *The Population Bomb*. New York: Ballantine, 1968.

Eizenstat, Stuart. "Stick with Kyoto: A Sound Start on Global Warming." *Foreign Affairs 77* (May/June 1998): 119–21.

Eldridge, E. F. *Industrial Waste Treatment Practice*. New York: McGraw-Hill, 1942.

Elements of Oil-Tanker Transportation. Tulsa, Okla.: PennWell Books, 1982.

Elkin, Harold F. "Petroleum Refining." In *Engineering Control of Air Pollution*, edited by Arthur Stern. Vol. 4, *Air Pollution*. New York: Academic Press, 1977.

Ellis, Carleton, and Joseph V. Meigs. *Gasoline and Other Motor Fuels*. New York: D. Van Nostrand Co., 1921.

Emmons, William Harvey. *Geology of Petroleum*. New York: McGraw-Hill, 1921.

Engler, Robert. *The Politics of Oil: A Study of Private Power and Democratic Directions*. Chicago: University of Chicago Press, 1961.

Enos, John L. *Petroleum Progress and Profits: A History of Process Innovations*. Cambridge, Mass.: MIT Press, 1962.

Epstein, Samuel S., Lester O. Brown, and Carl Pope. *Hazardous Waste in America*. San Francisco: Sierra Club Books, 1982.

Esposito, John, and Larry J. Silverman. *Vanishing Air: The Ralph Nader Study Group Report on Air Pollution*. New York: Grossman Publishers, 1970.

Evans, Ulick R. *The Corrosion of Metals*. London: Edward Arnold and Co., 1924.

Ezell, Samuel. *Innovations in Energy: The Story of Kerr-McGee*. Norman: University of Oklahoma, 1971.

Fairhall, David, and Phillip Jordan. *The Wreck of the Amoco Cadiz*. New York: Stein and Day, 1980.

Fanning, Leonard M. *The Story of the American Petroleum Institute: A Study and Report with Personal Reminiscences*. New York: World Petroleum Policies, 1959.

Ferguson, Jim G. *Minerals in Arkansas*. Little Rock, Ark.: State of Arkansas, 1922.

Ferrier, R.W. *The History of the British Petroleum Industry: The Developing Years, 1901–32*. New York: Cambridge University Press, 1982.

Fiorino, Daniel J. *Making Environmental Policy*. Berkeley: University of California Press, 1995.

Forbes, Gerald. *Flush Production: The Epic of Oil in the Gulf Southwest*. Norman: University of Oklahoma Press, 1942.

Fortuna, Richard C., and David J. Lennett. *Hazardous Waste Regulations, the New Era* New York: McGraw-Hill, 1987.

Fox, Stephen. *The Mirror Makers: A History of American Advertising and Its Creators*. New York: William Morrow and Co., 1984.

Franks, Kenny A. *The Oklahoma Petroleum Industry*. Norman: University of Oklahoma Press, 1980.

Franks, Kenny A. and Paul F. Lambert. *Early Louisiana and Arkansas Oil: A Photographic History, 1901–1946*. College Station, Tex.: Texas A&M University Press, 1982.

Fraser, Nicholas, Philip Jacobson, Mark Ottaway, and Lewis Chester. *Aristotle Onassis*. Philadelphia: J. B. Lippincott Co., 1977.

Freedman, Martin and Bikki Jaggi. *Air and Water Pollution Regulation: Accomplishments and Economic Consequences*. Westport, Conn.: Quorum Books, 1993.

Frey, John W. "The Interstate Oil Compact." In *Energy Resources and National Policy*, edited by National Resources Committee. Washington: Government Printing Office, 1939.

Frick, Thomas C., editor. *Petroleum Production Handbook*. New York: McGraw-Hill, 1962.

Garland, T. M., and Frank Parrish, Jr. *Sources of Water for Water Flooding*. Prepared for the North Texas Oil and Gas Association by the Wichita Falls Petroleum Field Office. Wichita Falls, Tex.: U.S. Bureau of Mines, 1958.

Garner, F. H. "Distillation." In *Petroleum: Twenty-five Years Retrospect, 1910–1935*, edited by Institute of Petroleum Technologists. London: Aldine House, 1935.

Gary, James H., and Glenn E. Handwerk. *Petroleum Refining: Technology and Economics*. New York: Marcel Dekker, 1994.

Gibb, D. E. W. *Lloyd's of London: A Study in Individualism*. London : Macmillan and Co., 1957.

Gibb, George S., and Evelyn H. Knowlton. *The Resurgent Years, 1911–1927: History of the Standard Oil Company (New Jersey)*. New York: Harper & Brothers, 1955.

Giddens, Paul H. *The Birth of the Oil Industry*. New York: Macmillan, 1938.

Giebelhaus, August W. *Business and Government in Industry: A Case Study of Sun Oil, 1876–1945*. Greenwich, Conn.: JAI Press, 1980.

Gorman, Hugh S. "Efficiency, Environmental Quality, and Oil Field Brines: The Success and Failure of Pollution Control by Self-Regulation." *Business History Review 73* (winter 1999): 601–40.

———. "Manufacturing Brownfields: The Case of Neville Township, Pennsylvania, 1899–1989." *Technology and Culture* 38 (July 1997): 539–74.

Graebner, William. "Hegemony Through Science: Information Engineering and Lead Toxicology, 1925–1965." In *Dying for Work: Workers' Safety and Health in Twentieth-Century America*, edited by David Rosner and Gerald Markowitz. Bloomington: Indiana University Press, 1987.

Gray, W. O. "Accidental Spills from Tankers and Other Vessels." In *The Prevention of Oil Pollution*, edited by J. Wardley-Smith. New York: Wiley, 1979.

Grinder, R. Dale. "The Battle for Clean Air: The Smoke Problem in Post-Civil War America." In *Pollution and Reform in American Cities, 1870–1930*, edited by Martin V. Melosi. Austin: University of Texas Press, 1980.

Green, Donald E. *Land of the Underground Rain: Irrigation in the Texas High Plains, 1910–1970*. Austin: University of Texas Press, 1973.

Haddad, Brent M. "Marketable Permits and Pollution Charges: Two Case Studies." In *The Environment Goes to Market: The Implementation of Economic Incentives for Pollution Control*. Washington, D.C.: National Academy of Public Administration, 1994.

Hager, Dorsey. *Fundamentals of the Petroleum Industry*. New York: McGraw-Hill, 1939.

Hamilton, Alice. *Exploring the Dangerous Trades: The Autobiography of Alice Hamilton, M.D.* Boston: Little, Brown, and Co., 1943.

Hardwicke, Robert E. *The Oilman's Barrel*. Norman: University of Oklahoma Press, 1958.

Harris, Kenneth. *The Wildcatter: A Portrait of Robert O. Anderson*. New York: Weidenfeld and Nicolson, 1987.

Hart, W. B. *Industrial Waste Disposal For Petroleum Refineries and Allied Plants*. Cleveland, Ohio: Petroleum Processing, 1947.

Harter, Harry. *East Texas Oil Parade*. San Antonio, Tex.: Naylor Co., 1934.

Hawley, Ellis W. "Herbert Hoover, the Commerce Secretariat, and the Vision of an 'Associative State,' 1921–1928." *Journal of American History* 62 (1974): 116–40.

Hays, Samuel P. *Beauty, Health, and Permanence: Environmental Politics in the United States, 1955–1985*. New York: Cambridge University Press, 1987.

———. *Conservation and the Gospel of Efficiency: The Progressive Conservation Movement, 1890–1920*. Cambridge, Mass.: Harvard University Press, 1959.

Henry, J. D. *Thirty Five Years of Oil Transport: Evolution of the Tank Steamer*. London: Bradbury, Agnew, and Co., 1907.

Henry, J. T. *Early and Later History of Petroleum*. Philadelphia: Jas. B. Rodgers, 1873.

Herold, Stanley C. *Oil Well Drainage*. Stanford, Calif.: Stanford University Press, 1941.

Hidy, Ralph W., and Muriel E. Hidy. *Pioneering in Big Business, 1882–1911: History of the Standard Oil Company (New Jersey)*. New York: Harper and Brothers, 1955.

Hill, E. F., W. A. Cannon, and C. E. Welling. "Single Cylinder Engine Testing of Hydrocarbon Oxidation Catalysts." Paper no. 174 presented at the SAE meeting, Seattle, August 1957.

Hirsch, Richard F. *Technology and Transformation in the American Electric Utility Industry*. New York: Cambridge University Press, 1989.

Hoover, Herbert. *The Memoirs of Herbert Hoover.* 3 vols. New York: Macmillan, 1951.

Hortig, F. J. "Jurisdictional, Administrative, and Technical Problems Related to the Establishment of California Coastal and Offshore Boundaries." In *The Law of the Sea: Offshore Boundaries and Zones,* edited by Lewis M. Alexander. Columbus: Ohio State University Press, 1967.

Hoult, David P., ed. *Oil on the Sea: Proceeding of a Symposium Sponsored by MIT and Woods Hole Oceanographic Institution.* New York: Plenum Press, 1969.

Hounshell, David A. *From the American System to Mass Production, 1800–1932: The Development of Manufacturing Technology in the United States.* Baltimore, Md.: Johns Hopkins University Press, 1984.

Hounshell, David A., and John Kenley Smith, Jr. *Science and Corporate Strategy: Dupont R&D, 1902–1980.* New York: Cambridge University Press, 1988.

Hughes, Thomas P. *Networks of Power: Electrification in Western Society, 1880–1930.* Baltimore, Md.: Johns Hopkins University Press, 1983.

Hunt, T. S. *Notes on the History of Petroleum or Rock Oil.* Washington, D.C.: Smithsonian Institution, 1862.

Hurley, Andrew. "Creating Ecological Wastelands: Oil Pollution in New York City, 1870–1900." *Journal of Urban History* 20 (May 1994): 340–64.

Hurley, William D. *Environmental Legislation.* Springfield, Ill.: Charles C. Thomas, 1971.

Ingersoll-Rand Company. *Oil Well Blowing Handbook.* New York: Ingersoll-Rand, 1927.

Inkley, F. A. *Oil Loss Control In the Petroleum Industry.* New York: John Wiley and Sons, 1985.

Institute of Petroleum Technologists. *Petroleum: Twenty-five Years Retrospect, 1910–1935.* London: Aldine House, 1935.

Inter-Governmental Maritime Consultative Organization. *Tanker Casualties Report.* London: IMCO, 1978.

Interstate Oil Compact Commission. *Production and Disposal of Oil Field Brines in the United States and Canada.* Oklahoma City, Okla.: Interstate Oil Compact Commission, 1960.

Ise, John. *The United States Oil Policy.* New Haven: Yale University Press, 1926.

James, H. G. *Refining Industry of the United State.* Oil City, Pa: Derrick Publishing, 1916.

James, Marquis. *The Texaco Story: The First Fifty Years, 1902–1952.* New York: Texas Company, 1953.

Jeffery, W. H. *Deep Well Drilling.* Houston: Gulf Publishing Company, 1931.

Johnson, Arthur M. *Petroleum Pipelines and Public Policy, 1906–1959.* Cambridge, Mass.: Harvard University Press, 1967.

———. *The Challenge of Change: The Sun Oil Company, 1945–1977.* Columbus: Ohio State University Press, 1983.

———. *The Development of American Petroleum Pipelines, 1862–1906.* Ithaca: Cornell University Press, 1956.

Johnson, Branden B. and Vincent Covello, editors. *The Social and Cultural Construction of Risk: Essays on Risk and Perception.* Boston: D. Reidel, 1987.

Johnson, Leland R. *The Headwaters District: A History of the Pittsburgh District, U.S. Army Corps of Engineers.* Washington, D.C.: U.S. Army Corps of Engineers, 1978.

Johnson, Roswell H. and Huntley, L. G. *Principles of Oil and Gas Production.* New York: John Wiley & Sons, 1916.

Jones, Charles O. *The Policies and Politics of Pollution Control.* Pittsburgh: University of Pittsburgh Press, 1975.

Jones, Harold R. *Pollution Control in the Petroleum Industry.* Park Ridge, N. J.: Noyes Data Corporation, 1973.

Jones, Ogden S. *Fresh Water Protection from Pollution Arising in the Oil Fields* Lawrence: University of Kansas Publications, 1950.

Jones, Park J. *Petroleum Production: Oil Production By Water.* New York: Reinhold Publishing, 1947.

Kahan, Archie M. *Acid Rain: Reign of Controversy.* Golden, Colorado: Fulcrum, 1986.

Kalichevsky, Vladimir A., and Bert A. Stagner. *Chemical Refining of Petroleum: The Action of Vari-*

ous *Refining Agents and Chemicals on Petroleum and Its Products*. New York: Chemical Catalog Co., 1933.

Katzman, Martin T. *Chemical Catastrophes, Regulating Environmental Risk Through Pollution Liability Insurance*. Homewood, Ill.: Richard D. Irwin, 1985.

Kennedy, Harold W., and Martin E. Weekes. "Control of Automobile Emission—California Experience and the Federal Legislation." In *Air Pollution Control*, edited by Clark C. Havighurst. Dobbs Ferry, N.Y.: Oceana Publications, 1969.

Kerlin, Gregg, and Daniel Rabovsky. *Cracking Down: Oil Refining and Pollution Control*. New York: Council on Economic Priorities, 1975.

King, Arthur T. *Oil Refinery Terms in Oklahoma*. American Dialect Society Publication no. 9. Greensboro, N.C.: Woman's College of the University of North Carolina, 1948.

King, G. A. B. *Tanker Practice: The Construction, Operation, and Maintenance of Tankers*. London: Maritime Press, 1969.

Kingston, Benson M. *Acidizing Handbook*. Houston, Tex.: Gulf Publishing, 1936.

Kneese, Allen V., Robert U. Ayres, and Ralph C. D'Arge. "Economics and the Environment: A Material Balance Approach." In *The Economics of Pollution*, edited by Harold Wolzin. Morristown, N.J.: General Learning Press, 1974.

Kostecki, Paul T. and Edward J. Calabrese, editors. *Hydrocarbon Contaminated Soil and Groundwater: Analysis, Fate, Environmental and Public Health Affects*. Chelsea, Michigan: Lewis Publishers, 1991.

Krier, James E., and Edmund Ursin. *Pollution and Policy: A Case Essay on California and Federal Experience with Motor Vehicle Air Pollution, 1940–1975*. Berkeley: University of California Press, 1977.

Lagerstrom, Larry R. "Constructing Uniformity : The Standardization of International Electromagnetic Measures, 1860–1912." Ph.D. diss., University of California, Berkeley, 1992.

Lambert, Paul F., and Kenny A. Franks, editors. *Voices from the Oil Field*. Norman: University of Oklahoma Press, 1959.

Larson, Erick. *Pipeline Corrosion and Coatings*. New York: American Gas Journal, 1938.

Larson, Henrietta M., Evelyn H. Knowlton, and Charles S. Popple. *New Horizons, 1927–1950: History of the Standard Oil Company, New Jersey*. New York: Harper and Row, 1971.

Larson, Henrietta M., and Kenneth W. Porter. *History of Humble Oil and Refining: A Study in Industrial Growth*. New York: Harper and Brothers, 1959.

Latour, Bruno. *Science in Action*. Cambridge, Ma.: Harvard University Press, 1987.

Layton, Edwin. *The Revolt of the Engineers: Social Responsibility and the American Engineering Profession*. Cleveland, Ohio: Case Western University Press, 1971.

Lents, J. M., and P. Leyden. "RECLAIM: Los Angeles' New Market-Based Smog Cleanup Program." *Journal of the Air and Waste Management Association* 46 (1996), 196–205.

Leslie, Eugene H. *Motor Fuels: Their Production and Technology*. New York: Chemical Catalog Company, 1923.

Leslie, Stuart W. *Boss Kettering: Wizard of General Motors*. New York: Columbia University Press, 1983.

Leven, David D. *Petroleum Encyclopedia: Done in Oil*. New York: Ranger Press, 1942.

Levy, Walter J. *Oil Strategy and Politics, 1941–1981*. Boulder: Westview Press, 1982.

Lewelling, Henry, and Monte Kaplan. "What to do about Salt Water." In *Frontiers in Petroleum Engineering*. Dallas: Petroleum Engineering Publishing Co., 1960.

Lilley, Ernest R. *The Oil Industry: Production, Transportation Resources, Refining, and Marketing*. New York: D. Van Nostrand, 1925.

Liroff, Richard A. *Air Pollution Offsets: Trading, Selling, and Banking*. Washington, D.C.: Conservation Foundation, 1986.

———. *Reforming Air Pollution Regulation: The Toil and Trouble of EPA's Bubble*. Washington, D.C. Conservation Foundation, 1986.

Livingston, Dennis. "Oil on the Seas: Two Cheers for a New Treaty." *Environment* 16 (September 1974): 38–43.

Logan, Leonard M., Jr. *Stabilization of the Petroleum Industry.* Norman: University of Oklahoma Press, 1930.

Logsdon, Jeanne Marie. "Organizational Responses to Environmental Issues: Oil Refining Companies and Air Pollution." Ph.D. diss., University of California, Berkeley, 1983.

Loos, John L. *Oil on Stream! A History of the Interstate Oil Pipe Line Company, 1909–1959.* Baton Rouge: Louisiana State University Press, 1959.

MacFarlane, John M. *Fishes: The Source of Petroleum.* New York: Macmillan, 1923.

Maler, Karl-Goran. *Environmental Economics: A Theoretical Inquiry.* Baltimore: Resources for the Future, 1974.

Manners, Ian R. *North Sea Oil and Environmental Planning: The United Kingdom Experience.* Austin: University of Texas Press, 1982.

Manning, Francis S., and Eric H. Snider. *Environmental Assessment Data Base For Petroleum Refining Wastewaters and Residuals.* Ada, Ok.: Robert S. Kerr Environmental Research Laboratory, 1983.

Marland, E. W. *Shall There Be Communism in Oil?* Ponca City, Okla.: published by the author, 1924.

Marshall, Hubert and Betty Zisk. *The Federal-State Struggle for Offshore Oil.* Indianapolis: Bobbs-Merrill Co. Inc., 1966.

Martin, Robert, and Lloyd Symington. "A Guide to the Air Quality Act of 1967." In *Air Pollution Control,* edited by Clark C. Havighurst. Dobbs Ferry, N.Y.: Oceana Publications, 1969.

Matthews, John Joseph. *Life and Death of an Oilman: The Career of E. W. Marland.* Norman: University of Oklahoma Press, 1951.

Maybourn, R. "Operational Pollution from Tankers and Other Vessels." In *The Prevention of Oil Pollution,* edited by J. Wardley-Smith. New York: Wiley, 1979.

McBryde, W. A. E. "Petroleum Deodorized: Early Canadian History of the 'Doctor Sweetening' Process." *Annals of Science* (Great Britain) 48 (1991): 103–11.

McCoy, Ruth H. "Halliburton Oil Well Cementing Company." M.A. thesis, Oklahoma Agricultural and Mechanical College, 1958.

McCraw, Thomas K. "Regulation in America: A Review Article." *Business History Review* 49 (summer 1975): 159–83.

McDaniel, Ruel R. *Some Ran Hot.* Dallas, Tex.: Regional Press, 1939.

McEvoy, Arthur F. *The Fisherman's Problem: Ecology and Law in California Fisheries, 1850–1980.* New York: Cambridge University Press, 1990.

McKee, Jack E. *Report on Oil Substances and their Effects on the Beneficial Uses of Water.* Sacramento: State Water Pollution Control Board, 1956.

McLaurin, John J. *Sketches in Crude Oil.* Harrisburg, Pa.: published by the author, 1896.

McMillion, L. G. "Hydrological Aspects of Disposal of Oil-Field Brines in Texas." *Ground Water* 3 (October 1965): 36–42.

McWane, R. C. *Pipe and the Public Welfare.* New York: Stirling Press, 1917.

Meadows, Donella, Dennis L. Meadows, Jorgen Randers, and William W. Behrens III. *The Limits to Growth: A Report for the Club of Rome's Project on the Predicament of Mankind.* New York: Universe Books, 1972.

M'Gonigle, R. M., and Mark W. Zacher. *Pollution, Politics, and International Law: Tankers at Sea.* Berkeley: University of California Press, 1979.

Milkovich, Barbara Ann. "A Study of the Impact of the Oil Industry on the Development of Huntington Beach, California, Prior to 1930." M.A. thesis, California State University, Long Beach, 1988.

Miller, David W., editor. *Waste Disposal Effects on Groundwater.* Berkeley, Calif.: Premier Press, 1980.

Miller, Keith L. "Plucking the Apple Without Rooting Up the Tree: Environmental Concern in the Production of Prairie State Petroleum." *Illinois Historical Journal* 84 (autumn 1991): 161–72.

Millikan, Charles Van Ormer. *Oil-Well Cementing Practices in the United States.* New York: API, 1959.

Mitchell, Ronald B. *Intentional Oil Pollution at Sea: Environmental Policy and Treaty Compliance.* Cambridge, Mass.: MIT Press, 1994.

Mohlman, F. W. "Twenty-Five Years of Activated Sludge." In *Modern Sewage Disposal,* edited by Langdon Pearse. Lancaster, Pa.: Lancaster Press, 1938.

Moore, Harold. *Liquid Fuels for Internal Combustion Engines: A Practical Guide for Engineers.* New York: D. Van Nostrand, 1917.

Morrel, Robert W. *Oil Tankers.* New York: Simmons-Bordman, 1931.

Moss, James E. *Character and Control of Sea Pollution by Oil.* Washington, D.C.: API, 1963.

Mostert, Noël. *Supership.* New York: Warner, 1974.

Murphy, Blakely M. *Conservation of Oil and Gas: A Legal History.* Chicago: Section of Mineral Law, American Bar Association, 1949.

Myres, Samuel D. *The Permian Basin: Era of Advancement.* El Paso: Permian Press, 1977.

————. *The Permian Basin: Era of Discovery.* El Paso: Permian Press, 1974.

Nash, Gerald D. *United States Oil Policy, 1890–1964: Business and Government in the Twentieth Century.* Pittsburgh, Pa.: University of Pittsburgh Press, 1968.

National Academy of Sciences. *Oil in the Sea.* Washington, D.C.: National Academy of Sciences, 1985.

National Academy of Sciences. *Petroleum in the Marine Environment.* Washington, D.C.: Government Printing Office, 1975.

National Research Council. Committee on Effectiveness of Oil Spill Dispersants. *Using Oil Dispersants on the Sea.* Washington, D.C.: National Academy Press, 1989.

National Research Council. *Double Hull Tanker Legislation: An Assessment of the Oil Pollution Act of 1990.* Washington, D.C.: National Academy Press, 1998.

National Research Council. *Tanker Spills: Prevention By Design.* Washington, D.C., National Academy Press, 1991.

National Research Council. Transportation Research Board. *Pipelines and Public Safety.* Special report 219. National Research Council, 1988.

Nelson, W. L. *Petroleum Refinery Engineering.* New York: McGraw-Hill, 1941.

Neustadt, Richard E., and Ernest R. May. *Thinking in Time: The Uses of History for Decision Makers.* New York: Macmillan, 1986.

Nevins, Allan. *Study in Power: John D. Rockefeller, Industrialist and Philanthropist.* 2 vols. New York: Charles Scribner's Sons, 1953.

Nicholson, Patrick J. *Mr. Jim: The Biography of James Smither Abercrombie.* Houston, Tex.: Gulf Publishing, 1983.

Noble, David F. "Social Choice in Machine Design: The Case of Automatically Controlled Machine Tools." In *Case Studies on the Labor Process,* edited by Andrew Zimbalist. New York: Monthly Review Press, 1979.

Noggle, Burl. *Teapot Dome: Oil and Politics in the 1920s.* New York: W. W. Norton and Co., 1962.

Nordhauser, Norman E. *The Quest for Stability: Domestic Oil Regulation, 1917–1935.* New York: Garland, 1979.

Nussbaum, Barry D. "Phasing Down Lead in Gasoline in the U.S.: Mandates, Incentives, Trading and Banking." In *Climate Change: Designing a Tradeable Permit System,* edited by Organization for Economic Co-Operation and Development. Paris: Organization for Economic Co-Operation and Development, 1992.

O'Brien, Hubbert L. *Petroleum Tankage and Transmission.* East Chicago, Ind.: Graver Tank and Mfg. Co., 1951.

O'Fallon, John E. "Deficiencies in the Air Quality Act of 1967." In *Air Pollution Control,* edited by Clark C. Havighurst. Dobbs Ferry, N.Y.: Oceana Publications, 1969.

Ohio River Valley Water Sanitation Commission. *Preventing Stream Pollution from Oil Pipeline Breaks.* Cincinnati, Ohio: Ohio River Valley Water Sanitation Commission, 1950.

Oil Compact Commission, Research Committee. *Underground Storage of Liquid Petroleum Hydrocarbons in the United States.* Oklahoma City, Okla.: Interstate Oil Compact Commission, 1956.

Oil Well Supply Company. *Oilwell.* Pittsburgh: Oil Well Supply Company, 1916.

Olien, Roger M., and Diana Davids Olien. *Oil and Ideology: The Cultural Creation of the American Petroleum Industry.* Chapel Hill: University of North Carolina Press, 2000.

———. *Wildcatters: Texas Independent Oilmen.* Austin, Tex.: Texas Monthly Press, 1984.

Olson, Donel R. "The Control of Motor Vehicle Emissions." In *Engineering Control of Air Pollution,* edited by Arthur Stern. Vol. 4, *Air Pollution.* New York: Academic Press, 1977.

Opie, John. *Ogallala: Water for a Dry Land.* Lincoln: University of Nebraska Press, 1993.

Organisation for Economic Co-operation and Development. *Pipeline and Tankers: A Report On the Use of Pipelines and the Transport of Oil By Tankers.* Paris: OCED, 1961.

Osgood, Wentworth H. *Increasing the Recovery of Petroleum.* 2 vols. New York: McGraw-Hill, 1930.

Ostroff, A. G. *Introduction to Oilfield Water Technology.* Englewood Cliffs, N.J.: Prentice-Hall, 1965.

Page, R. C., and A. Ward Gardner. *Petroleum Tankship Safety.* London: Maritime Press, 1971.

Page, Talbot. *Conservation and Economic Efficiency: An Approach to Materials Policy.* Baltimore: Johns Hopkins Press, 1977.

Panyity, Louis S. *Prospecting for Oil and Gas.* New York: John Wiley and Sons, 1920.

Parker, Marshall. *Pipe Line Corrosion and Cathodic Protection.* Houston, Tex.: Gulf Publishing, 1954.

Patterson, D. R. "Petroleum Automation." In *Frontiers in Petroleum Engineering.* Dallas: Petroleum Engineering Publishing Co., 1960.

Patton, Charles C. *Oilfield Water Systems.* Norman, Oklahoma: Campbell Petroleum Series, 1977.

Payne, Darwin. *Initiative in Energy: The Story of Dresser Industries, 1880–1978.* New York: Simon and Schuster, 1979.

Peckham, S. F. *Report on the Production, Technology, and Uses of Petroleum and Its Byproducts.* Washington, D.C.: U.S. Dept. of the Interior, Census Bureau, 1885.

Perrow, Charles. *Normal Accidents: Living with High Risk Technologies.* New York: Basic Books, 1984.

Peterson, Alfred. *Oil and Gas: Be Your Own Geologist.* Kansas City, Mo.: Franklin Hudson, 1921.

Petroleum Extension Service. *Prevention and Control of Blowouts.* Austin: University of Texas, 1958.

Petrow, Richard. *In the Wake of the Torrey Canyon.* New York: David McKay Co., 1968.

Pettengill, Samuel B. *Hot Oil: The Problem of Petroleum.* New York: Economic Forum Co., 1936.

Philips, Elizabeth. "Managing Underground Storage Tanks." In *Hazardous Materials Management,* edited by Doy Cox. Rockville, Md.: Institute of Hazardous Materials Management, 1995.

Pigou, Arthur C. *The Economics of Welfare.* New York: St. Martin's Press, 1932.

Polind, Norman. "Effects of Torrey Canyon Pollution on Marine Life." In *Oil on the Sea: Proceeding of a Symposium Sponsored by MIT and Woods Hole Oceanographic Institution,* edited by David P. Hoult. New York: Plenum Press, 1969.

Portney, Paul R., editor. *Public Policies for Environmental Protection.* Washington, D.C.: Resources for the Future, 1990.

Potter, Jeffrey. *Disaster by Oil: Oil Spills, Why they Happen, What They Do, and How We Can End Them.* New York: Macmillan, 1973.

Powell, Fred W. *The Bureau of Mines: Its History, Activities, and Organizations.* Service Monographs of the United States Government, no. 3. New York: D. Appleton and Co., 1922.

Pratt, Joseph A. "Creating Coordination in the Modern Petroleum Industry: The American Petroleum Institute and the Emergence of Secondary Organizations in Oil." *Research in Economic History* 8 (1983): 179–215.

———. "Growth or a Clean Environment? Responses to Petroleum-related Pollution in the Gulf Coast Refining Region." *Business History Review* 52 (spring 1980): 1–29.

———. "Letting the Grandchildren Do It: Environmental Planning During the Ascent of Oil as a Major Energy Source." *Public Historian* 2 (1980): 28–61.

———. *The Growth of a Refining Region.* Greenwich, Conn.: JAI Press, 1980.

Presley, James. *Never In Doubt: A History of the Delta Drilling Company.* Houston: Gulf Publishing Co., 1981.

Prindle, David F. *Petroleum Politics and the Texas Railroad Commission.* Austin: University of Texas Press, 1981.

Pritchard, Sonia Zaide. *Oil Pollution Control.* London: Croom Helm, 1987.

Quam-Wickham, Nancy. "Cities Sacrificed on the Altar of Oil: Popular Opposition to Oil Development in 1920s Los Angeles." *Environmental History* 3 (April 1998): 189–209.

Quebedeaux, W. A. "Air and Stream Pollution Control in Harris County, Texas." *Public Health Reports* 69 (September 1954): 836–40.

Randle, Russell V. "The Oil Pollution Act of 1990: Its Provisions, Intents, and Effects." In *Oil Pollution Deskbook.* Washington, D.C.: Environmental Law Reporter, 1991.

Raymond, Jack. *Robert O. Anderson: Oil Man/Environmentalist.* Aspen: Aspen Institute for Humanistic Studies, 1988.

Riegel, Emil R. *Industrial Chemistry.* New York: Reinhold Publishing, 1942.

Ripley, Randall B. "Congress and Clean Air: The Issue of Enforcement, 1963." In *Congress and Urban Problems,* edited by F. N. Cleaveland. Washington, D.C.: Brookings Institute, 1963.

Rister, Carl Coke. *Oil! Titan of the Southwest.* Norman: University of Oklahoma Press, 1949.

Robert, Joseph C. *Ethyl: A History of the Corporation and the People Who Made It.* Charlottesville: University Press of Virginia, 1983.

Robertson, Jerome B. *Oil Slanguage: An Explanation of Terms and Slang of Oil Fields From Pennsylvania to California, Texas to Montana—and Around the World.* Evansville, Ind.: Petroleum Publishers, 1954.

Rogers, William H. "Where Environmental Law and Biology Meet: Of Panda's Thumbs, Statutory Sleepers, and Effective Law." *University of Colorado Law Review* 65 (1993): 25–75.

Rosen, Christine M. "Noisome, Noxious, and Offensive Vapors, Fumes, and Stenches in American Towns and Cities, 1840–1865." *Historical Geography* 25 (1997): 49–82.

Rosenberg, Nathan. *Perspectives on Technology.* New York: Cambridge University Press, 1976.

Rosner, David, and Gerald Markowitz. "'A Gift of God'?: The Public Health Controversy over Leaded Gasoline During the 1920s." In *Dying for Work: Workers' Safety and Health in Twentieth-Century America,* edited by David Rosner and Gerald Markowitz. Bloomington: Indiana University Press, 1987.

Ross, Steven J. *Workers on the Edge: Work, Leisure, and Politics in Industrializing Cincinnati, 1788–1890.* New York: Columbia University Press, 1985.

Rostow, Eugene V. *A National Policy for the Oil Industry.* New Haven: Yale University Press, 1948.

Russell, Clifford S. *Residual Management in Industry: A Case Study of Petroleum Refining.* Baltimore: Resources for the Future, Johns Hopkins University Press, 1973.

Sabin, Paul. "Searching for Middle Ground: Native Communities and Oil Extraction in the Northern and Central Ecuadorian Amazon, 1967–1993. *Environmental History* 3 (April 1998): 144–68.

Sampson, Anthony. *The Seven Sisters: The Great Oil Companies and the World They Made.* New York: Viking Press, 1975.

Santiago, Myrna. "Rejecting Progress in Paradise: Huastecs, the Environment, and the Oil Industry in Veracruz, Mexico, 1900–1935." *Environmental History* 3 (April 1998): 169–88.

Sawyer, L. A., and W. H. Mitchell. *Tankers.* New York: Doubleday and Co., 1967.

Schelling, Thomas C. "The Cost of Combating Global Warming: Facing the Tradeoffs." *Foreign Affairs* 76 (November/December 1997): 8–13.

Schumacher, E. F. *Small Is Beautiful: A Study of Economics As If People Mattered.* London: Blond and Briggs, 1973.

Scott, Otto J. *The Exception: The Story of Ashland Oil and Refining Company.* New York: McGraw-Hill, 1968.

Scoville, John, and Noel Sargent, editors. *Fact and Fancy in the T.N.E.C. Monographs.* New York: National Association of Manufacturers, 1942.

Seely, Bruce E. *Building the American Highway System: Engineers as Policy Makers.* Philadelphia: Temple University Press, 1987.

Sellers, Christopher. "Factory as Environment: Industrial Hygiene, Professional Rivalries and the Modern Sciences of Pollution." *Environmental History Review* 18 (spring 1994): 55–83.

———. "The Public Health Service's Office of Industrial Hygiene and the Transformation of Industrial Medicine." *Bulletin of the History of Medicine* 65 (1992): 42–73.

Sharrer, G. Terry. "Naval Stores, 1781–1881." In *Material Culture of the Wooden Age,* edited by Brooke Hindle. Tarrytown, N. Y.: Sleepy Hollow Press, 1981.

Sheaffer, John R., and Leonard A. Stevens. *Future Water.* New York: William Morrow and Co., 1983.

Sheridan, Richard B. "John Ise, 1885–1969: Economist, Conservationist, Prophet of the Energy Crisis." *Kansas Hist.* 5 (1982): 83–106.

Sherrill, Robert. *The Oil Follies of 1970–1980: How the Petroleum Industry Stole the Show (and Much More Besides).* Garden City, N.Y.: Anchor Press, 1983.

Sinclair, Bruce. *Centennial History of the American Society of Mechanical Engineers, 1880–1980.* Toronto: University of Toronto Press, 1980.

Sittig, Marshall. *Petroleum Refining Industry: Energy Saving and Environmental Control.* Park Ridge, N. J.: Noyes Data Corporation, 1978.

Skillern, Frank F. *Environmental Protection: The Legal Framework.* New York: McGraw-Hill, 1981.

Smith, A. D. "Refining." In *A Handbook of Petroleum Industry,* edited by David T. Day. New York: John Wiley and Sons, 1922.

Smith, J. E., editor. *"Torrey Canyon" Pollution and Marine Life: A Report by the Plymouth Laboratory of the Marine Biological Association of the United Kingdom.* Cambridge: Cambridge University Press: 1968.

Smith, Merritt Roe, and Leo Marx, editors. *Does Technology Drive History: The Dilemma of Technological Determinism.* Cambridge, Mass.: MIT Press, 1994.

Smith, V. Kerry and William H. Desvousges. *Measuring Water Quality Benefits.* Boston: Kluwer, 1986.

Snyder, Lynne P. "The Death-Dealing Smog over Donora, Pennsylvania: Demanding a Legal Response: Industrial Air Pollution, Public Health Policy, and the Politics of Expertise, 1948–1949." *Environmental History Review* 18 (spring 1994): 117–41.

Solberg, Carl. *Oil Power.* Mason: Charter, 1976.

Solomon, Barry D., and Hugh S. Gorman. "State-Level Air Emissions Trading: The Michigan and Illinois Models." *Journal of the Air and Waste Management Association* 48 (1998): 1156–65.

Southwestern Legal Foundation. *Economics of the Petroleum Industry: New Ideas, New Methods, New Developments.* Houston, Tex.: Gulf Publishing Company, 1963–89.

Speight, James G. *The Chemistry and Technology of Petroleum.* New York: Marcel Dekker, 1991.

Speller, Frank N. *Corrosion—Causes and Prevention.* New York: McGraw-Hill, 1926 and 1951.

Spence, Hartzell. *Portrait in Oil: How the Ohio Oil Company Became Marathon.* New York: McGraw-Hill, 1962.

Spitz, Peter. H. *Petrochemicals: The Rise of an Industry.* New York: Wiley, 1988.

Stanford Research Institute. *The Smog Problem in Los Angeles County.* Los Angeles: Western Oil and Gas Association, 1954.

Steinhart, Carol, and John Steinhart. *Blowout: A Case Study of the Santa Barbara Oil Spill.* Belmont, Calif.: Duxbury, 1972.

Stillwagon, C. K. *Rope Chokers: A Collection of Human Interest Stories, Anecdotes, Historical Fragments, and Pictures of the Oil Fields.* Houston, Texas: The Rein Company, 1945.

Stine, Jeffrey K., and Joel A. Tarr. "At the Intersection of Histories: Technology and the Environment." *Technology and Culture* 39 (October 1998): 601–40.

Stocking, George Ward. *The Oil Industry and the Competitive System: A Study in Waste.* Westport, Conn.: Hyperion Press, 1925.

Stoff, Michael B. *Oil, War, and American Security: The Search for a National Policy on Foreign Oil.* New Haven: Yale University Press, 1980.

Stradling, David. *Smokestacks and Progressives: Environmentalists, Engineers, and Air Quality in America, 1881–1951.* Baltimore: Johns Hopkins University Press, 1999.

Suman, John R. *Petroleum Production Methods.* Houston, Texas: Gulf Publishing Company, 1922.

Super Ocean Carrier Conference. *The Million Ton Tanker: Proceedings of the Super Ocean Carrier Conference.* San Pedro, Ca.: Super Ocean Carrier Conference, 1974.

Sweet, George Elliot. *The History of Geophysical Prospecting.* Los Angeles: Science Press, 1978.

Talbot, Frederick A. *The Oil Conquest of the World.* Philadelphia: J. B. Lippincott, 1914.

Tarbell, Ida M. *The History of the Standard Oil Company.* New York: McClure, Phillips and Co., 1904.

Tarr, Joel A. "Industrial Wastes and Public Health: Some Historical Notes." *American Journal of Public Health* 75 (September 1985): 1059–67.

———. *The Search for the Ultimate Sink: Urban Pollution in Historical Perspective.* Akron, Ohio: The University of Akron Press, 1996.

Tarr, Joel A. and Kenneth E. Koons. "Railroad Smoke Control." In *Energy Transport: Historical Perspectives on Policy Issues,* edited by George H. Daniels and Mark H. Rose. Beverly Hills, Calif.: Sage Publications, 1982.

Taylor, Frank J. and Welty, Earl M. *Black Bonanza: How an Oil Hunt Grew into the Union Oil Company.* New York: Whittlesey House, 1950.

Teleky, Ludwig. *History of Factory and Mine Hygiene.* New York: Columbia University Press, 1948.

The Texas Company. *Texaco.* Houston, Tex.: Texas Company, 1931.

Thompson, Craig. *Gulf: A Human Story of Gulf's First Half Century.* Pittsburgh, Pa.: Gulf Oil Company, 1951.

Thornhill, Jerry T. "Pollution Control Activities of the Water Pollution Control Board and Texas Water Commission." In *West Texas Oil Lifting Short Course.* Lubbock, Tex.: Texas Tech University, 1962.

Tinkle, Lon. *Mr. De: A Biography of Everette Lee DeGolyer.* Boston: Little, Brown, and Co., 1970.

Todd, David Keith, and Daniel E. Orren McNulty. *Polluted Groundwater: A Review of Significant Literature.* Port Washington, N.Y.: Water Information Center, Inc., 1976.

Tomany, James P. *Air Pollution: The Emissions, the Regulations, and the Controls.* New York: American Elsevier, 1975.

Tower, Walter D. *The Story of Oil.* New York: Appleton, 1920.

Tucker, Elton B. "Removal of Sulfur Compounds From Petroleum Naphthas By Catalytic Decomposition with Bauxite." Ph.D. diss., Columbia University, 1928.

Turner, Alvin O. "The Regulation of the Oklahoma Oil Industry." Ph.D. diss., Oklahoma State University, 1977.

U.S. Steel. *The Making, Shaping, and Treating of Steel.* 7th ed. Pittsburgh, Pa.: U.S. Steel, 1957.

United Nations, Industry and Environment Office. *Environmental Management Practices in Oil Refineries and Terminals: An Overview.* Paris: United Nations Environment Program, 1987.

University of Oklahoma. *Proceedings of the Petroleum Fluid Metering Conference.* Norman: University of Oklahoma Press, 1938.

Uren, Lester C. *Petroleum Production Engineering.* New York: McGraw-Hill, 1924 and 1946.

Usselman, Steven. "Running the Machine: Technological Change in the Railroad Industry, 1865–1910." Ph.D. diss., University of Delaware, 1986.

Ver Weibe, Walter A. *Oil Fields in the United States.* New York: McGraw-Hill, 1937.

Vietor, Richard H. K. "The Evolution of Public Environmental Policy: The Case of 'No Significant Deterioration'." *Environmental Review* 3 (Winter 1979): 2–18.

———. *Energy Policy in America Since 1945: A Study of Business-Government Relations.* Cambridge: Cambridge University Press, 1984.

———. *Environmental Politics and the Coal Coalition.* College Station, Tex.: Texas A&M University Press, 1980.

Vincenti, Walter. *What Engineers Know and How They Know It: Analytical Studies from Aeronautical History.* Baltimore: John Hopkins University Press, 1990.

Vlachos, William, and C. A. Vlachos. *The Fire and Explosion Hazards of Commercial Oils.* Philadelphia, Pa.: Vlachos and Co., 1921.

Vogel, David. *National Styles of Regulation: Environmental Policy in Great Britain and the United States.* Ithaca: Cornell University Press, 1986.

Waddell, Paul R., and Robert F. Niven. *Sign of the 76: The Fabulous Life and Times of the Union Oil Company of California.* Los Angeles: Union Oil Co., 1976.

Walker, George R., and Leo F. Edison. *National Forum on Growth with Environmental Quality: Proceedings of a National Forum.* Tulsa, Oklahoma: Mid-continent Environmental Center Association, Inc., 1974.

Wall, Bennett H. *Growth in a Changing Environment: A History of Standard Oil Company (New Jersey) 1950–1972 and Exxon Corporation 1972–1975.* New York: McGraw-Hill, 1988.

Wasson, Theron. "Creole Field, Gulf of Mexico, Coast of Louisiana." In *Structure of Typical American Oil Fields,* vol. 3., edited by J. V. Howell. Tulsa, Okla.: American Association of Petroleum Geologists, 1948.

Waters, W. G., II, T. D. Heaver, and T. Verrier. *Oil Pollution From Tanker Operations: Causes, Costs, and Controls.* Vancouver: Centre For Transportation Studies, 1980.

Weaver, E. E. "Effects of Tetraethyl Lead on Catalyst Life and Efficiency in Customer Type Vehicle Operation." Paper presented at the International Automotive Engineering Congress, Detroit, Mich., January 13–17, 1969.

Wenk, Edward, Jr. *Making Waves: Engineering, Politics, and the Social Management of Technology.* Urbana: University of Illinois Press, 1995.

Werey, R. B. *Instrumentation and Automatic Control in the Oil Refining Industry.* Philadelphia: Brown Instrument Company, 1941.

Wescott, James H. *Oil: Its Conservation and Waste.* New York: Beacon, 1930.

Westbrook, Robert B. "Tribune of the Technostructure: The Popular Economics of Stuart Chase." *American Quarterly* 32 (1980): 387–408.

Weston, Roy F., Robert G. Merman, and Joseph G. DeMann. "Waste Disposal Problems in the Petroleum Industry." In *Industrial Wastes: Their Disposal and Treatment,* edited by Willem Rudolfs. New York: Reinhold, 1953.

White, Captain Herbert John. *Oil Tank Steamers: Their Working and Pumping Arrangements Thoroughly Explained.* Glasgow: James Brown and Sons, 1920.

Whiteshot, Charles A. *The Oil-Well Driller: A History of the World's Greatest Enterprise, The Oil Industry.* Mannington, W. Va.: published by the author, 1905.

Whorton, James. *Before Silent Spring: Pesticides and Public Health in Pre-DDT America.* Princeton: Princeton University Press, 1974.

Wiebe, A. H., J. G. Burr, and H. E. Faubion. "The Problem of Stream Pollution in Texas with Special Reference to Salt Water from the Oil Fields." *Transactions of the American Fisheries Society* 64 (1934): 81–86.

Williams, Charles C., and Charles K. Bayne. *Ground-Water Conditions in Elm Creek Valley, Barber County, Kansas.* Bulletin 64. Topeka, Kans.: State Geological Survey of Kansas, 1946.

Williams, Howard R. and Charles J. Meyers. *Oil and Gas Terms: Annotated Manual of Legal, Engineering, and Tax Words and Phrases.* New York: Banks & Company, 1957.

Williams, James C. *Energy and the Making of Modern California.* Akron, Ohio: University of Akron Press, 1997.

Williams, T. Harry. *Huey Long.* New York: Knopf, 1970.

Williamson, Harold F., and Arnold R. Daum. *The American Petroleum Industry: The Age of Illumination, 1859–1899.* Evanston, Ill.: Northwestern University Press, 1959.

Williamson, Harold F., Ralph L. Andreano, Arnold R. Daum, and Gilbert C. Klose. *The American Petroleum Industry: The Age of Energy, 1899–1959.* Evanston, Ill.: Northwestern University Press, 1963.

Wilson, Charles Morrow. *Oil Across the World: The American Saga of Pipelines.* New York: Longmans, Green, and Co., 1946.

Winkler, John K. *John D.: A Portrait in Oils.* New York: Blue Ribbon Books, 1929.

Winslow, Ron. *Hard Aground: The Story of the Argo Merchant Oil Spill.* New York: Norton, 1978.

Wolpert, George S., Jr. *U.S. Oil Pipelines: An Examination of How Oil Pipe Lines Operate and the Current Public Policy Issues Concerning Their Ownership.* Washington, D.C.: API, 1979.

Working Group on the Study of Critical Environmental Problems. *Man's Impact on the Global*

Environment: Report of Critical Environmental Problems. Cambridge, Mass.: MIT, 1970.

Worster, Donald. *Rivers of Empire: Water, Aridity, and the Growth of the American West.* New York: Pantheon Books, 1986.

Wright, William. *The Oil Regions of Pennsylvania.* New York: Harper and Brother, 1865.

Yancey, J. C. *Why and Where Oil is Found.* New York: published by the author, 1919.

Yergin, Daniel. *The Prize: The Epic Quest for Oil, Money, and Power.* New York: Simon and Schuster, 1991.

Zachariasen, Fredrik. *Oil Pollution in the Sea: Problems for Future Work.* Arlington, Va.: Institute for Defense Analysis, 1968.

Zimmerman, Erich. *Conservation in the Production of Petroleum.* New Haven: Yale University Press, 1957.

Index

References to tables are printed in boldface type. References to illustrations are printed in italic type.

131, 169; —, refinery emissions, 239; —, vapor losses, 79–85; work in El Dorado, Arkansas, 43, 54
Bureau of Navigation, U.S., 23
Burr, J. G., 16
Burton, William, 99
Butler County, Pennsylvania, 69

Caddo oil field (Louisiana), 36
California: disposal of brine in, 57; oil production in, 41, 51, 186, 260–63, 281
California South Coast Air Quality Management District, 318
California State Mining Bureau, 51
Canada: attendance at conference to prevent oil pollution, 132; earliest production of oil, 369n. 4
carbon dioxide: emissions of, 355–56
Carbosand, 213
Carll, John, 177
Carpenter Steel Company, 200
Carpoloy brand steel, 200
catalytic converters, 275, 311
cathodic protection, 197, 199, 205
Cayo del Oso Bay, 350
Celanese Chemical Co., 349
Champion Paper Company, 242
Chemical Construction Company, 219
Chesapeake and Ohio Railroad, 69
Chester, Pennsylvania, 220
Chevron Oil Company, 13
Cincinnati, Ohio, 232, 276
Cities Service Oil and Gas Company, 155, 176, 178, 203, 206, 241, 274
Claus, C. F., 235
Clean Air Act Amendments of 1970, 302, 311
Clean Air Act Amendments of 1977, 314
Clean Air Act Amendments of 1990: acid rain program, 317; need for oxygenated fuels, 345
Clean Air Act of 1963, 246
Cleveland, Ohio: as center of refining, 68
coal: sulfur dioxide from burning of, 273; vs. fuel oil, 127, 409n. 53
coal oil, 39
Coast Guard, U.S. , 256; cleanup and prevention of oil spills, 283, 287, 324–29, 338–39; Oil Pollution Panel, 251
Colonial Pipeline Company, 347
Columbia Conduit Company, 69
Commoner, Barry, 299, 327
Comprehensive Environmental Response, Compensation, and Liability Act (CERCLA) of 1980, 342, 344; See also superfund
Coney Island Board of Trade: concerns about oil pollution of, 16
Connally Act (Hot Oil Act), 166
Conroe oil field (Texas), 166

conservation: of oil, 59, 154, 156; scientific, 139, 358; utilitarian ethic, and the efficient use of resources, 26, 29, 32, 34, 154, 159; —, as a pollution control ethic, 2–3, 29, 88, 137, 168, 176, 182, 186, 233, 269; —, as compared to an environmental ethic, 137, 141, 182, 269; —, costs and benefits, use of, 4, 22–26, 84, 152, 279, 306, 317; —, economic vs. physical waste, 165; —, failure as a pollution control ethic, 8, 194, 241, 270, 299–300, 322, 359; —, power to shape policy, 167; —, regulations based on, 50, 51, 160, 163, 165–66; —, rhetorical power of, 138, 167
Consolidated Edison, 273, 274
Constant, Edward, 186
Constantin Oil and Refining Company, 36
continental shelf: dispute over ownership, 262, 265
Coordinating Research Council, 271, 275
Corpus Christi, Texas, 263, 350
corrosion: causes of, 201; chemistry of, 196–97; in refineries, 108, 202; of galvanized steel, 199, 201; of oil tankers, 252; of pipelines, 76–77; of stainless steel, 200–201; of storage tanks, 79; of tankers, 337; prevention of, 195–205; symposium on, 201–2
Corrosion, 205
Council on Economic Priorities, 303
crude oil, See petroleum
Cushing oil field (Oklahoma), 36
cyclones: use in refineries, 228, 229–30
Cywin, Alleb, 285

Darst Creek oil field, 172
Darst Creek Salt Water Disposal Company, 172–73
Davis, Dwight, 158
Davison, G. S., 15, 105
De Laval Separator Company, 219
deadweight tons (dwt), 259
deepwell injection: for disposal of industrial wastes, 342, 349; See also injection wells
DeGolyer, E. L., 180
Delaware River, 205
Delaware River Valley Compact Commission, 232
Dennis, John, 255–56
Department of Commerce, U.S., 23, 116
Department of Health, Education, and Welfare (HEW), U.S., 244, 272, 275, 316
Department of the Interior, U.S., 271, 292, 293; report on oil pollution, 280
Department of Justice, U.S., 311
Department of Transportation, U.S., 298; report on oil pollution, 280
Doheny, Edward, 157
Doherty, Henry L., 155–56, 158, 159, 160

Mexico: oil production in, 156
Mid-Continent Cathodic Protection
    Association, 205
Midgley, Thomas, Jr., 139
Midland, Michigan, 233
Miller, Keith, 291
Mining Leasing Act of 1920, 291–92, 294
Mississippi River, 206, 299, 305
Mobil Oil Company, 13, 233, 281, 298
Moore, H., 25–26
Moore, Joseph G., Jr., 308
Moore, W. W. , 18–19
Moran, Eugene, 26
Mostert, Nöel, 331
"mother hubbard suit," 245

Nantucket, Massachusetts, 249
naphtha: as a constituent of petroleum, **93**;
    distillation of, 95–96; used in:
    illuminating oil, 149; —, motor fuel, 96
Narragansett Bay: oil spill in, 16
National Academy of Sciences, 146
National Association of Corrosion Engineers,
    205
National Board of Fire Underwriters, 83–84;
    concerns about oil pollution of, 16–17
National Bureau of Standards, U.S., 147;
    origins of, 146; study of corrosion, 197,
    204
National Coast Anti-Pollution League, 23, 27,
    112
National Commission on Water Quality, 308
National Environmental Policy Act, 288, 292,
    294
National Fire Protection Association, 82
National Lead Company, 219
National Multi-Agency Oil and Hazardous
    Material Pollution Contingency Plan, 280,
    291
National Oil Pipe Line Company, 70
National Oil Recovery Corporation, 286
National Park Service, U.S., 255
National Petroleum Mutual Fire Insurance
    Association, 151
National Petroleum War Services Committee,
    30–31
National Physical Laboratory (Great Britain),
    146
National Pollutant Discharge Elimination
    System, 310
National Transportation Board, U.S., 346
Native Alaskans: role in Trans-Alaska
    pipeline controversy, 290–91
Natural Resources Defense Council, 309
Neches River, 206
Nelson, Thurlow, 22
Netherlands, The, 251
Neuberger, David, 20, 22–23, 27–28
New Brunswick, Canada, 297

New Orleans, Louisiana, 128, 204; oil-related
    fire in, 17
New York: brine disposal concerns in, 352; oil
    production in, 35
New York City, 272; concerns about oil
    pollution in, 17–18, 20–21
New York Dept. of Health: pollution-related
    investigations, 17–18
New York Harbor: pollution in, 18, 26
Newark Bay: pipeline under, 70
Newport, Rhode Island: oil spill in, 129
Niarchos, Stavros, 254
Nicholas, L. V., 161
Nixon, Richard M., 288, 294, 301
nuclear power, 355–56

Occidental Petroleum, 297
Ocean Grove Camp Association, 21
octane, 138, 275
offsets, of air emissions, *See* emissions
    reduction credits
offshore wells, 258, *264*; as incentive toward
    automation, 211; blowouts, 263, 282,
    283–84, 332–33; origins, 260–62;
    perception of discharges from, 266
Ogallala aquifer, 190, 354
Ohio: oil production in, 35
Ohio River, 231
Ohio River Valley Water Sanitation
    Commission, 232
Oil and Gas Journal, 244, 353
oil exploration, 35, 157; dry holes, 49;
    geophysical knowledge and instruments,
    182–83
Oil in Navigable Waters Act (Great Britain),
    132
Oil Pollution Act of 1924, 132, 358; alternative
    bills, **114**; amendment to, 286; British
    equivalent, 132; congressional hearings on,
    20–28, 112–17, 129–32; efforts to amend,
    152, 154, 215; enforcement of, 250;
    requirements of, 117, 133, 151, 169;
    treatment of accidents, 249
Oil Pollution Act of 1990, 324, 334, 336, 339
Oil Pollution Committee of New York, 129
oil production: conditions in "flush" fields,
    40–*44*, 53–54; domestic vs. foreign, 156–57,
    255; maintenance of reservoir pressure,
    53–54, 178, 180; oil left unrecoverable, 59,
    154–55; pollution concerns, 55, 58–**61**, **170**,
    173–75, 343; regulation of, 60, 154–67,184;
    secondary and tertiary recovery, 179, 191,
    194, 353; storage methods, 43–*44*, 66; use
    of geophysical knowledge and
    instruments, 48–49, 184; *See also* brine;
    petroleum; rule of capture
oil refineries, *See* refineries
oil spills: ecological effects of, 16, 76–79, 105,
    213, 256, 278, 297; fingerprinting of, 330;

from offshore wells, 281–*84*, 332–33; from
tankers, 129, 249, 277–79, 330–31, 334–**35**;
from pipelines, 19, 77–79, 347; in the
Arctic Ocean, 292; in western
Pennsylvania, 66; liability for, 249–50,
280, 286–87, 324, 334; prevention and
cleanup of, 213, 278–81, 287, 324–29,
335–39; reporting of, 324; *See also*
pollution
oil wells: abandoned, 186, 188, 191, 353;
acidizing of, 203; corrosion problems in,
203; crooked, 48–49, 185; directional
drilling of, 186, 261–62, 263; drilling of,
cable tool, 44–45, 185; drilling of, rotary
43–*47*, 52, 185; logging of, 47–49; use of
cement in, 49–52; water in, 49–52; *See also*
blowouts; injection wells; offshore wells;
oil production
oil-water separation, *See* brine; refineries:
pollution control; tankers: equipment
Oklahoma: conservation statutes, 60; disposal
of brine in, 57; oil production in, 36,
161–*64*, 354
Oklahoma City oil field, 162
Oklahoma Pipeline Company, 71
Oklahoma Water Resources Board, 354
Onassis, Aristotle, 254, 257
Organization of Petroleum Exporting
Countries, 294
Outer Continental Shelf Lands Act of 1953, 266
oyster beds: damage from pollution, 16, 22, 70

Panama: as flag of convenience, 254
Pasadena, California, 275
Pendleton, 256
Pennsylvania: disposal of brine in, 57, 352; oil
production in, 35, 66
Pennsylvania Dept. of Labor and Industry,
25
Pennsylvania Geological Survey, 177
Pennsylvania Railroad, conflict with pipeline
companies, 69; need for industrial
standards, 145
Pennsylvania State Water Board, 220
Pensacola, Florida, 281
Perkins, Almond A., 51
permafrost: concern for, 290
Peru: search for oil offshore, 260
petroleum: constituents of, 91–92, 94; flow
through formations, 37, 54, 59, 156; impurities in, 99, 100, 272, 274; measuring
quality of, 92–93, 211; origins of, *38–39*, 40;
price of, 92–93, 164; transport by barrel,
66, 120; *See also* brine; gas, natural; oil
production
petroleum engineers: as a new type of
engineer, 47–48, 51; education of, 184; support for regulating the rule of capture, 155,
159

petroleum industry: charges of monopolistic
practices, 142, 158, 245; effects of
overproduction, 155, 161–62, *164*–65; *See
also* American Petroleum Institute;
pipeline companies; refineries; tankers
Pew, J. Edgar, 147–48, 159, 161, 172
Pew, J. Howard, 84
Pew, J. N., 185
phenols, 225, 233, 307; in refinery effluent, 103
Philadelphia, Pennsylvania, 229, 273, 276
Phillips Petroleum Company, 176
Phinney, Sedley Hopkins, 112
Physikalische-Technische Reichsanstalt
(Germany), 146
pipeline companies: as common carriers,
70–72, 208; conflict with railroads, 69, 70;
efforts to prevent corrosion, 76–78,
195–205, 295; monitoring flows, 65, 210;
need for storage, 66; nuisance and
damage suits faced by, 70, 77–78, 79, 87,
212, 348; purchasing of oil, 55, 65–66;
securing rights of way, 71, 291–92, 294
pipelines, petroleum, 69; automation of,
210–11; breaks and leaks, 16, 19, 69, 70,
75–76, 87, 213, 346, 347; corrosion of, 63,
201–2, 347; design and construction of, 70,
72–73, 294–95, 348; for refined products,
205–6; for transporting gas, 36–37, 75, 208;
inspection and maintenance of, 75–76,
206, 207, 211, 348; leak detection in, 347;
list of pollution-related concerns, **86**;
mileage of, 63; one-call systems, 346;
pumping stations, 72–73; replacing
teamsters with, 67; shutoff valves, 213, 295,
348; size of, 208; threaded vs. welded,
74–75, 201, 205; underwater, 19, 75, 206,
298; vapor losses from, 63, 205; *See also*
Trans-Alaska pipeline
Pithole City, Pennsylvania, 67
Pittsburgh, Pennsylvania, 205, 231; as center
of refining, 68; first pipeline to, 69
Plimsoll Mark, 125
Point Barrow, Alaska, 288
Point Breeze refinery (Pennsylvania), 97
pollution: as a symptom of inefficiency,
29–30; damage and nuisance suits, 70,
77–78, 79, 87, 103, 172–73, 175, 187–88;
ecological effects of, 16, 76–79, 105, 152,
278, 297, 324; from municipal sewers,
20–21, 25, 232; from oil, 14–*15*, 17, 28, 255,
256, 283–*84*, **325**; liability for, 21, 22,
249–50, 280, 286–87, 324; measuring and
monitoring of, 18, 90, 119, 220, 238, **325**;
perceptions of, 1–2, 7–8, 17, 21–22, 25, 67,
70, 77–79, 102–5, 132, 138–41, 143, 172–73,
176, 213, 214, 219, 232, 233, 255, 270; *See also*
air quality; oil spills; refineries: emissions
and wastes; tankers: oily discharges; water
quality

regulations on, 7–8, 305, 319–22; evolution of tankers, 335–40.
Temporary National Economic Committee, 245
tetraethyl lead, 311, 358; effect on catalytic converters, 275; initial debate over, 138–43; phase-out of, 312–13
Texaco, 281, 302; *See also* Texas Company
Texas: brine disposal in, 180–82, 188–89, 191, 350; disposal of brine in, 57; oil production in, 36, 43, 120, 122, 127, 174–80, 262–63
Texas Company, 31, 66, 120, 121, 206, 235
Texas Dept. of Agriculture, 352
Texas Market Demand Act, 165
Texas Parks and Wildlife Department, 350
Texas Pipeline Company, 77
Texas Railroad Commission, 165–66, 174, 178, 180, 188, 242, 349, 351
Texas State Land Commission, 262–63
Texas Water Pollution Control Board, 189–90
Texas Water Quality Board, 327
Texas, Fish, Game, and Oyster Commission, 16, 173, 175, 263
Tide Water Pipeline Company, 69–70
Tonsina, 330
Torrey Canyon, 277, 283, 302, 334–**35**
torsion balances, 183
Toxic Substances Control Act, 345
Trans-Alaska pipeline, 288, 290–95, 302; breaks and leaks, 295; design and construction of, 294–95; environmental impact statement, 293
Trans-Alaska Pipeline System (TAPS), 290–92
Tulsa, Oklahoma, 201; as pipeline destination, 73
tundra: concern for, 290

Udall, Stewart L., 279, 291
ultralarge crude carriers (ULCCs), *See* tankers
Union Oil Company, 51, 85, 235, 277, 281, 302
United Kingdom Ministry of Transport, 251
United Nations, 255; study of oil pollution, 251–52
Uren, Lester C., 161
utilitarian ethic, *See* conservation: utilitarian ethic

Vacuum Oil Company, 27
Valdez, Alaska, 290, 293, 296

Van Dyke, J. W., 96–97
Van Syckel, Samuel, 67
Venezuela, 255, 260, 263, 323
very large crude carriers (VLCCs), *See* tankers

War Department, U.S., 22; comments on Trans-Alaska pipeline, 292; enforcement of Oil Pollution Act of 1924, 116
water flooding: of oil fields, 177–78; promotion of, 191; use of fresh water in, 194; use of radioactive isotopes in testing, 193; *See also* brine; oil production: secondary recovery
water quality: concerns and debates, 1–2, 14–18, 20–28, 112–17, 129–32, 152, 215–17, 270, 350; measuring and monitoring of, 3, 8, 220, 232–33, 242–44, 286, 307, 326–27, 345; regulation of, legitimacy, 360–61; regulation of, compared to air quality regulations, 358; state vs. federal responsibility for, 20, 231–32, 271, 334, 340, 351; *See also* groundwater; pollution
Water Quality Act of 1965, 244, 246, 270, 271, 301
Water Quality Improvement Act of 1970, 286
Waterway League of America, 22; concerns about oil pollution, 16
West Virginia: oil production in, 35
Western Oil and Gas Association, 236, 240
Weston, Roy F., 229
Wheeling, West Virginia, 232
White, David, 40
White, Herbert John, 123
Wiggins, J. H., 80–82
Wilbur, Curtis, 158
Wild Mary: blowout of, 162–64
wildcatters, 35
Wilderness Society, 292
Work, Hubert, 158
Wood River, Illinois, 305
Woods Hole Oceanographic Institute, 297
World War I: demand for oil, 157; oil exploration after, 35; oil pollution concerns after, 14; origins of API during, 30; sale of tankers after, 253
World War II: expansion of refining capacity during, 224; pipeline construction during, 208–9; role of federal oil coordinator during, 180; sale of tankers after, 253
Wright, James C., Jr., 276

ABOUT THE AUTHOR

HUGH S. GORMAN is assistant professor of environmental history and policy at Michigan Technological University. He holds a Ph.D. in History and Policy from Carnegie Mellon University.